P9-CAG-627

Neutrinos–1974

(Philadelphia)

AIP Conference Proceedings
Series Editor: Hugh C. Wolfe
Number 22
Particles and Fields Subseries No. 9

Neutrinos–1974

(Philadelphia)

Editor
Charles Baltay
Columbia University

American Institute of Physics
New York 1974

Copyright © 1974 American Institute of Physics, Inc.

This book, or parts thereof, may not be
reproduced in any form without permission

L.C. Catalog Card No. 74-82413
ISBN 0-88318-121-5
AEC CONF-740418

American Institute of Physics
335 East 45th Street
New York, N.Y. 10017

Printed in the United States of America

This volume is dedicated to

Professor Henry Primakoff

on his 60th birthday

PREFACE

The neutrino is the only known particle that interacts exclusively via the weak interactions. It is therefore a unique tool for studying the weak interactions at high energies and large momentum transfers. In addition, the neutrino can be used as a sensitive probe of the basic structure of the proton and neutron, and the properties of the neutrinos have fundamental implications in branches of physics other then particle physics, such as astrophysics and cosmology.

In the past decade condiserable progress has been made in understanding the properties and interactions of the neutrinos. Therefore a series of International Conferences on Neutrino Physics and Astrophysics has been initiated in 1968. The fourth conference in this series was held in Downingtown, Pennsylvania, on April 26-28, 1974. This volume contains a collection of articles based on the invited papers presented at this conference.

The organizing committee of the conference consisted of S. Bludman (chairman), C. Baltay, R. Davis, W. Fry, M. Goldhaber, A.Lagarrigue, K. Lande, A.K. Mann, G. Marx, M.G.K. Menon, H. Primakoff, F. Reines, A. Salam, S. Weinberg, A.W. Wolfendale, and Ya. B. Zeldovich. The organizing committee extends its warm gratitude to the University of Pennsylvania, as well as the U.S. Atomic Energy Commission and the National Science Foundation, for undertaking the sponsorship of this conference.

We would also like to thank the speakers at the conference who, in spite of many other commitments, prepared these articles on a tight schedule to make the speedy publication of this volume possible.

Charles Baltay

Columbia University
New York, N.Y.
June 1974

TABLE OF CONTENTS

INTRODUCTORY REMARKS

Welcome to the Fourth International Conference on Neutrino Physics and Astrophysics! This is the fourth conference in a series which began with Moscow in 1968, Cortona in 1970, Balaton in 1972. The University of Pennsylvania, with the support of the U. S. Atomic Energy Commission and the National Science Foundation, is especially happy to sponsor the first appearance of this important conference in the Americas.

We feel especially favored in having you here at this time because we feel that, in the last two years, the field of neutrino physics and astrophysics has matured. Since the Balaton Conference we have seen an accelerating pace of theoretical and experimental developments: unified theories of weak and electromagnetic interactions, the earliest decisive laboratory experiments on neutral currents and using high energy neutrinos as a probe of nucleonic structure, exciting observations on extraterrestial neutrinos. It is remarkable indeed that an entire conference devoted strictly to one massless, neutral, weakly interacting particle can be scheduled for three days through Sunday, and draw so distinguished a group of theoretical and experimental physicists and astrophysicists.

These neutrino conferences appear to now be an established biennial institution. I hope that in the future we will still be successful in maintaining our unified focus on one particle which is apparently central to physics on the very smallest to the very largest scale.

Sidney A. Bludman
Chairman, Conference
Organizing Committee

INTRINSIC AND ELECTROMAGNETIC PROPERTIES OF NEUTRINOS

Maurice Goldhaber
Brookhaven National Laboratory, Upton, L.I., N.Y. 11973

ABSTRACT

Empirical limits for the intrinsic and electromagnetic properties of neutrinos are discussed.

It was very appropriate that this conference was introduced by a neurophysiologist because a good part of what we know about the neutrino is the brainchild of physicists. This is a field where the ratio of fact to fiction, measured in suitable units, is very small.

The title of the conference has its usual ambiguity because we have no more neutral term than neutrinos for the neutral members of the lepton family but, as you know, the term "neutrinos" is often restricted to mean the antiparticles of the antineutrinos. (Delayed laughter.) Neutrinos are remarkable particles: They induce courage in theoreticians and perseverance in experimenters.

Most of the properties of neutrinos are widely considered to be either 0, +1, -1 or ∞. Of course, we must remember that physics is an empirical science, and empirically you don't measure 0, you don't measure 1, or ∞, so you ought to give some limits, or at least you should remember that such numbers are generalizations. Some or all of these generalizations may of course be true. If you can reduce them to some more general principles, to some more generally accepted conservation laws, you may come out with some of these generalizations; it is often forgotten that these numbers are inputs to a theory which are then taken to be eternal truth.

What are the properties of neutrinos? First of all, in spite of its name, the neutrino could conceivably have a small electric charge, which might differ for ν_e and ν_μ. Of course, it would probably not be too hard from general principles to argue that it should be zero. But let us say we take the empirical point of view. What do we know about these charges? First, for the e-neutrino. You know of course from Ray Davis's experiment that it has a very large charge because it cannot cross the sun—it never arrives here! If ^8B neutrinos were produced in the center of the sun and if we assume that they lost sufficient energy by ionization in crossing the sun, going, say, from 8 MeV to below 2 MeV, thus becoming pretty useless for producing ^{37}A from ^{37}Cl, a <u>lower</u> limit for the electric charge, $q_{\nu_e} > 10^{-6}\, q_e$, would follow.

One can do a more sensitive experiment on earth and obtain a smaller <u>upper</u> limit. L. J. Frazer, E. R. Carlson and V. W. Hughes of Yale have searched for net electric charges of the neutral atoms ^{39}K and ^{133}Cs. With δq defined as the net charge of a hydrogen atom ($\delta q = q_p + q_e$) and q_n as the electric charge of a neutron, they find

$$Q(^{39}K) = 19\ \delta q + 20\ q_n = (0.84 \pm 0.78) \times 10^{-18}\ q_e\ ,$$

$$Q(^{133}Cs) = 55\ \delta q + 78\ q_n = (1.62 \pm 0.70) \times 10^{-18}\ q_e\ ,$$

$$\delta q = (0.9 \pm 2.0) \times 10^{-19}\ q_e\ ,$$

$$q_n = (0.4 \pm 1.5) \times 10^{-19}\ q_e\ .$$

Charge conservation in neutron decay, $n \rightarrow p + e^- + \bar{\nu}_e$, yields

$$q_{\bar{\nu}_e} = -(0.5 \pm 3.5) \times 10^{-19}\ q_e.$$

Now, what are the other properties which are supposedly zero for a neutrino? One property usually considered to be zero is the mass of the neutrino. It is widely believed that unless both ν_e and ν_μ have mass zero their masses are likely to be different. The upper limit for m_{ν_e} has been gradually reduced over many years by measuring the beta spectrum of ^3H, which, because of its low β-ray energy, is the most suitable nucleus for this search. The shape of the high energy end of the spectrum is sensitive to the mass of the neutrino. The last few years have seen a number of experiments which have given limits between 35 and 100 eV/c^2. For m_{ν_μ} the limit is not as good. The most accurate measurement involves a three-body decay process, e.g., $K_L^o \rightarrow \pi\mu\nu$. Recent results obtained in Berkeley by Clarke, Ellioff, Frisch, Johnson, Kerth, Shen and Wenzel, of which Dr. Rol Johnson kindly informed me, yield $m_{\nu_\mu} < 650$ keV/c^2. It would be very interesting if one could find in some deeper theory a relation between the masses of ν_e and ν_μ but at present we do not even have a promising speculation. Of course it is still possible that both ν_e and ν_μ have mass zero, but remember, if that should be the final answer, it will not be an experimental statement.

Other properties of neutrinos which may be zero, or close to zero, for which speculations exist are, e.g.,

charge radius,

possible magnetic moment
(if neutrinos have a finite mass),

electric dipole moment
(since time reversal invariance need not be assumed, a neutrino could possess an electric dipole moment if it has a finite mass).

I believe that these possible properties which could have been looked for experimentally and which sometimes people thought they had looked for already are at present being overwhelmed by the apparent discovery of neutral currents that would produce similar effects. However, it is quite possible that as the experiments yield detailed angular distributions or detailed cross section dependence on energy, we might ultimately also get an insight into these "zero" properties, or at least obtain upper limits for them.

What neutrino properties are "+1" or "-1"? For all leptons one can assign lepton numbers ℓ_e and ℓ_μ. By definition we put $\ell = +1$ for the particles and $\ell = -1$ for the antiparticles. These lepton numbers are believed to be absolutely conserved in all weak interactions. The lack of evidence for double beta decay without neutrino emission permits the conclusion that lepton number is conserved to ~1%. Such a conclusion depends on somewhat uncertain estimates of nuclear matrix elements. A comparable limit can be obtained from interactions of neutrinos where there are no doubts about matrix elements, such as the absence of the reaction

$$^{37}Cl + \overline{\nu}_e \rightarrow {}^{37}A + e^-$$

for reactor antineutrinos and the absence of μ^+ production in interactions of energetic ν_μ's with nuclei.

What other properties are either +1 or -1? An important one is the helicity of neutrinos. Direct evidence from the γ helicity in the ^{152m}Eu experiment and from the μ helicity in $\pi \rightarrow \mu + \nu$ decay gives the sign of the helicity and a lower limit of ~0.6. These experiments could be further improved. Indirect evidence from nuclear β-decay and from μ-decay shows that the helicity is within a few percent equal to +1 for $\overline{\nu}$ and -1 for ν.

Which property of neutrinos may be ∞? It is the lifetime of neutrinos. Again, because of Ray Davis's lack of success in his search for solar neutrinos a way out has been thought up by postulating that neutrinos decay on the way here from the sun. This is extremely speculative. One should use astrophysical arguments with extreme caution when they lead to such radical solutions. So, for the time being let us assume neutrinos to be stable.

Normally, when we teach neutrino physics to students we say $m_\nu = 0$ or $q_\nu = 0$, etc. To show an open mind, I would propose that we use a new equality sign for <u>empirical</u> equality to distinguish it from <u>mathematical</u> equality. The new equality sign is this: [eye symbol], meaning: it looks to us equal to zero, thus m_ν [eye symbol] 0, etc.

I have said that the helicity of neutrinos is equal to either +1 or -1. However, this could arise from the fact that the only interactions we have studied thoroughly produce either left-handed neutrinos or right-handed antineutrinos. Assume there were an interaction which could produce neutrinos of the opposite helicities; then this would allow us without apparently contradicting fundamental notions to have neutrinos with positive helicities and antineutrinos with negative helicities. Such a theory is being proposed by Fischbach, Garvey, Kayser and Rosen, to be discussed in more detail by Dr. Rosen in his talk. They considered the following: Assume the neutral current has really been observed. How do you know that that interaction conserves the helicity of the neutrinos? Assume the neutral current were due to an interaction which flips helicities of neutrinos. A neutrino of a certain helicity hitting a nucleus, say of spin zero, could be reflected and change helicity. Therefore, purely empirically, if you see forward recoils from such a nuclear target, a flip of neutrino helicity would be implied, but if you see only very low momentum transfer then the neutrino could retain its helicity. This is a conceptionally easy but actually very hard experiment which could distinguish between neutral current theories with or without conservation of helicity.

While we must continuously test the assumption of "simple numbers", we have to admit that they have often turned out to be remarkably resilient.

SYMMETRIES AND CONSERVATION LAWS IN NEUTRINO PHYSICS*

S. P. Rosen
Purdue University, West Lafayette, Indiana 47907

ABSTRACT

The basic conservation laws of neutrino physics are described, and the experimental limits on their validity are discussed. The best limit comes from double beta decay, and it shows that lepton number and helicity are conserved in charged weak current processes to order 3 x 10^{-4}. No such limits are available for the recently discovered neutral currents, and so the speculation is made that there may be a large breakdown of the usual helicity rule for neutrinos in neutral current interactions. Experiments to test this speculation are suggested in neutrino scattering off various targets, in neutral beta decay, and in pseudoscalar meson decay. The roles of isospin and strangeness in neutral currents are also discussed. Predictions based upon lepton conservation and helicity are made for the decay distributions of heavy leptons.

INTRODUCTION

Today I want to discuss the basic conservation laws of neutrino physics and the extent to which they are satisfied. For charged current processes the best experimental limits come from double beta decay, and they indicate that the laws are valid to the order of three parts in ten thousand. In the case of the recently discovered weak neutral currents, no limits are available as yet, and so I am going to consider the possibility of large breakdowns. My point of view is that, even though the impetus to look for neutral currents came from the Salam-Weinberg gauge theories, the determination of their properties is nevertheless an experimental problem. If nothing else, entertaining the possibility of a large breakdown in some well-known conservation law is a dramatic way for a theorist to suggest experiments to his experimental colleagues.

The topics I shall cover are: Basic Conservation Laws; Double Beta Decay; Muon Conservation; the Konopinski-Mahmoud Scheme, Muon-Electron Universality; Neutrino Helicity Flip in Neutral Currents; Isospin in Neutral Currents; Strange Neutrinos; and Heavy Leptons.

BASIC CONSERVATION LAWS

Apart from electric charge the basic quantum numbers assigned

---

* Supported in part by the U.S. Atomic Energy Commission.

to neutrinos and other leptons are lepton number, helicity, and muon number. Lepton number L is used to distinguish particles, or leptons, from anti-particles, or anti-leptons, and the conventional assignment of values is:

$$L = \begin{cases} +1 & \text{for } e^-, \nu_e, \mu, \nu_\mu \\ -1 & \text{for } e^+, \bar{\nu}_e, \mu^+, \bar{\nu}_\mu \\ 0 & \text{for all other presently known particles.} \end{cases} \qquad (1)$$

Notice the phrase "presently known" in the third line of eq$^{\underline{n}}$ (1): I have used it because there are some theories[1] that postulate the existence of particles which carry both lepton number <u>and</u> baryon number at the same time. Notice also that muons and electrons with the same electric charge are given the same lepton number: there is, as we shall see below, an alternative scheme in which muons and electrons of opposite charges are treated as leptons.

Conservation of lepton number means that the algebraic sum of the lepton numbers of particles in the final state of any weak process must equal the corresponding sum for the particles in the initial state. Thus L is an additive quantum number like electric charge and baryon number.

From our studies of ordinary weak interactions[2], we know that, within an accuracy of some five to ten per cent, the helicities of electrons, positrons, neutrinos, and anti-neutrinos are given by

$$\mathcal{H}(e^{\pm}) = \pm\, v/c \qquad ; \quad \mathcal{H}(\nu, \bar{\nu}) = -1, +1 \qquad (2)$$

This corresponds to the so-called γ_5-invariance of the theory, according to which every lepton field ψ appears in the weak interaction Lagrangian in the combination $\frac{1}{\sqrt{2}}(1+\gamma_5)\psi$. In its turn, γ_5-invariance gives rise to the V-A form of weak charged currents.

With the assignment of lepton number given in eq$^{\underline{n}}$ (1), we can re-express γ_5-invariance by saying that the helicity of a particle with lepton number L follows the general rule:

$$\mathcal{H}(p_L) = -(L)\; v/c \qquad (3)$$

In the discussion below I am going to consider the possibility of both very small deviations from this rule and very large ones.

The concept of muon number was introduced to account for the absence of the decay mode $\mu \to e + \gamma$, and for the failure of neutrinos from $\pi_{\mu 2}$ decay to stimulate inverse β-decay[3] ($\nu_\mu + n \nrightarrow e^- + p$). Two types of quantum number have been proposed; one, M, is conserved in an additive manner and has the values:

$$M = \begin{cases} +1 & \text{for } \mu^-, \nu_\mu \\ -1 & \text{for } \mu^+, \bar{\nu}_\mu \\ 0 & \text{for all other particles} \end{cases} \qquad (3);$$

and the other M′, is conserved as a multiplicative quantum number like parity[4]. It has the values

$$M' = \begin{cases} -1 & \text{for } \mu^-, \mu^+, \nu_\mu, \bar{\nu}_\mu \\ +1 & \text{for all other particles} \end{cases} \qquad (4)$$

Conservation of either M or M′ is enough to forbid the unobserved processes I have mentioned, but whereas M′ conservation allows the transition of muonium into anti-muonium ($\mu^+e^- \to \mu^-e^+$), M forbids it. Unfortunately this transition is exceedingly hard to detect, and so no choice as to the correct muon number can yet be made on this basis. Other differences between M and M′ show up in the final state of muon decay; as we shall see in due course, what meagre evidence there is seems to be in favor of M rather than M′. The matter, however, is far from being settled.

As an alternative to the schemes involving two quantum numbers, namely (L and M) or (L and M′), we can consider the single quantum number scheme of Konopinski and Mahmoud[5]. It was in fact the first lepton number scheme to be proposed and it treats μ^+ and e^- as particles, rather than μ^- and e^-. The quantum number, which for the sake of clarity I shall call K, is conserved additively and it has the values

$$K = \begin{cases} +1 & \text{for } \mu^+, \bar{\nu}_\mu, e^-, \nu_e \\ -1 & \text{for } \mu^-, \nu_\mu, e^+, \bar{\nu}_e \end{cases} \qquad (5)$$

The analogue of the helicity rule of eq$^{\underline{n}}$ (3) does not hold in this scheme, and so if we do take helicity (or γ_5-invariance) into account in the Konopinski-Mahmoud scheme, we must distinguish between muons and electrons. In other words we must really bring in another quantum number[6].

In the limit of exact conservation of leptons, muons and helicity the Konopinski-Mahmoud scheme is equivalent to the other schemes we have described. As an example consider

$$\nu_\mu + p \to e^+ + n \qquad (6)$$

which is forbidden in the conventional scheme because it requires a change of lepton number $\Delta L = 2$. In the Konopinski-Mahmoud scheme reaction (6) conserves K, but, if we insist on the helicity assignments of eq$^{\underline{n}}$ (2), it can proceed only through scalar, pseudoscalar, or tensor couplings in the four-fermion Lagrangian $(\bar{e}^+ 0\ \nu_\mu)(\bar{n}\ 0\ p)$; thus it violates γ_5-invariance. If we use only vector and axial vector couplings in the Lagrangian, then the positron helicity of eq$^{\underline{n}}$ (2) does not hold, and we again violate the simple γ_5 invariance of the ordinary theory. Thus eq$^{\underline{n}}$ (6) is forbidden as long as γ_5-invariance is an exact symmetry[7].

Should there be a breakdown of γ_5-invariance, reaction (6) would become an allowed process with respect to K-conservation, but it would remain forbidden by L conservation. In this case, the observation of (6), even with a cross-section reduced by two or three orders of magnitude from normal weak interaction cross-sections, would indicate that the Konopinski-Mahmoud scheme gives a better account of the properties of leptons than do the other schemes I have described. Failure to detect (6) beyond a certain level (see below) would indicate the reverse.

What we are saying here is that, even though the different schemes are equivalent to one another when all quantum numbers are conserved exactly, they do develop differences as soon as the slightest breakdown occurs. Let us now turn to experiment to see how large this breakdown can be.

DOUBLE BETA DECAY AND THE LIMITS OF L AND $\mathcal{H}$ CONSERVATION

In principle, limits on the validity of lepton conservation laws can be obtained from electron helicity measurements, and from cross-sections for $\Delta L = 2$ reactions like[8]

$$\bar{\nu}_e + C\ell^{37} \rightarrow e^- + Ar^{37} \tag{7}$$

but in practice the best limit is obtained from double beta decay. The process[9]

$$(A,Z) \rightarrow (A, Z+2) + 2e^- + 2\bar{\nu}_e \tag{8}$$

can occur as a second-order effect of the usual beta decay Hamiltonian whenever the single beta decay $(A,Z) \rightarrow (A, Z+1) + e + \bar{\nu}_e$ is either energetically forbidden, or strongly inhibited by a large change in nuclear spin. The half-life of this two-neutrino double beta decay is estimated to be of order 10^{21} years.

Another conceivable transition is one in which the virtual neutrino emitted at the first stage of the second-order process is reabsorbed at the second stage to yield[9]

$$(A, Z) \rightarrow (A, Z+2) + 2e^- \tag{9}$$

For this no-neutrino double beta decay to occur, both L conservation and the helcity rule of eq$\underline{^n}$ (2) for neutrinos must be violated[6,10]. In the event that these violations are maximal, the half-life of no-neutrino double beta decay is expected to be of order 10^{15} years, that is some six orders of magnitude shorter than the half-life for two-neutrino decay. It is this large difference that enables us to set a sharp limit on the conservation of L and $\mathcal{H}$.

Experiments on double beta decay have been going on for many years, and the most successful ones to date have made use of geochemical and mass spectroscopic techniques. Tellurium ores of known age are subjected to stepwise heating so as to drive out the Xenon occluded in the ore. The isotopic abundance of the occluded Xenon is compared with that of atmospheric Xenon, and any excess of a given isotope is attributed to the transition Te → Xe. Its half-life can be computed from the amount of the excess and the known age of the ore, but before the transition can be interpreted as double beta decay, all competing mechanisms for producing Xenon must be shown to be negligible.

The first experiment of this kind was reported by Ingraham and Reynolds in 1949 and 1950[11]. They obtained a half-life of 1.4×10^{21} years for the transition $Te^{130} \to Xe^{130}$, but because of uncertainties in the energetics of nuclear levels at that time, they could not rule out entirely the possibility that the transition was a succession of real single beta decays, $Te^{130} \to I^{130} + e^- + \bar{\nu}_e$, $I^{130} \to Xe^{130} + e^- + \bar{\nu}_e$ rather than a genuine double beta decay[9]. By 1968 the problem of the energetics was cleared up, and Kirsten, Schaeffer, Norton, and Stoenner[12] carried out another experiment in which they were able to eliminate all the alternatives to double beta. Moreover, their value for the half-life of Te^{130}, namely $10^{21.34 \pm 0.12}$ years, is in very good agreement with the earlier result of Ingraham and Reynolds[11].

Although this observed half-life is consistent with theoretical estimates for two-neutrino double beta decay, it does not constitute definitive evidence because the final state electrons are not detected in mass spectroscopic experiments. There is always the possibility that the no-neutrino decay of eq$\underline{\text{n}}$ (9), while strongly suppressed relative to its maximal rate of 10^{-15} year^{-1}, may nevertheless compete on roughly equal terms with the two-neutrino process of eq$\underline{\text{n}}$ (8). Very recently, however, Hennecke, Manuel, and Sabu[13] reported an experiment which provides good evidence against this possibility.

The idea behind their experiment was first suggested by Pontecorvo[14]. He pointed out that if the double beta decays of two different isotopes of tellurium are compared with one another, the nuclear matrix elements appearing in the theoretical formulae for the decay rates ought to be approximately the same, and so the ratio of the rates should be equal to the ratio of the phase space available to each isotope. Now the phase space for the four-lepton final state of eq$\underline{\text{n}}$ (8) is proportional to somewhere between the eighth and tenth power of the total kinetic energy release, ϵ, while the phase space for the two-lepton process of eq$\underline{\text{n}}$ (9) varies as the fourth to the sixth power of ϵ. (The exact behaviour can be found in the literature[15], but the power law gives a reasonable approximation to it.) Furthermore, it so happens that the energy release in Te^{130} ($\epsilon^{130} \approx 2.5$ Mev) is roughly three times the energy release in Te^{128} (0.86 Mev), and so the ratio of their half-lives depends very strongly upon whether there are two leptons or four leptons in the final state.

If we define the ratio of half-lives as

$$R = t_{\frac{1}{2}}(Te^{128} \to Xe^{128})/(t_{\frac{1}{2}}(Te^{130} \to Xe^{130}) \qquad (10)$$

then the predicted values of R for two-neutrino and no-neutrino double beta decay are[15]

$$R(2\bar{\nu}_e) \cdot = 8 \times 10^3$$

and

$$R(0\,\bar{\nu}_e) = 1.4 \times 10^2 \tag{11}$$

respectively. In their experiment, Hennecke, Manuel, and Sabu[13] measure R directly and obtain a value

$$R_{expt} = (1.59 \pm 0.05) \times 10^3 \tag{12}$$

which is closer to $R(2\bar{\nu}_e)$ than to $R(0\,\bar{\nu}_e)$. Their half-life for Te^{128}, $t_{\frac{1}{2}}^{128} = 1.5 \times 10^{24}$ years, is somewhat longer than an earlier measurement of Takaoka and Ogata[16], but they are able to do a much better job of ruling out mechanisms which might compete with double beta decay.

On the basis of the measurements of $t_{\frac{1}{2}}^{130}$ and R, we may reasonably conclude that double beta decay occurs as a two-neutrino process; but in order to set sharp limits on the breakdown of L and $\mathcal{H}$ conservation, we shall assume that it is an admixture of the two-neutrino and the no-neutrino varieties. To generate the no-neutrino decay we take the lepton current in the conventional weak Hamiltonian to be[15,17]

$$L_\lambda = \{\bar{\psi}_e\, \gamma_\lambda[(1+\gamma_5) + \eta(1-\gamma_5)](\psi_\nu + \xi\psi_{\bar{\nu}})\} \tag{13}$$

where η measures a breakdown in the helicity rule of eq$^{\underline{n}}$ (2) (γ_5 invariance) and ξ is an L-nonconserving parameter. (Ideally we might take $\xi = 1$, in which case the neutrino would be described by a Majorana field). Fitting η and ξ to the data on tellurium, we find that

$$|\xi\eta| \approx 3 \times 10^{-4} \tag{14}$$

This, then, is the extent to which L and $\mathcal{H}$ are presently known to be conserved.

(Notice that a non-zero mass for the neutrino would also give rise to a breakdown of $\mathcal{H}$ conservation. However, the present upper limit of $m_\nu < 60$ ev is too small by a factor of 250 to account for the observed half-life of Te^{130}, and so we can safely respect its effects. The "weak-interaction-induced" rest mass of the neutrino ν_e is approximately $m_\nu \approx \eta(G\Lambda^2)m_e \approx 0.15$ ev for $\xi = 1$ and a cut-off $\Lambda \approx 10$ Gev[18].)

One way of deciding whether or not no-neutrino double beta decay really does occur is to observe the electrons themselves. In the two-neutrino case the sum of electron energies will exhibit a continuous spectrum with a broad peak at about[19] $\frac{1}{2}\epsilon$, whereas in the no-neutrino case the sum of electron energies is always equal to ϵ, the energy release of the process. Thus if no-neutrino decay is present, the spectrum will consist of a continuous distribution with a sharp spike at the high-energy end.

Several attempts to observe the spectrum have been made, the most recent ones being down a salt mine[20] and in the Mont Blanc Tunnel[21], but none of them have been successful so far. In 1150 hours of running time with a 10.6 gms source of Ca^{48}, Bardin, Gollon, Ullman, and Wu[20] saw one possible event in which the sum of electron energies was very close to the maximum for $Ca^{48} \rightarrow Ti^{48}$, and they set a lower limit of 2×10^{21} years on the half-life for no-neutrino double beta decay. Fiorini, Pullia, Bertolini, Capellani, and Restelli[21] ran for 4400 hours with an effective source of 28 gms of Ge^{76} but they did not see any signal in their Ge($\mathcal{L}$i) source-detector; they set a lower limit of 5×10^{21} years for $Ge^{76} \rightarrow Se^{76} + 2e$.

The principal problem of these experiments is, of course, the background, and it may come to pass that with some more improvements in detection techniques, the electrons from double beta decay will become visible. The recent geochemical determination[22] of the half-life for $Se^{82} \rightarrow Kr^{82}$ may be a great help in this regard: the half-life is 2×10^{20} years, or ten times shorter than the lower limit on Ca^{148}. Given the one possible event in the Ca^{48} experiment and a 100 gms of Se^{82}, this factor of ten means that we may see as many as 50 events of $Se^{82} \rightarrow Kr^{82}$ in 1000 hours of running time!

On that optimistic note, I now turn to another subject.

MUON CONSERVATION

Two processes which would enable us to determine whether muon number is the additive quantum number M of eq$^{\underline{n}}$ (3) or the multiplicative one M′ of eq$^{\underline{n}}$ (4) are:

$$\mu^+ e^- \rightarrow \mu^- e^+$$

and

$$e^- e^- \rightarrow \mu^- \mu^- \tag{15}$$

Both are forbidden by M conservation, but allowed by M′ conservation. Unfortunately they are extremely difficult to detect, and the present limits on them are much too large for us to choose between M and M′: for muonium → anti-muonium the effective coupling constant is only less than 5800 times G_W[23], and for the two-electron process it is less than 600 G_W[24]. We must therefore look to muon decay itself as a means for learning about the nature of muon number.

In the decay of the positive muon, the final state consists of a positron and two neutrinos which, for the sake of discussion, I shall denote by ν_1 and $\bar{\nu}_2$:

$$\mu^+ \rightarrow e^+ + \nu_1 + \bar{\nu}_2 \tag{16}$$

From the helicities of the μ^+ and e^+ in the $\pi - \mu - e$ decay sequence, together with the angular distribution of the e^+ and the value of the Michel parameter ($\rho = 0.752 \pm 0.003$) in the e^+ energy spectrum[2], we know that ν_1 is left-handed and $\bar{\nu}_2$ is right-handed.[25] We do not know, however, what kind of neutrinos they are.

If M is the correct muon number, then $(\nu_e, \bar{\nu}_\mu)$ is the only possibility, but if M′ is the correct one, then both $(\nu_e, \bar{\nu}_\mu)$ and $(\nu_\mu, \bar{\nu}_e)$ are allowed. (Notice that I am now assuming L conservation.) We can describe this situation by writing the neutrino final state as an admixture[10]

$$|\nu_1, \bar{\nu}_2\rangle \equiv \alpha|\nu_e, \bar{\nu}_\mu\rangle + \beta|\nu_\mu, \bar{\nu}_e\rangle \tag{17}$$

and by noting that

$$M \rightarrow \beta = 0 \ ; \quad M' \rightarrow \quad \alpha = \beta = 1/\sqrt{2} \tag{18}$$

To distinguish between these two cases, we must obviously "see" the neutrinos from μ^+ decay. When the muon decays in a detector, at rest or in flight, the resulting e-type neutrinos can interact with the material of the detector and produce electrons and positrons[26]. By L-conservation, ν_e must give rise to an e^- and $\bar{\nu}_e$ to an e^+, and so the ratio of electrons to positrons created by neutrinos from muon decay gives us a direct measure of $|\alpha|^2/|\beta|^2$.

An experiment of this type has been carried out in the Gargamelle bubble chamber at CERN, and it gives a result, namely[27]

$$|\beta|^2/(|\alpha|^2 + |\beta|^2) < 0.25 \tag{19}$$

which favors the additive quantum number M. The statistics, however, are very poor, and so the issue of M versus M′ cannot be regarded as settled at this time. Other experiments are in the works at the Los Alamos Meson Facility, and hopefully they will provide us with more definitive results.

THE KONOPINSKI-MAHMOUD SCHEME

I have already pointed out that there are processes that conserve the Konopinski-Mahmoud lepton number[5] K of eq$^{\underline{n}}$ (5) but not L, M, and M′ of eq$^{\underline{ns}}$ (1), (3), and (4). They do violate the helicity rule of eq$^{\underline{n}}$ (2) (γ_5-invariance), and so their amplitudes are expected to be suppressed relative to those of processes allowed in all lepton number schemes.[7] If the limit on helicity from double beta decay (eq$^{\underline{n}}$ (14)) is any guide, the suppression factor lies between 10^{-3} and 10^{-4}.

The example I gave before was the production of positrons by muon meutrinos incident upon a proton target (eq$^{\underline{n}}$ (6)). This is, of course, an exceptional case because ν_μ cannot, collide with a proton and create a μ^- unless it has enough energy to knock the proton into an excited state such as Δ^{++} $(\nu_\mu + p \rightarrow \mu^- + \Delta^{++} \rightarrow \mu^- + p + \pi^+)$. When the neutrino collides with a nucleus of charge Z, however, it will be capable of producing both positrons and negative muons, and on the basis of the above suppression factor we expect a branching ratio[7]

$$B_1 \equiv \frac{\sigma(\nu_\mu + Z \rightarrow e^+ + (Z-1))}{\sigma(\nu_\mu + Z \rightarrow \mu^- + (Z+1))} \approx 10^{-7} \tag{20}$$

The present experimental limit[28], $B_1 < 4 \times 10^{-3}$, is still well above this expectation.

Another example put forward by Kisslinger[29] is μ^--capture followed by the emission of a positron instead of a neutrino:

$$\mu^- + Z \rightarrow e^+ + (Z - 2) \qquad (21)$$

Because the electric charge of the leptons changes by two units, this process either occurs as a second order effect of $\Delta Q = 1$ currents, as is the case with double beta decay, or it is a manifestation of some hitherto undetected $\Delta Q = 2$ current $L_{(d)}$. The first alternative predicts an extremely small branching ratio ($\leqslant 10^{-16}$) relative to ordinary muon capture, while the second depends upon a new coupling constant[29] and could give rise to a much larger branching ratio. I shall therefore consider some of the consequences of a leptonic current L(d).

At the present time there exists one event[30] which fits the description of eq$^{\underline{n}}$ (21). If it is indeed a true example of the process then the branching ratio to ordinary muon capture is

$$B_2 \equiv \frac{\sigma(\mu^- + Z \rightarrow e^+ + (Z - 2))}{\sigma(\mu^- + Z \rightarrow \nu_\mu + (Z - 1))} = 2.6 \times 10^{-8} \qquad (21)$$

and the effective coupling constant G(d) for the $\Delta Q = 2$ current is[7,31]

$$G_{(d)} \approx 10^{-3}\, G_W , \qquad (22)$$

where G_W is the usual weak interaction coupling constant ($\approx 10^{-5}/m_p^2$). If the event proves to be spurious, then eq$^{\underline{ns}}$ (21) and (22) would serve as upper limits on B_2 and $G_{(d)}$.

Another process engendered by the current $L_{(d)}$ is $A \rightarrow B + \mu^{\pm} + e^{\pm}$. Because of the energy required to create a μ^+, the most likely examples of this decay are to be found amongst strange particles[7]:

$$K^+ \rightarrow \pi^- + \mu^+ + e^+$$
$$\Sigma^- \rightarrow p + \mu^- + e^- \qquad (23)$$
$$\Xi^- \rightarrow \Sigma^+ + \mu^- + e^-$$

Assuming that the coupling of strangeness changing $\Delta S = 1$ hadronic terms to $L_{(d)}$ is suppressed relative to strangeness conserving terms by about the same factor (namely $\sin\theta_c$) as occurs in ordinary semi-leptonic decay, we estimate the branching ratios relative to semi-leptonic decay to be of order $(G_{(d)}/G_W)^2$, and from eq$^{\underline{n}}$ (22) their numerical values are about one part in a million. Thus the absolute branching ratio for the K^+ decay should be

$$B_3 = \frac{\Gamma(K^+ \rightarrow \pi^- \mu^+ e^+)}{\Gamma(K^+ \rightarrow \text{all})} \approx 3 \times 10^{-8} \qquad (24)$$

This value is very close to the present experimental upper limit,[32] and so a further research for $K^+ \to \pi^- \mu^+ e^+$ would be of considerable interest.

In the Konopinski-Mahmoud scheme the decay $\mu^+ \to e^+ + \gamma$ satisfies $\Delta K = 2$ and it could occur as a second-order weak effect provided that there be a breakdown of lepton conservation. Its branching ratio is

$$B_4 \equiv \frac{\Gamma(\mu^+ \to e^+ \gamma)}{\Gamma(\mu^+ \to e^+ \nu_e \bar{\nu}_\mu)} \approx \alpha (G_W \Lambda^2)^2 |\xi|^2 \approx 10^{-15} \qquad (25)$$

for a cut-off $\Lambda \approx 10$ Gev and $|\xi| \approx 3 \times 10^{-4}$. In this estimate I have used a value of ξ consistent with the double beta decay limits: now no-neutrino double beta decay is essentially a low energy process with a natural cut-off of $m_{nucleon}$, whereas $\mu^+ \to e^+ + \gamma$ can have much higher energies entering into its potentially divergent amplitude[7]. Furthermore, there is no a priori why a parameter which is small at low to moderate energies need remain small at high energies: indeed various gauge theories suggest that muon conservation will ultimately break down, and that as energies grow, the breakdown will become more severe[33]. We may therefore speculate that the lepton-nonconserving parameter ξ is energy dependent, with very low values ($\sim 3 \times 10^{-4}$) for energies in the region of the nucleon mass, and larger values ($\sim 10^{-1}$) at 10 Gev; ultimately it should reach its limiting value $\xi = 1$[7]. This implies that the branching ratio for $\mu^+ \to e^+ \gamma$ could be as large as 10^{-10}, a value about two orders of magnitude below the present upper limit. Further attempts to see this decay are therefore well worthwhile.

Other $\Delta K = 2$ processes in which this energy dependence might show up are

$$\begin{gathered} \nu_\mu + e \to \nu_e + e \\ \nu_\mu + Z \to \mu^+ + e^- + \nu_e + Z \end{gathered} \qquad (26)$$

At low energies their branching ratios to the $\Delta K = 0$ processes

$$\begin{gathered} \nu_\mu + e^- \to \nu_e + \mu^- \\ \nu_\mu + Z \to \mu^- + e^+ + \nu_e + Z \end{gathered} \qquad (27)$$

should be small, but at high energies they could become as large as 25%[7]. Furthermore, if the helicity violating parameter η of eq^n (13) is also energy dependent and attains its maximum value ($\eta = 1$) at high energy, then all parity violating effects should disappear.

As a final point on the Konopinski-Mahmoud scheme, we note that these $\Delta K = 2$ processes provide a mechanism for Pontecorvo's[34] "neutrino oscillations". The basic oscillation is $\nu_e \leftrightarrow \nu_\mu$ and its estimated oscillation length is $2E_\nu \times 10^3$km where the neutrino energy E_ν is measured in Gev[7].

MUON-ELECTRON UNIVERSALITY

Next to the existence of the muon, the greatest mystery in lepton physics is muon-electron universality[2]. As far as we can tell at present, muons, electrons, and their respective neutrinos enter into the weak charged current in a completely symmetrical way

$$L_\lambda = (\bar{\mu}\ \gamma_\lambda(1+\gamma_5)\nu_\mu) + (\bar{e}\ \gamma_\lambda(1+\gamma_5)\nu_e) \tag{28}$$

and the only difference in their electromagnetic properties comes from their different masses. We do not yet know whether this universality extends to weak neutral currents, but it is generally assumed to do so.

In many gauge theories, muon-electron universality is put in by hand. The terms in the Lagrangian involving the electron and its neutrino are constructed according to the requirements of the specific theory, and the corresponding terms for the muon and its neutrino are obtained by substituting

$$e \to \mu\ ,\quad \nu_e \to \nu_\mu \tag{29}$$

in the electronic ones. In the Weinberg model[35], for example, the electronic terms are required to be

$$g\ \bar{L}_e \underline{t}\ L_e \cdot \underline{W} + \frac{1}{2}\ g'(\bar{L}_e L_e + 2\bar{R}_e R_e)B_o \tag{30}$$

where L_e denotes the left-handed doublet (ν_e, e_L), R_e is the right-handed electron e_R, and $\underline{W}$ and B_o are isotriplet and isosinglet gauge bosons respectively; note that space-time indices have been suppressed in eq$\underline{\text{n}}$ (30). From the substitution rule of eq$\underline{\text{n}}$ (29) the full leptonic Lagrangian

$$\mathcal{L}_\ell = g(\bar{L}_e \underline{t}\ L_e + \bar{L}_\mu \underline{t}\ L_\mu)\cdot\underline{W} + \frac{1}{2}\ g'(\bar{L}_e L_e + \bar{L}_\mu L_\mu + 2\bar{R}_e R_e + 2\bar{R}_\mu R_\mu)B_o \tag{31}$$

It is not difficult to see that, in addition to the substitutional symmetry of eq$\underline{\text{n}}$ (29), this Lagrangian possesses a stronger SU(2) symmetry. If we treat (ν_e, ν_μ) and (e^-, μ^-) as two doublets under the action of M-spin[36,37] and regard the gauge bosons as singlets, then $\mathcal{L}_\ell$ conserves M-spin. An immediate consequence of this conservation is that the amplitudes for neutrino-electron scattering satisfy the sum-rule[36]

$$A(\nu_e e^- \to \nu_e e^-) - A(\nu_\mu e^- \to \nu_\mu e^-) = A(\nu_\mu e^- \to \nu_e \mu^-) \tag{32}$$

The sum rule cannot be satisfied exactly because the Higgs scalars couple to muons and electrons unsymmetrically; however, since their coupling strength is of order[38] $(g\, m_{lepton}/M_W)$, their effects in neutrino-electron scattering may be small.

The conservation of M-spin will be valid in other gauge models[39] in which muon-electron universality is inserted by substitution (see eq$^{\underline{n}}$ (29)) as long as leptons appear bilinearly in the Lagrangian. If there are direct quadrilinear couplings of leptons, then symmetry under $e \leftrightarrow \mu$ does not, by itself, imply M-spin conservation. Such models will not predict the sum rule of eq$^{\underline{n}}$ (32), and so it will be interesting to see the extent to which it is satisfied when sufficient data become available.

NEUTRAL CURRENTS: CAN THE NEUTRINO FLIP ITS HELICITY?

Having shown that in weak interactions engendered by charged currents the basic conservation laws are valid to about three parts in ten thousand, I now want to consider the possibility that there may be large violations of these laws in the recently discovered neutral currents. My point of view is that the neutral current events observed at CERN[40], NAL[41], and Argonne[42] represent a new phenomenon, and we should ask the same questions about its space-time and internal symmetry properties as we once did of beta decay. This does not necessarily bring us into conflict with Salam-Weinberg gauge theories,[35,39,43] but it does allow us to examine a wider range of possibilities.

One of these possibilities concerns the helicity of the neutrino. In charged current interactions the neutrino always follows the helicity rule of eq$^{\underline{n}}$ (2); but we may ask, does it also do so in neutral current phenomena? Is it possible that these processes can give rise to right-handed ν_e and ν_μ instead of the left-handed ones of the charged currents? These are questions that my colleagues E. Fischbach, G. Garvey, B. Kayser and I[44] have been asking, and I would like to discuss some of the ways in which we think the answers may be found.

Consider a beam of neutrinos ν_μ scattering off some target by means of a neutral current interaction. The incident neutrino, having been created in pion decay $\pi^+ \rightarrow \mu^+ + \nu_\mu$, is left-handed, and so the question of producing a right-handed ν_μ in the scattering process amounts to asking whether the ν_μ can flip its helicity. Now the five covariant combinations of spin ½ fields have well-defined properties with respect to helicity[2]: the vector and axial vector terms

$$(\bar{\nu}\,\gamma_\lambda \nu) \text{ and } (\bar{\nu}\, i\gamma_\lambda \gamma_5 \nu) \qquad (33)$$

conserve the helcity of the incident neutrino whereas the scalar, pseudoscalar, and tensor interactions

$$(\bar{\nu}\,\nu),\ (\bar{\nu}\,\gamma_5 \nu), \text{ and } (\bar{\nu}\,\sigma_{\alpha\beta}\nu) \qquad (34)$$

flip the helicity of the neutrino. Therefore I shall consider the possibility that the so-called neutral "current" is, in fact, an admixture of S, P, and T.

The possibility can be tested in neutrino scattering off various targets, $\nu + A \rightarrow \nu' + B$; in the neutral beta decay of excited states of nuclei, $Z^* \rightarrow Z + \nu + \bar{\nu}$; and in the decay of the pseudoscalar mesons π^o and η^o into neutrino-anti-neutrino pairs. If the interaction is an admixture of S and P alone, and contains no tensor (T), then we can make some statements about the scattering process which are independent of the nature of the target A and the recoil B:

(i) the differential cross-section always contains a factor $(k \cdot k') = -EE'(1 - \cos\theta)$ where k, and k′ are the four-momenta of the incident and scattered neutrinos, E and E′ are their energies, and θ is the scattering angle; and

(ii) the differential cross-section for anti-neutrino scattering ($\bar{\nu} + A \rightarrow \bar{\nu}' + B$) is equal to the differential cross-section for neutrino scattering ($\nu + A \rightarrow \nu' + B$) whenever A is unpolarized and the spin state of B is not observed. The first point follows from the fact that, in the limit of zero mass for the neutrinos, $(k \cdot k')$ is the only scalar quantity that can be formed from purely leptonic variables, and the second point comes about because there can be no interference terms between the scalar and pseudoscalar matrix elements for $A \rightarrow B$ when the spin states of A and B are undetected.

An immediate consequence of points (i) and (ii) is that forward scattering of neutrinos and anti-neutrinos is forbidden. When the mass of the target is light compared with the energy of the incident neutrino or anti-neutrino, as is the case for

$$\bar{\nu}_\mu + e \rightarrow \bar{\nu}_\mu + e \tag{35}$$

this forbiddenness will show itself as a tendency for the incident particle go give up a large fraction of its energy to the target. Consequently we anticipate that a significant number of fast forward electrons will be produced in both ν_μ- e and $\bar{\nu}_\mu$- e scattering if the interaction is S and P.

This situation is almost the opposite of the one predicted by a V-A interaction. In that case, forward scattering is allowed for both neutrinos and anti-neutrinos, but backward scattering is forbidden in an anti-neutrino process like the one in eq$^{\underline{n}}$ (35). Thus the neutrinos and anti-neutrinos will show a tendency (slightly more pronounced in $\bar{\nu}$ scattering) to transfer only a small fraction of their energy to the electron, and the production of energetic electrons in the forward direction will be very rare. Another difference between the S and P interaction and the V-A one is that, as a result of point (ii) above, the total cross-sections for anti-neutrino and neutrino scattering off electrons are equal to one another in the S, P case:

$$\sigma^{\bar{\nu}} = \sigma^{\nu} \quad \text{for S and P} \tag{36}$$

but they will be in the well-known ratio[1]

$$\sigma^{\bar{\nu}} = \frac{1}{3}\sigma^{\nu} \tag{37}$$

if the interaction is V-A.

The properties of these interactions can be understood in terms of helicity arguments. In the S. P case an incident left-handed ν_μ becomes an outgoing right-handed ν_μ, and in order for the neutrino to continue in the same direction, the interaction must be capable of inducing a spin change, $\Delta S_z = 1$. Obviously, scalar and pseudo-scalar interaction cannot do this, and so the forward scattering is forbidden. Exactly the same argument holds for anti-neutrino scattering, except that a right-handed $\bar{\nu}_\mu$ comes in, and a left-handed one goes out.

In the case of the V-A interaction, the helicity of particles is conserved, and the assignment of helicities follows the rule of eq$^{\underline{n}}$ (2). Therefore, in the limit of vanishing electron mass, backward $\bar{\nu}_\mu$ - e scattering requires a total spin change of two units, $\Delta S_z = 2$ - a requirement beyond the capabilities of vector interactions.

The tensor interaction, which we have neglected until now, does flip the helicity of the neutrino, but it is also capable of inducing a spin change $\Delta S_z = 1$. Consequently it allows forward scattering, and in the centre-of-mass of the ν_μ - e system it gives an angular distribution of the form $(3 + \cos\theta)^2$. However, if the neutrinos are incident upon a spin zero nucleus, and if the nucleus remains in a J = 0 state after collision, then the tensor interaction cannot contribute to the scattering amplitude. In such a case the choice lies between S and V, and it can be made on the basis of the differences I have already described.

Neutral beta decay $Z^* \rightarrow Z + \nu + \bar{\nu}$ can be analyzed in much the same way as ordinary beta decay[2]. Allowed Fermi transitions satisfy the selection rules $\Delta J = 0$, "no" parity change and are engendered by the scalar S, or vector V interactions; allowed Gamow-Teller transitions have selection rules $\Delta J = 1, 0$ (no $0 \rightarrow 0$ transitions) "no" parity change, and they are engendered by tensor T and axial vector A. The differences between these interactions show up in the angular correlations between ν and $\bar{\nu}$, or equivalently the energy distribution of the recoiling nucleus.

Consider a pure Fermi transition from a spin zero, positive parity (0^+) excited state to a 0^+ ground state. If the interaction is S, then the ν and $\bar{\nu}$ are emitted with the same helicity, and angular momentum conservation does not allow them to travel in the same direction. The angular distribution is therefore proportional to[45]

$$d\Gamma_S \sim (1 - \cos\theta)\, d(\cos\theta) \tag{38}$$

and the ν and $\bar{\nu}$ tend to travel in opposite directions. Consequently the recoiling nucleus does not have much kinetic energy.[46]

If the interaction is V, then ν and $\bar{\nu}$ have opposite helicities, and so travel in anti-parallel directions is now forbidden. The angular correlation,

$$d\Gamma_V \sim (1 + \cos\theta)\, d(\cos\theta) \tag{39}$$

is such that ν and $\bar{\nu}$ tend to travel in the same direction, and so the nucleus must have a relatively large recoil in the opposite direction in order to conserve linear lomentum.

By similar arguments, we can show that in a pure Gamow-Teller transition of the type $1^+ \rightarrow 0^+$, the angular distribution due to the tensor interaction T is[45]

$$d\Gamma_T \sim (1 + \frac{1}{3}\cos\theta)\, d\,(\cos\theta) \tag{40}$$

and the nucleus has a large recoil energy[46]. On the other hand the axial vector A angular distribution is

$$d\Gamma_A \sim (1 - \frac{1}{3}\cos\theta)\, d\,(\cos\theta) \tag{41}$$

and there is very little nuclear recoil. In summary we see that the S and T distributions are exactly the reverse of those for V and A.

The third situation in which we might look for S, T, and P interactions is the decay of neutral pseudoscalar mesons into $\nu\bar{\nu}$ pairs. If the neutral current interaction is of V, A type then the decays π^o, $\eta \rightarrow \nu\bar{\nu}$ are absolutely forbidden because ν and $\bar{\nu}$ have opposite helicities and the vacuum does not recoil. If, however, the interaction is S and P, the neutrinos have the same helicity and the decay is allowed. Furthermore my colleagues Fischbach, Garver, Kayser, and I[44] estimate the branching ratio to be

$$\frac{\Gamma(\pi^o \rightarrow \nu\bar{\nu})}{\Gamma(\pi^o \rightarrow \gamma\gamma)} = 3 \times 10^{-8} \qquad (S,\ P) \tag{42}$$

Given this branching ratio and the relatively short mean life of the neutral pion, any reasonable person world immediately conclude that a search for $\pi^o \rightarrow \nu\bar{\nu}$ is beyond the realm of possibility. However, much to their surprise, my colleagues Fischbach and Kayser discovered in conversations with experimentalists at Berkeley that it may be feasible to look for the sequence

$$K^+ \rightarrow \pi^+\pi^o \quad \hookrightarrow \nu + \bar{\nu} \tag{43}$$

In other words, the efficiencies for detecting the γ's from π^o decay may be so high, that should you miss one or two in a hundred million, you can be confident that the pion has decayed through some channel other than $\pi^o \rightarrow 2\gamma$! Amazing!

ISOSPIN AND NEUTRAL CURRENTS

In studying the effects of neutral current interactions with nuclei, we encounter one problem that does not appear in beta decay and related processes. The phenomenon of nuclear beta decay is generally viewed as one in which any one of the neutrons inside the nucleus may undergo the basic transformation into a proton, $n \rightarrow p + e^- + \bar{\nu}_e$. Consequently the operator that transforms the initial nuclear state into the final state is a coherent sum of operators for the individual neutrons, e.g.,[2,45] $\langle \Psi_f | \Sigma \tau_k^+ \sigma_k | \Psi_i \rangle$. In neutral current processes, neutrons and protons can both undergo the basic interaction,

$$\begin{aligned} \nu + n &\rightarrow \nu' + n \\ \nu + p &\rightarrow \nu' + p \end{aligned} \tag{44}$$

and so the transition operator now consists of a sum of neutron operators together with a sum of proton operators:

$$\langle \Psi_f | \Sigma\, 0_k^{(n)} + \Sigma\, 0_j^{(p)} | \Psi_i \rangle \tag{45}$$

The question then arises whether these two sums interfere with one another constructively, or destructively.

Complete constructive interference ($0^{(n)} = 0^{(p)}$) means that the neutral current is charge independent and conserves isospin ($\Delta T = 0$). Complete destructive interference ($0^{(n)} = -0^{(p)}$) means that the neutrral current behaves as an isovector with selection rule $\Delta T = 1$. If neither of these extremes holds then the current will be an admixture of isoscalar and isovector parts.

One way of studying this question is to scatter moderate energy neutrinos off an isoscalar nucleus like Ne^{20} and concentrate on those events in which the nucleus remains intact. If a single pion is produced in the collision, then the isospin of the final state is unity, and the neutral current will be, at least in part, an isovector. On the other hand if pions are produced only in pairs, then the current is most likely to be an isoscalar. The production of both single pions and pairs would indicate that the current is an admixture of isoscalar and isovector, and the ratio of singles to pairs would provide a measure of the relative amounts of each ΔT. The experiment may be summarized as

$$\nu + Ne^{20} \rightarrow \nu' + Ne^{20} + \begin{matrix} 1\pi & \Delta T = 1 \\ 2\pi & \Delta T = 0 \end{matrix} \tag{46}$$

Another was has been suggested by Sakurai[47] who has a theory in which the neutral leptonic current couples to baryon number. The neutral hadronic current is therefore an isoscalar, and to test this selection rule, Sakurai proposed a study of pion production by neutrinos on protons. For an isoscalar current, the final pion-nucleon system will have $T = \frac{1}{2}$ and the ratio of final states

$$\nu + p \rightarrow \nu + p + \pi^o \tag{47a}$$
$$\rightarrow \nu + n + \pi^+ \tag{47b}$$

will be 2:1 in favor of (47b). On the other hand, should the final state be dominated by the (3,3) resonance, then the ratio of final states will be reversed,

$$\frac{N(\pi^o)}{N(\pi^+)} = \begin{cases} \frac{1}{2} & \text{for } \Delta T = 0 \\ 2/1 & \text{for } \Delta T = 1 \quad (T = 3/2 \text{ final state}) \end{cases} \tag{48}$$

and the current will contain an isovector part.

Over the years, a succession of authors[36,48,49] have speculated that the leptons themselves carry isospin, and that this isospin may have the same origin as hadronic isospin. Thus (ν_e, e^-) and (p, n) may behave as doublets with respect to a single SU(2). The way is then open to consider such possibilities as the conservation of the combined hadronic and leptonic isospins. While I do not wish to discuss these possibilities in any detail, I would like to point out that they do yield testable predictions. For example, in "elastic" lepton-nucleon scattering, isospin conservation implies several sum rules for scattering amplitudes: one of them is[36]

$$A(\nu_\mu p \to \nu_\mu p) - A(\nu_\mu n \to \nu_\mu n) = A(\nu_\mu n \to \mu^- p) \tag{49}$$

STRANGE NEUTRINOS

Just as we try to apply the concept of isospin to leptons, so we can also think of applying strangeness to them[49]. To each of the known doublets (ν_e, e^-) and (ν_μ, μ^-) we add a third member which is often, but not always taken to be a neutral particle, and which is light enough to take part in the same processes as the regular neutrino. We also postulate a conservation law to the effect that ν_μ and ν_e are emitted only in strangeness conserving semi-leptonic processes, and the "strange" neutrinos λ_μ and λ_e only in $\Delta S = 1$ semi-leptonic processes. Thus the neutrino from π^+ decay is not the same as the neutrino emitted in K^+-decay.

Depending upon the particular theory, the difference between, say, λ_μ and ν_μ can range from partial identity[50] ($1 > \langle\lambda_\mu|\nu_\mu\rangle > 0$) to complete orthogonality[49] ($\langle\lambda_\mu|\nu_\mu\rangle = 0$). Present experimental data[51] seem to rule out complete orthogonality, but they are not good enough to exclude the case of partial identity. Setting aside the resolution of this question, I want to make a point here about muon decay and the Cabibbo angle.

From the known lifetimes of the muon and the neutron and from the observed rate for Λ-hyperon beta decay, we know that the neutron- and Λ-decay amplitudes are proportional to to G $\cos\theta$ and G $\sin\theta$ respectively[2]. If there is only one mode for muon decay, namely $\mu^- \to e^- \bar{\nu}_e \nu_\mu$, then from its lifetime we can conclude[2] that the appropriate coupling constant is G. However, should there be additional decay modes like $\mu^- \to e^- \bar{\lambda}_e \lambda_\mu$, then this coupling constant will have to be "shared" between them and the conventional mode. In turn this will mean that we will have to re-tailor the conventional charged current x current theory to include the new modes without upsetting the universality of the lifetimes. Georgi and Glashow[50] have discussed this problem in the context of gauge models, and I[52] have constructed a model Lagrangian in which the Cabibbo angle can no longer be regarded as a rotation parameter in the SU(3) space of strong interactions. The re-interpretation of the Cabibbo angle could well be the most important outcome of the existence of strange neutrinos.

HEAVY LEPTONS

To conclude this review I would like to come back to the basic conservation laws of lepton number and helicity, and use them to make predictions about the decay spectra of heavy leptons. I shall adopt the conventional quantum numbers L and M (eq$^{\underline{ns}}$ (1) and (3)), and assume that they are conserved absolutely.

Many gauge models of electromagnetic and weak interactions postulate the existence of positively charged leptons E^+ and M^+ with the same lepton and muon numbers as e^- and μ^- respectively.[39] These leptons are sufficiently heavy that they can decay into one lepton plus hadrons, as well as into three leptons. Estimates of branching ratios[39,53] indicate that the purely leptonic modes occur about 30% of the time, and so it may be helpful to study their decay distributions.

Because M^+ and E^+ have opposite lepton numbers from e^+ and μ^+, the only decay modes that conserve L, M, and electric charge are:

$$\begin{aligned} E^+ &\to e^+ \nu_e \nu_e & \qquad M^+ &\to \mu^+ \nu_\mu \nu_\mu \\ &\to \mu^+ \nu_\mu \nu_e & &\to e^+ \nu_e \nu_\mu \end{aligned} \tag{50}$$

The presence of identical neutrinos in the first two processes is reminiscent of the early days when the Michel parameter ρ was invented as a tool to help decide whether the neutrinos emitted in muon decay are identical particles, or distinct from one another[54]. The decay distribution $d^2\Gamma/dE\, d(\cos\theta)$ for the processes in eq$^{\underline{n}}$ (50) is proportional to[2,55]

$$\begin{aligned} pE\{3(W - E) &+ 2\rho(\tfrac{4}{3}E - W - \tfrac{m^2}{3E}) + 3\eta \tfrac{m}{E} (W - E) \\ &- \tfrac{p}{E} \cos\theta\ \xi[(W - E) + 2\delta(\tfrac{4}{3} E - W - \tfrac{m^2}{3E})]\} \end{aligned} \tag{51}$$

where p and E are the energy and momentum of the charged lepton in the rest-frame of E^+ or M^+, and m is its mass; W is its maximum energy, and θ is the angle between its momentum and the spin of the parent lepton. The parameters ρ, η, ξ, and δ all depend on the coupling constants in the four-fermion interaction, and ρ and δ lie between zero and 3/4 for identical neutrinos in the final state, and between 0 and 1 for distinct neutrinos.

The parameter ρ is a measure of the probability that the charged lepton be emitted with its maximum energy: if ρ vanishes, the probability vanishes; and if ρ is large, the probability is large. Let us see what this probability is when all the particles in eq$^{\underline{n}}$ (50) satisfy the helicity rule of eq$^{\underline{n}}$ (2): that is, when E^+, M^+ and the neutrinos are left-handed, and e^+ and μ^+ are right-handed.

If the e^+ in the first decay mode of eq$^{\underline{n}}$ (50) had its maximum energy in one direction, then the neutrinos would both travel in the opposite direction. Because e^+ and ν_e have opposite helicities, the spins of all three leptons would point in the same direction and would add up to a total of 3/2. The parent lepton has spin of only ½, and so it cannot possibly decay into a configuration in which e^+ has its maximum energy. Therefore the ρ parameter for $E^+ \to e^+\nu_e\nu_e$ must be zero. Similarly the ρ parameters for the other three decays in eq$^{\underline{n}}$ (50) are zero.

It is not difficult to show that in the charge-retention order, the only four-fermion interaction consistent with the helicity rule of eq$^{\underline{n}}$ (2) is of the S and P variety:

$$(\bar{e}^+ (1 + \gamma_5)E^+)(\bar{\nu}_e(1 - \gamma_5)\nu_e{}^c) \qquad (52)$$

where $\nu_e{}^c$ denotes the charge conjugate of ν_e. The interactions for the other processes in eq$^{\underline{n}}$ (50) are of the same form, and they all give a decay distribution for which

$$\rho = \delta = \eta = 0; \quad \xi = -3 \qquad (53)$$

in eq$^{\underline{n}}$ (51). The value of ξ can be understood from the fact that the scalar combination of neutrino fields makes them tend to travel in opposite directions (our old friend $(k_1 \cdot k_2)$ again!) and so their spins cancel out. Thus the momentum of the e^+, being parallel to its spin, must tend to be parallel to the spin of the E^+.

We can examine the decays of neutral heavy leptons E^o and M^o in much the same way. If E^o and M^o have the same lepton numbers as e^- and μ^- respectively, then the decay distribution of the e^+ in $E^o \to e^-e^+\nu_e$ will have the same parameters as in eq$^{\underline{n}}$ (53), but the distribution of the e^- will be exactly as in μ^- decay, namely[2]:

$$\rho = \delta = 3/4, \quad \xi = 1 \qquad (54)$$

As long as the masses of charged leptons in the final state are neglected, similar results hold for the other decay modes of E^o and M^o.

Should it turn out that these predictions are not fulfilled, then the helicity rule of eq$^{\underline{n}}$ (2) will not be valid for heavy lepton interactions.

CONCLUSION

In conclusion, I want to re-iterate what I said at the beginning. As far as charged current processes are concerned, the basic conservation laws appear to be valid to order 3×10^{-4}. Neutral currents, on the other hand, are entirely new phenomena and, while we may hope that their properties fall in line with Salam-Weinberg theories, we must not assume that they do without experimental proof. I have tried to suggest experiments by considering very large deviations from what we might expect, and I hope I have convinced some of you to try them.

REFERENCES

1. See for example G. Kalbfleisch and E. C. Fowler, Nuovo Cimento 19A, 173 (1974); H. Georgi and S.L. Glashow, Phys. Rev. Letters 32, 438 (1974); J. C. Pati and A. Salam, Phys. Rev. Letters 32, 1083 (1974).

2. For reviews of weak interactions see: E. J. Konopinski Ann. Rev. Nuclear Sci. 9, 99 (1959); T. D. Lee and C. S. Wu, Ann., Rev. Nuclear Sci. 15, 381 (1965) and 16, 471 (1966); D. Bailin, Rep. Prog. Phys. 34, 491 (1971); R. E. Marshak, Riazuddin, and C. P. Ryan, Theory of Weak Interactions in Particle Physics (Wiley-Interscience, New York 1969).

3. G. Danby, J-M. Gaillard, K. Goulianos, L. M. Lederman, N. Mistry, M. Schwartz, and J. Steinberger, Phys. Rev. Letters 9, 36 (1962).

4. G. Feinberg and S. Weinberg, Phys. Rev. 123, 1439 (1961).

5. E. J. Konopinski and H. M. Mahmoud, Phys. Rev. 92, 1045 (1953).

6. S. A. Bludman, Nuovo Cimento 27, 751 (1963); P. K. Kabir, Nuovo Cimento 28, 165 (1963).

7. H. Primakoff and S. P. Rosen, Phys. Rev. D5, 1784 (1972).

8. R. David, Phys. Rev. 97, 766 (1955).

9. For general reviews see: H. Primakoff and S. P. Rosen, Rep. Prog. Phys. 22, 121 (1959), and Alpha-, Beta-, and Gamma-Ray Spectroscopy, edited by K. Siegbahn (North-Holland Publishing Co., Amsterdam, 1965) Vol. 2, p. 1499; G. F. Dell'Antonio and E. Fiorini, Sup. Nuovo, Cim. 17, 132 (1960); and V. R. Lazarenko Usp. Fiz. Nauk 90, 601 (1966).

10. S. P. Rosen, Particles and Fields - 1971, edited by A. C. Melissinos and P. F. Slattery (A.I.P., New York 1971) p. 226; E. Fiorini, Revista del Nuovo Cimento 2, 1 (1972).

11. M. G. Inghram and J. H. Reynolds, Phys. Rev. 76, 1265 (1949), and ibid 78, 822 (1950).

12. T. Kirsten, O. A. Schaeffer, E. Norton, and R. W. Stoenner, Phys. Rev. Letters 20, 1300 (1968).

13. E. W. Hennecke, O. K. Manuel, and D. D. Sabu, Phys. Rev. Letters, (to be published).

14. B. Pontecorvo, Phys. Letters 26B, 630 (1968).

15. H. Primakoff and S. P. Rosen, Phys. Rev. 184, 1925 (1969).

16. N. Takaoka and K. Ogata, Z. Naturforsch. 21a, 84 (1966).

17. Pontecorvo (reference 14), and D. Smith, C. Picciotto and D. Bryman, Nuclear Phys. B64, 525 (1973) have considered the possibility that no-neutrino double beta decay is engendered not as a second-order weak interaction, but rather as the first-order effect of some new interaction. The value of R in eq^n (12) (see, reference 13) is not consistent with this alternative.

18. This estimate of m_ν is based upon the virtual dissociation of a neutrino into an electron and intermediate vector boson. Should there be no such boson, m_ν would be a second-order weak effect and the estimate would be smaller by a factor $G\Lambda^2 \approx 10^{-3}$.

19. S. P. Rosen, Proc. Phys. Soc. (London) 74, 350 (1959).

20. R. K. Bardin, P. J. Gollon, J. D. Ullman, and C. S. Wu, Phys. Letters 26B, 112 (1967), and Nuclear Phys. A158, 337 (1970).

21. E. Fiorini, A. Pullia, G. Bertolini, F. Cappellani, and G. Restelli, Nuovo Cimento 13A, 747 (1973).

22. T. Kirsten and H. W. Müller, Earth Planet Sci. Letters 6, 271 (1969); B. Srinivasan, E. C. Alexander, R. D. Beaty, D. E. Sinclair, and O. K. Manuel, Economic Geology 68, 252 (1973).

23. J. J. Amato, P. Crane, V. W. Hughes, J. E. Rothberg, and P. A. Thompson, Phys. Rev. Letters 21, 1709 (1968).

24. W. C. Barber, B. Gittelman, D. C. Cheng, and G. K. O'Neill, Phys. Rev. Letters 22, 902 (1969).

25. S. P. Rosen, Phys. Rev. Letters 4, 613 (1960); G. Backenstoss, B. D. Hyams, G. Knop, P. C. Marin and U. Stierlin, Phys. Rev. Letters 6, 415 (1961); M. Bardon, P. Franzini, and J. Lee, 7, 23 (1967).

26. G. Kalbfleisch, Nuclear Phys. B25, 197 (1970).

27. T. Eichten et al., Physics Letters 46B, 281 (1973).

28. D. Perkins, Proceedings of the Topical Conference on Weak Interactions, CERN, 1969 (CERN, Geneva, 1969) p. 1.

29. L. S. Kisslinger, Phys. Rev. Letters 26, 998 (1971).

30. D. A. Bryman, M. Blecher, K. Gotow, and R. J. Powers, Phys. Rev. Letters 28, 1469 (1972).

31. M. D. Shuster and M. Rho, Phys. Letters 42B, 45 (1972).

32. E. W. Beier, D. A. Buchholz, A. K. Mann, and S. H. Parker, Phys. Rev. Letters 29, 678 (1972).

33. P.G.O. Freund, Nuclear. Phys. B47, 200 (1972).

34. B. Pontecorvo, Soviet Phys. JETP 26, 984 (1968); V. Gribov and B. Pontecorvo, Physics Letters 28B, 493 (1969).

35. A. Salam and J. C. Ward, Physics Letters 13, 168 (1964); S. Weinberg, Phys. Rev. Letters 19 264 (1967). For a review of the subject see Professor B. W. Lee's talk at this conference.

36. S. Meshkov and S. P. Rosen, Phys. Rev. Letters 29, 1764 (1972).

37. F. Gursey and G. Feinberg, Phys. Rev. 128, 378 (1962); T. D. Lee, Nuovo Cimento 35, 915 (1965); T. Appelquist and J. D. Bjorken, Phys. Rev. D4, 3726 (1971).

38. See for example C.H Llewellyn Smith, CERN Report No. TH. 1710-CERN, (July, 1973).

39. See for example H. Georgi and S. L. Glashow, Phys. Rev. Letters 28, 1494 (1972); J. D. Bjorken and C. H. Llewellyn Smith, Phys. Rev. D7, 887 (1973).

40. F. Hasert et al, Physics Letters 46B, 121 (1973), and ibid 46B, 138 (1969).

41. A. Benvenuti et al, Phys. Rev. Letters 32, 800 (1974).

42. S. J. Barish et al, Phys. Rev. Letters (to be published).

43. J. C. Pati and A. Salam, Phys. Rev. Letters 31, 661 (1973).

44. E. Fischbach, B. Kayser, G. Garvey, and S. P. Rosen (to be published).

45. See for example, E. J. Konopinski (reference 2) for a discussion of helicity and the choice of interactions in beta decay.

46. J. S. Allen et al, Phys. Rev. 116, 134 (1959).

47. J. J. Sakurai, Phys. Rev. D9, 250 (1974).

48. R. L. Ingraham and M. A. Melvin, Nuovo Cimento 29, 1034 (1963);

R. Gatto, Nuovo Cimento, 27, 313 (1963); R. E. Marshak, C. Ryan, T. K. Radha and K. Raman, Phys. Rev. Letters 11, 396 (1963); S. Okubo, C. Ryan, and R. E. Marshak, Nuovo Cimento 34, 753 and 759 (1964).

49. V. G. Gupta, Phys, Rev. B135, 783 (1964); T. Das and K. Mahanthappa, Nuovo Cimento 44, 469 (1966); G. Kalbfleisch and E. Fowler (Reference 1); S. Pakvasa and K. Tennakone (to be published); and R. M. Weiner Phys. Rev. Letters 20, 396 (1968).

50. H. Georgi and S. L. Glashow, Physics Letters 44B, 191 (1973).

51. D. H. Perkins, Proceedings of the Sixteenth Conference on High Energy Physics, Chicago - Batavia (National Accelerator Laboratory, 1972), Volume 4, p. 189.

52. S. P. Rosen, Nuovo Cimento 20A, 288 (1974).

53. Y. S. Tsai, Phys. Rev. D4, 2821 (1971).

54. L. Michel, Proc. Phys. Soc. (London) A63, 514 (1950).

55. T. Kinoshita and A. Sirlin, Phys. Rev. 107, 593 (1957), and ibid 108, 844 (1957); C. Bouchiat and L. Michel, Phys. Rev. 106, 170 (1957); C. Jarlskog, Nuclear Phys. 75, 659 (1966).

NEUTRINO PHYSICS--THEORETICAL CONSIDERATIONS

Benjamin W. Lee
National Accelerator Laboratory, Batavia, IL. 60510

O. PROLOGUE

Neutrino physics is of interest not only for the study of intrinsic characteristics of the elusive neutrinos, but also as diagnostics of hadronic structures, and as a probe of weak interactions.

What follows is a more or less faithful reproduction of the transparencies used in the oral presentation. In preparing this talk, I relied heavily on the following excellent reviews:

S. Adler, "Accelerator Neutrino Physics, Present and Future," NAL-Conf-74/39-THY,
C.H. Llewellyn-Smith, Physics Reports 3C, No. 5 (1972),
E.A. Paschos, NAL-Conf-73/65-THY: Lectures delivered at the "Ettore Majorana" Summer School, 1973.

For comprehensive bibliographies, the above references should be consulted.

I have used a number of figures and illustrations from published papers available before this Conference. More up-to-date data have been presented at the Conference and the reader should consult other contributions in these Proceedings for these.

I. KINEMATICS AND STANDARD V-A THEORY

1. Kinematics.

$$\nu(k_1) + A(p) \rightarrow \ell(k_2) + B(p+q),$$

where $q = k_1 - k_2$ is the momentum transfer.

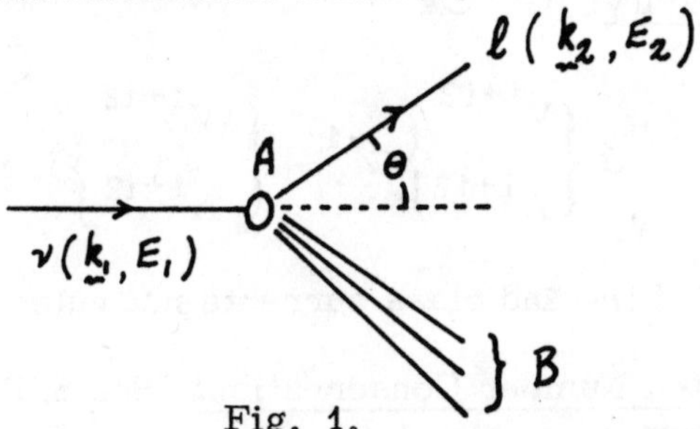

Fig. 1.
Kinematics of neutrino-induced reactions.

Invariant variables:

$$\nu = p \cdot q = m_A(E_1 - E_2),$$

$$q^2 = (k_1 - k_2)^2 = -4E_1 E_2 \sin^2\frac{\theta}{2} \text{ (neglecting } \ell \text{ mass).}$$

2. V-A Theory.

Local Current × Current Interaction:

$$\mathcal{L}_{eff} = \frac{G_F}{\sqrt{2}} \mathcal{J}^\dagger_\mu \mathcal{J}^\mu , \quad G_F \simeq 10^{-5}\, m_p^{-2} .$$

$$\mathcal{J}_\mu = J_\mu + \ell_\mu ;$$

$$J_\mu = (V_\mu^{1+i2} - A_\mu^{1+i2}) \cos\theta_c + (V_\mu^{4+i5} - A_\mu^{4+i5}) \sin\theta_c ,$$

$$\ell_\mu = \bar{\nu}\gamma_\mu(1-\gamma_5)e + \bar{\nu}'\gamma_\mu(1-\gamma_5)\mu ,$$

and $$\theta_c \cong 15° = \pi/12.$$

Current Algebra:

Normalization of V^i_μ and A^i_μ is fixed by Gell-Mann's SU(3) × SU(3) algebra:

$$[A^i_0(\vec{x},0), A^j_0(\vec{y},0)] = if^{ijk}\, V^k_0(\vec{x},0)\,\delta^3(\underset{\sim}{x}-\underset{\sim}{y}) ,$$

etc.

G-parity: $G = C e^{i\pi I_2}$.

$$G \begin{Bmatrix} V^{1+i2} \\ A^{1+i2} \end{Bmatrix} G^{-1} = \begin{Bmatrix} +V^{1+i2} \\ -A^{1+i2} \end{Bmatrix} \text{ i.e., 1st class.}$$

Existence of the 2nd class currents not ruled out .

Lepton Number Conservation: See S. P. Rosen's discussion at this conference.

II. EXCLUSIVE REACTIONS

1. Quasi-elastic: A = N, B = N.

$$\nu_\mu + n \to \mu^- + p .$$

Hadronic matrix element:

$$<p(p_2)|J_\mu|n(p_1)> = \cos\theta_c \bar{u}_p(p_2)\Gamma_\mu u_n(p_1),$$

where

$$\Gamma_\mu = \gamma_\mu F_V^1(q^2) + \frac{i\sigma_{\mu\nu}q^\nu}{2m_N} F_V^2(q^2) + \frac{q_\mu}{2m_N} F_V^3(q^2)$$

$$-\gamma_\mu\gamma_5 G_A(q^2) - \gamma_5 q_\mu H_A(q^2) + \frac{1}{m_N}\gamma_5(p_1+p_2)_\mu F_A^3(q^2).$$

Absence of 2nd class current implies $F^3_{V,A} = 0$; CVC implies $F_V^3 = 0$, and $F_V^{1,2}$ are as in ep scattering. Since $q^\lambda\bar{\mu}\gamma_\lambda(1-\gamma_5)\nu \propto m_\mu$, H_A may be neglected. Parameterize $G_A = 1.24(1-q^2/M_A{}^2)^{-2}$.

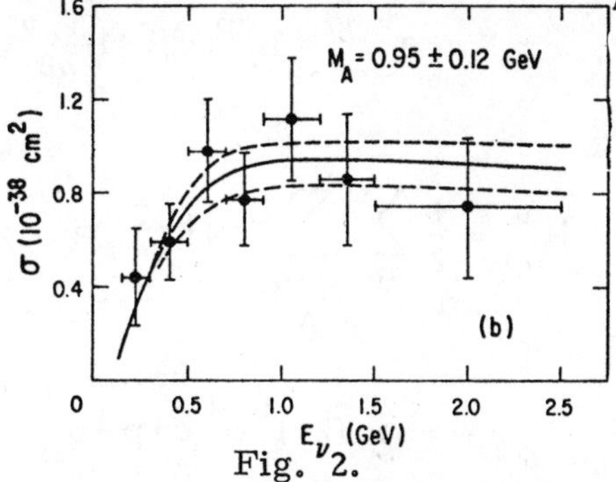

Fig. 2.

Cross Section for $\nu_\mu + n \to \mu^- + p$ and fit with $M_A = 0.95 \pm 0.12$ GeV. From W.A. Mann, et al., Phys. Rev. Letters 31, 844 (1973). See also P. Schreiner's contribution.

2. $\Delta(1236)$ Production: A=N, B=Δ (1236, $P_{3/2}$, 3/2).
$\quad\quad\hookrightarrow$ N+π.

Adler's theory [Ann. Phys., 50, 189 (1968)]: Relativistic version of the static model. Non-resonant amplitudes =

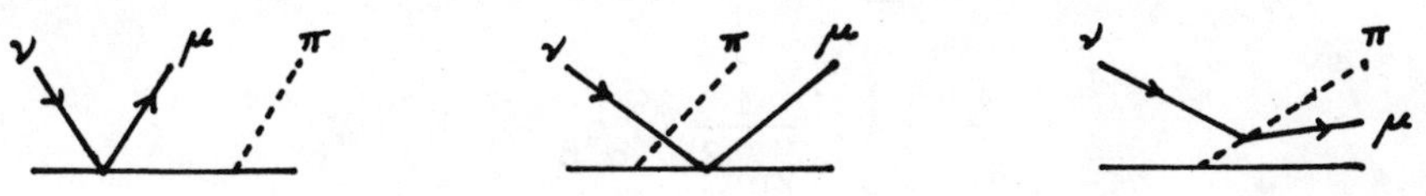

Fig. 3.
Born diagrams in Adler's theory.

Born approximations; Resonant multipoles = Born approximation + rescattering corrections via πN phase shifts. Complete and essentially unique predictions in terms of $\delta_{\ell\pm}$, $F_V^{1,2}$, G_A and H_A. Works o.k. for photoproduction (CGLN); for electroproduction ($q^2 \lesssim 1$ GeV) and also for ν production. For comparison with experiment, see J. Campbell et al., Phys. Rev. Letters 8, 335 (1973); P. Schreiner et al., ibid., 8, 339 (1973); P. Schreiner, in these Proceedings.

III. INCLUSIVE REACTIONS

1. Kinematics.

A=N, B= sum over all final states for fixed m_B. The cross section is

$$\sigma^{\nu,\bar{\nu}} = L^{\alpha\beta}(\nu,\bar{\nu})\, H^{\nu,\bar{\nu}}_{\alpha\beta}$$

where

$$H^{\nu}_{\alpha\beta} = \frac{1}{2}\sum_{B}\sum_{\substack{\text{spin}\\ A}} \langle N|J^{\dagger}|B\rangle\langle B|J_{\beta}|N\rangle$$

$$\times\, (2\pi)^3\, \delta^4(q+p-p_B)$$

$$(\nu \to \bar{\nu},\; J \longleftrightarrow J^{\dagger})$$

$$= -q_{\alpha\beta} W_1 + \frac{p_\alpha p_\beta}{m_N^2} W_2 - i\,\frac{1}{2m_N^2}\,\epsilon_{\alpha\beta\rho\sigma}\, p^\rho q^\sigma W_3$$

$$+ \frac{q_\alpha q_\beta}{m_N^2} W_4 + \frac{1}{2m_N^2}(p_\alpha q_\beta + p_\beta q_\alpha) W_5$$

$$+ i\,\frac{1}{2m_N^2}(p_\alpha q_\beta - p_\beta q_\alpha) W_6$$

$$\frac{d^2\sigma}{d|q^2|\,d\nu} = \frac{G_F^2}{2\pi m_N^2}\left(\frac{E_\mu}{E_\nu}\right)\left[\cos^2\frac{\theta}{2} W_2 + 2\sin^2\frac{\theta}{2} W_1\right.$$

$$\left.\mp \frac{E_\mu + E_\nu}{m_N}\sin^2\frac{\theta}{2} W_3 + \mathcal{O}(m_\mu^2)\right] \text{ for } \frac{\nu}{\bar{\nu}} .$$

Charge symmetry constraints ($\theta_c = 0$):

$$e^{i\pi I_2} J_\mu e^{-i\pi I_2} = -J_\mu^\dagger \Rightarrow W_i^{\nu n} = W_i^{\bar{\nu}p},$$

$$W_i^{\nu p} = W_i^{\bar{\nu}n} .$$

2. Current algebra prediction.

$$\left[\int d^3x\, J^0(\vec{x}, 0)e^{iq\cdot x}, \int d^3y\, J^{0\dagger}(\vec{y}, 0)\right]$$

$$= 4\, I_3 \cos^2\theta_c + (3y + 2I_3)\sin^2\theta_c + \ldots$$

Take spin averaged matrix element between nucleons and go to infinite momentum frame $\Rightarrow$

Adler sum rule:

$$\frac{1}{m_N^2}\int_0^\infty d\nu\left[W_2^{\bar{\nu}}(\nu, q^2) - W_2^{\nu}(\nu, q^2)\right]$$

$$= \langle 4\cos^2\theta_c I_3 + (3y + 2I_3)\sin^2\theta_c\rangle_N$$

$$\simeq \left.\begin{array}{ll} 2 & \text{for } N = p \\ -2 & \text{for } N = n \end{array}\right\} \text{ independent of } q^2!!$$

Alternatively

$$\lim_{E_\nu\to\infty}\left[\frac{d\sigma^{\nu p}}{d|q^2|}-\frac{d\sigma^{\bar{\nu} p}}{d|q^2|}\right]=-\frac{G_F}{\pi}(\cos^2\theta_c+2\sin^2\theta_c)$$

3. Scaling

Dimensional analysis gives

$$W_1(\nu,q^2)=G_1(x,|q|^2/m_N^2),$$

$$\frac{\nu}{m_N^2}W_2(\nu,q^2)=G_2(x,|q^2|/m_N^2),$$

$$\frac{\nu}{m_N^2}W_3(\nu,q^2)=G_3(x,|q^2|/m_N^2).$$

Convenient scaling variables are

$$x=-g^2/2\nu\equiv\omega^{-1},\quad y=\frac{\nu}{m_N E_\nu}=\left(1-\frac{E_\mu}{E_\nu}\right)\cong\frac{E_B}{E_\nu},$$

$$0\le x,\ y\le 1.$$

[Note: $z=xy=|q|^2/2m_N E_\nu=2\left(\frac{E_\mu}{m_N}\right)\sin^2\frac{\theta}{2}$ is scale-invariant.]

$$\frac{d^2\sigma}{dxdy}=\frac{G^2 m_N E_\nu}{\pi}\Big[\Big(1-y-\frac{1}{2}xy\frac{m_N}{E_\nu}\Big)G_2$$

$$+xy^2G_1\mp xy\Big(1-\frac{1}{2}y\Big)G_3\Big].$$

<u>Bjorken scaling hypothesis</u>:

$$G_i(x,|q^2|/m_N^2)\xrightarrow[\substack{\nu,q^2\to\infty\\ x\text{ fixed}}]{}F_i(x).$$

In Bj limit,

$$\left(\frac{\pi}{G_F^2 m_N E_\nu}\right)\frac{d^2\sigma}{dx\,dy} = xy^2 F_1(x) + (1-y)F_2 \mp xy\left(1-\frac{1}{2}y\right)F_3 .$$

Positivity considerations, $\epsilon^\alpha \epsilon^{\beta *} H_{\alpha\beta} \geq 0$, give

$$0 \leq \sigma_s \propto W_2\left(\frac{\nu}{-q^2 m_N^2} + 1\right) - W_1 \Rightarrow \frac{F_2}{2x} - F_1 \geq 0$$

$$0 \leq \sigma_{\substack{R\\L}} \propto W_1 \pm \frac{1}{2}\left(\frac{\nu^2}{m_N^4} - \frac{q^2}{m_N^2}\right)^{\frac{1}{2}} W_3 \Rightarrow F_1 + \frac{1}{2}F_3 \geq 0.$$

Thus

$$F_2 \geq 2xF_1, \quad F_1 \geq \frac{1}{2}|F_3| .$$

Integrate over y:

$$\frac{\pi}{G_F^2 m_N E_\nu}\frac{d\sigma}{dx}\binom{\nu}{\bar{\nu}} = a_s^{\binom{\nu}{\bar{\nu}}} + \begin{pmatrix}1\\ \frac{1}{3}\end{pmatrix} xa_L^{\binom{\nu}{\bar{\nu}}} + \begin{pmatrix}\frac{1}{3}\\ 1\end{pmatrix} xa_R^{\binom{\nu}{\bar{\nu}}}$$

where $a_s = \frac{1}{2}F_2 - xF_1$, $a_{\substack{L\\R}} = F_1 \mp \frac{1}{2}F_3$.

4. Simple consequences of scaling:

i) Adler sum rule

$$\int_1^\infty \frac{d\,\omega}{\omega}\left[F_2^{\bar{\nu}} - F_2^{\nu}\right] = \int_0^1 \frac{dx}{x}\left[F_2^{\bar{\nu}} - F_2^{\nu}\right](x)$$

$$= \langle 4I_3 \cos^2\theta_c + (3y + 2I_3)\sin^2\theta_c \rangle_N .$$

In the scaling region, the left hand side is automatically q^2-independent.

ii) Total cross sections increase linearly with E_ν:

$$\sigma^{\nu,\bar{\nu}} = C_{\nu,\bar{\nu}}\, E_1$$

where

$$C_{\nu,\bar{\nu}} = \frac{G^2 m_N}{\pi} \int dx \left[a_3^{\nu,\bar{\nu}}(x) + \begin{pmatrix} 1 \\ \frac{1}{3} \end{pmatrix} x\, a_L^{\nu,\bar{\nu}}(x) + \begin{pmatrix} \frac{1}{3} \\ 1 \end{pmatrix} x\, a_R^{\nu,\bar{\nu}}(x) \right].$$

$$[\text{Also } \langle q^2 \rangle \propto \nu].$$

Experimentally

$$c_{\bar{\nu}}/c_\nu \simeq 1/3 \begin{pmatrix} 0.38 \pm 0.02 & \text{CERN} \\ 0.35 \pm 0.18 & \text{NAL} \end{pmatrix}.$$

To understand this, set $\theta_c = 0$ and consider an isosinglet target. From charge symmetry, get $F_i^{\nu p} = F_i^{\bar{\nu} n}$, $F_i^{\nu n} = F_i^{\bar{\nu} p}$, or

$$F_i^{\nu} = (F_i^{\nu p} + F_i^{\nu n})/2 = (F_i^{\bar{\nu} p} + F_i^{\bar{\nu} n})/2 = F_i^{\bar{\nu}} = F_i$$

$$\frac{\sigma^{\bar{\nu}}}{\sigma^{\nu}} = \frac{\int a_s + \frac{1}{3}\int x a_L + \int x a_R}{\int a_s + \int x a_L + \frac{1}{3}\int x a_R}.$$

so

$$1/3 \le \sigma^{\bar{\nu}}/\sigma^{\nu} \le 3,$$

and exp $\Rightarrow \int a_s = 0, \int x a_R = 0 \Rightarrow a_s \simeq 0,\ a_R \simeq 0.$

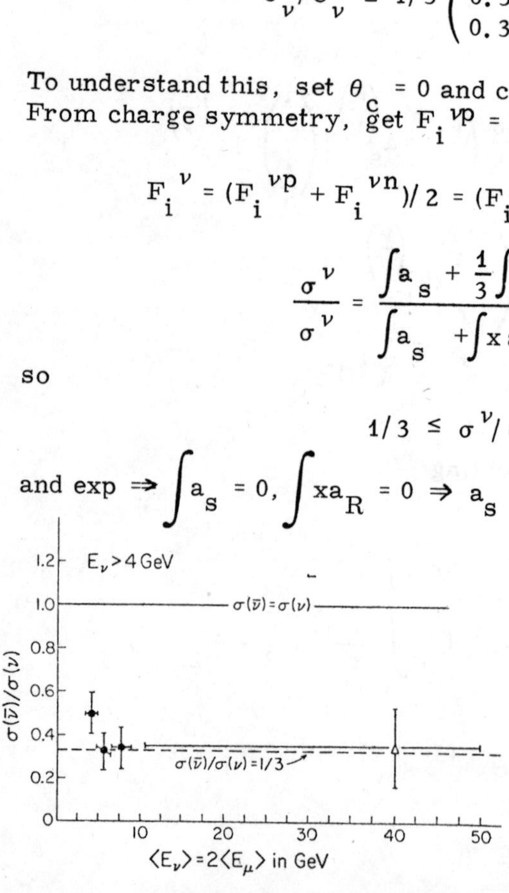

Fig. 4.
Plot of the ratio $\sigma(\bar{\nu})/\sigma(\nu)$ as a function of $\langle E_\nu \rangle$. The value of 1/3 is expected in the scattering of neutrinos and antineutrinos by fundamental fermions such as electrons and muons. From Benvenuti et al., Phys. Rev. Letters 30, 1084 (1973).

$$a_s \simeq 0 \text{ means } F_2 \simeq 2x F_1 .$$

(Callan-Gross relation: spin 1/2 constituents)

$$a_R \simeq 0 \text{ means } F_3 \simeq -2F_1$$

(Maximal VA interference).

iii) y-distribution; above considerations predict y-distribution.

$$\frac{d^2\sigma}{dxdy} = \frac{G_F^2 m_N E_1}{\pi} F_2(x) \qquad \text{for } \nu,$$

$$= \frac{G_F^2 m_N E_1}{\pi} F_2(x)(1-y)^2 \qquad \text{for } \bar{\nu}.$$

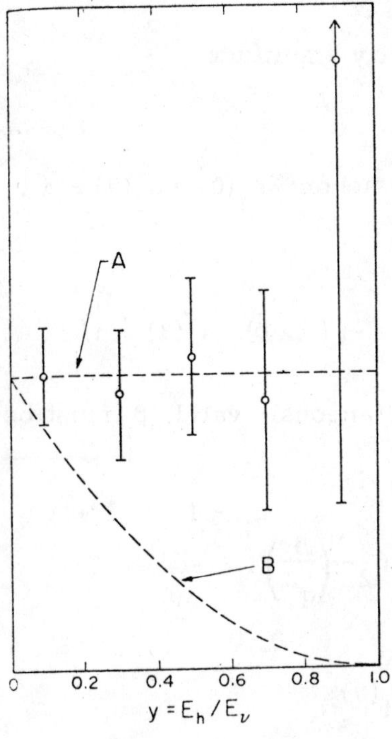

Fig. 5.
y-distribution in the inclusive ν process. From B. C. Barish et al., Phys. Rev. Letters 8, 565 (1973).

iv) Mean muon energy:

$$\left\langle \frac{E_2}{E_1} \right\rangle = \langle 1 - y \rangle = \begin{cases} 1/2 & \text{for } \nu \\ 3/4 & \text{for } \bar{\nu} . \end{cases}$$

This relation is flux independent.

v) Distribution in z = xy:

$$\frac{1}{N}\frac{dN}{dz} = \frac{d\sigma}{dz}\frac{1}{\sigma}$$

$$= \frac{\int_v^1 \frac{dx}{x} F_2(x)}{\int_0^1 dx\, F_2(x)}$$

This is flux, ν energy independent.

5. Regge Asymptotics.

From the Regge analysis of helicity amplitudes,

$$\left.\begin{array}{l} W_1 \xrightarrow[\nu\to\infty]{} \beta_1(q^2)\,\nu^{\alpha_1(0)} \\ W_2 \longrightarrow \beta_2(q^2)\,\nu^{\alpha_2(0)-2} \end{array}\right\} \quad \text{Pomeron: } \alpha_1(0) = \alpha_2(0) \simeq 1 .$$

$$W_3 \longrightarrow \beta_3(q^2)\,\nu^{\alpha_3(0)-1} \qquad G = -1\ (\omega, \phi):\ \alpha_3(0) \simeq 1/2 .$$

If Regge and Bjorken limits are simultaneously valid, β_i must be power behaved:

$$\beta_{1,3} \sim \left(\frac{1}{q^2}\right)^{\alpha_1, \alpha_3}, \quad \beta_2 \sim \left(\frac{1}{q^2}\right)^{\alpha_2 - 1},$$

and

$$F_1 \to \beta_1\,\omega^{\alpha_1(0)},$$

$$F_2 \to \beta_2\,\omega^{\alpha_1(0)-1} \quad : \text{ const as } \omega \to \infty ,$$

$$F_3 \to \beta_3\,\omega^{\alpha_3(0)} .$$

Also

$$F_2^{\nu p} - F_2^{\bar{\nu} p} \sim \omega^{\alpha_\rho(0)-1} \sim \frac{1}{\sqrt{\omega}} \text{ as } \omega \to \infty, \text{ etc.}$$

IV. SCALING AND QUARK PARTON MODEL

1. Assumptions: Infinite momentum frame description of hadrons in terms of almost free constituents of light mass.

i) Leptons scatter incoherently from partons.

ii) Partons are almost free: They are near their mass shell before and after interaction.

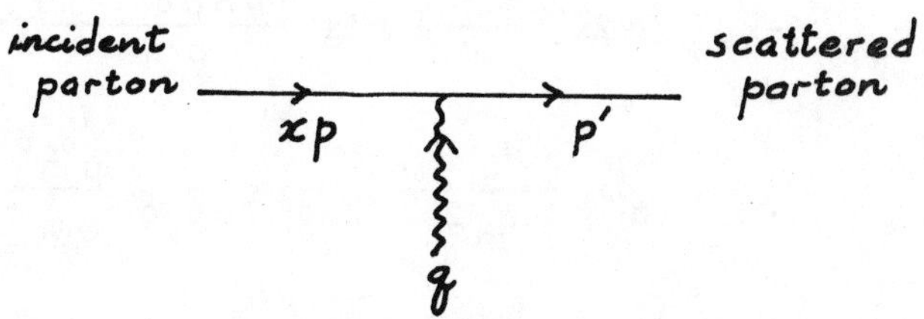

Fig. 6.
Parton kinematics. x is the fraction of longitudinal momentum carried by parton.

$$0 \simeq (m_{parton})^2 = p'^2 = (xp+q)^2 \simeq 2xp\cdot q + q^2,$$

or

$$x \simeq -q^2/2p\cdot q$$

i.e., for given q^2 and ν, deep inelastic scattering probes the parton distribution with longitudinal momentum fraction x. [References: R. P. Feynman, "Hadron-Photon Interactions", (Benjamin, NY., (1972); D. H. Perkins, in Proceedings of Chicago-Batavia Conference (1972), Vol. 4, p. 189.]

2. Structure function in the parton model.

$$H_{\alpha\beta}^{\nu,\bar{\nu}} = \sum_i u_i(x) h_{\alpha\beta}^{\nu,\bar{\nu}}$$

$u_i(x)$: probability that a parton (anti-parton) of type i be found with x.

$h_{\alpha\beta}$: structure function for a parton calculated in the Born approximation.

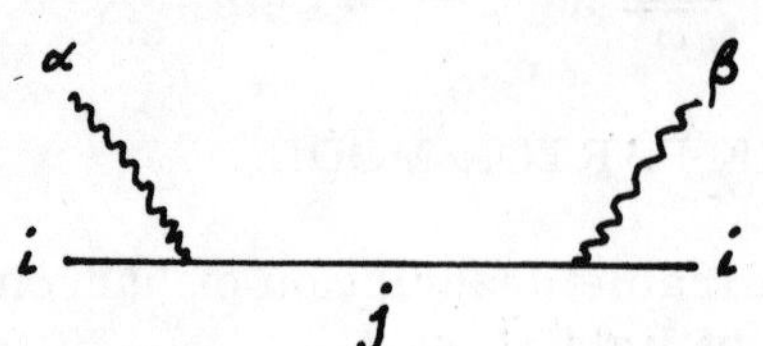

Fig. 7
Born diagram for the parton structure function $h_{\alpha\beta}$.

$$J_\mu = \cos\theta_c \underbrace{\bar{p}\gamma_\mu(1-\gamma_5)n}_{\text{pure V-A}}$$

$$h_{\alpha\beta} = -x g_{\alpha\beta} + 2x^2 \frac{p_\alpha p_\beta}{p\cdot q} \pm ix \frac{\epsilon_{\alpha\beta\rho\sigma} p^\rho q^\sigma}{p\cdot q} \text{ for } \begin{matrix}\text{parton}\\ \text{anti-parton}\end{matrix}$$

$$= -g_{\alpha\beta}\omega_1 + \frac{p_\alpha p_\beta}{m_N^2}\omega_2 - i\epsilon_{\alpha\beta\rho\sigma}\frac{p^\rho q^\sigma}{2m_N^2}\omega_3 .$$

So

$$\left.\begin{aligned} f_1 &= \omega_1 = x \\ f_2 &= p\cdot q\ \omega_2/m_N^2 = 2x^2 \\ f_3 &= p\cdot q\ \omega_3/m_N^2 = \mp 2x \end{aligned}\right\} \quad \begin{aligned} f_2 &= 2x f_1 \Longrightarrow F_2 - 2x F_1 = 0 \\ &\quad \text{(Callan-Gross)} \\ f_3 &= -2f_1 \text{ (partons)} \\ &= +2 f_1 \text{ (antipartons)} \end{aligned}$$

Experimentally $F_3 \simeq -2F_1 \Rightarrow u_{\bar{p}}, u_{\bar{n}}, u_{\bar{\lambda}} \simeq 0$.

With Callan-Gross relation we can write.

$$\frac{d^2\sigma}{dxdy} = \frac{G_F^2 m_N E_1}{\pi} x\left\{\begin{bmatrix}1\\(1-y)^2\end{bmatrix} F_L(x) + \begin{bmatrix}(1-y)^2\\1\end{bmatrix} F_R(x)\right\}$$

$$\text{for}\begin{pmatrix}\nu\\ \bar{\nu}\end{pmatrix}.$$

where

$$F_L^k = \sum_{i=p,n,\lambda} a_i^k u_i(x) ,$$

$$F_R^k = \sum_{i=\bar{p},\bar{n},\bar{\lambda}} b_i^k \bar{u}_i(x),$$

k denotes different processes, νp, $\bar{\nu} p$, $\bar{\nu} n$, νn, ep, and en.

3. Equality: From the above linear relations follow

$$12(F_1^{ep} - F_1^{en}) = F_3^{\nu p} - F_3^{\nu n} \text{ (not tested)}$$

4. Sum Rules:

$$\langle S \rangle = \int_0^1 dx\,[u_\lambda - u_{\bar{\lambda}}]$$

$$\langle I_3 \rangle = \int_0^1 dx\,\frac{1}{2}\,[u_p - u_{\bar{p}} - u_n + u_{\bar{n}}]$$

$$\langle B \rangle .= \int_0^1 dx\,\frac{1}{3}\,[u_p + u_n + u_\lambda - u_p - u_n - u_\lambda]$$

From these

a. Adler sum rule $\Leftarrow I_3$.

b. Llewellyn Smith-Gross: $-\int_1^\infty \frac{d\omega}{\omega}(F_3^\nu + F_3^{\bar{\nu}}) =$

$$= \langle 4B + Y(2 - 3\sin^2\theta_c) + 2I_3 \sin^2\theta_c \rangle \ .$$

5. Inequalities: $u_i \geq 0$.

$$F_2^{ep} + F_2^{en} - \frac{5}{18}(F_2^{\nu p} + F_2^{\nu n}) \propto \text{pos. const } (u_\lambda + u_{\bar{\lambda}}) \geq 0.$$

Experimentally

$$\int_0^1 dx\,(F_2^{ep} + F_2^{en}) \simeq \frac{5}{18}\int_0^1 dx\,(F_2^{\nu p} + F_2^{\nu n})$$

implying $u_\lambda \approx u_{\bar{\lambda}} \approx 0$, and $\frac{18}{5}\,F_2^{ed}(x) \simeq F_2^{\nu d}(x)$.

This is a test for fractionally charged quark partons.

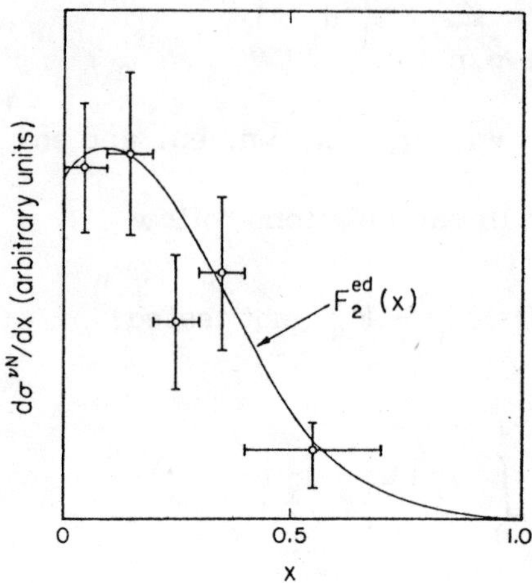

Fig. 8
The distribution $d\sigma^{\nu N}/dx$ (arbitrary units) versus $x = Q^2/2m\nu$ (the scaling variable). For comparison, the fit for $F_2^{ed}(x)$ is shown. From B. C. Barish et al., Phys. Rev. Letters 8, 565 (1973). See also F. Sciulli's contribution.

6. Overall picture

$$\frac{d\sigma}{dxdy} = \frac{G_F^2 m_N E_\nu}{\pi} 2x \begin{cases} u_n(x) & \nu p \\ (1-y)^2 u_p(x) & \bar{\nu} p \\ u_p(x) & \nu n \\ (1-y)^2 u_n(x) & \bar{\nu} n \end{cases}$$

V. SCALING AND LIGHT CONE

1. Light Cone Analysis

$$T_{\mu\nu}(p, q) = \int d^4x \, e^{iq\cdot x} \langle p | [j_\mu\left(\frac{x}{2}\right), j_\nu^\dagger\left(-\frac{x}{2}\right)] | p \rangle$$

$$= \int d^2x_\perp \int dt\, dz\, e^{i\frac{\nu}{m}(t-z) - imx(t+z)} \langle p | [\quad] | p \rangle$$

Dominant contribution comes from $x^2 \simeq 0$, because $t - z \simeq 0 \left(\frac{m}{\nu}\right)$ and because of the commutativity of two currents at space-like separation.

Wilson's operator product expansion [K. Wilson, Phys. Rev. 179, 1499 (1969)].

$$\mathscr{J}_\mu(x)\,\mathscr{J}_\nu(0) = \sum_i \mathscr{C}_i(x^2)\,N[\,\mathscr{O}^i_{\mu\nu}(0)]$$

$\mathscr{C}_i(x^2)$ → c number function of x^2

Gell-Mann, Fritsche postulate that leading light cone singularity structure of $\mathscr{J}_\mu(x)\mathscr{J}^\dagger_\nu(0)$ same as that of the free quark model → reproduce predictions of the quark parton model, and scaling. [See, for example, Gross and Treiman, Phys. Rev. D4, 1059 (1971).]

More rigorously [see, for example, D. Gross, "Scaling in Quantum Field Theory," unpublished],

$$T[J_\alpha \tfrac{x}{2}\, J_\beta -\tfrac{x}{2}\,] = \sum_{n=0}^{\infty} \mathscr{C}^{(n)}(x^2)\,\mathscr{O}_{\alpha\beta\mu_1\ldots\mu_n}(0)X^{\mu_1}\ldots X^{\mu_n}$$

$$+ \text{ terms contributing to } W_1, W_3$$

$$+ \text{ operators of "twist"} > 2,$$

where

$$\mathscr{O}_{\alpha\beta\mu_1\ldots\mu_n} \simeq \underbrace{\bar\psi\gamma_\alpha\,\partial_\beta\,\gamma_{\mu_1}}\ldots\psi, \text{ etc.}$$

sym. and traceless in $\alpha\beta, \mu_1, \ldots \mu_n$

$$\begin{aligned}\text{twist} &= \text{dimensions} - \text{spin}\\ &= (3+n+1) - (n+2) = 2.\end{aligned}$$

Twist > 2 operators contribute to W's asymptotically down by full powers of $(q^2)^{-1}$.

Take the matrix element of $\mathscr{O}_{\alpha\beta\ldots}$ and spin average:

$$\bar\Sigma \langle p|\,\mathscr{O}_{\alpha\beta\mu_1\ldots\mu_n}(0)\,|p\rangle \sim p_\alpha p_\beta p_{\mu_1}\ldots p_{\mu_n}.$$

The n-th term contains n+2 factors of p. Compare this to the forward Compton amplitude:

$$p_\alpha p_\beta \int d\nu' \,\frac{W_2(\nu', q^2)}{\nu' - \nu} \cong p_\alpha p_\beta \sum_n (p\cdot q)^n \int dx\, x^n\,[\nu' W_2(\nu, q^2)]$$

Thus

$$\int_0^1 dx\, x^N \left[\frac{\nu W_2(\nu, q^2)}{m_N^2}\right] = C^{(N)}(q^2)$$

$$\tilde{C}^{(N)}(q^2) \simeq (q^2)^{n+1}\left(\frac{\partial}{\partial q^2}\right)^n \int d^4x\, e^{iq\cdot x} \mathscr{C}^n(x^2)$$

For free fields, $\tilde{C}^{(N)}(q^2)$ is independent of $q^2 \Rightarrow \nu W_2(\nu, q^2)$ a function only of x.

2. Interacting Fields and Renormalization Groups.

Coefficient functions $\tilde{C}^{(N)i}(q^2)$ satisfy the renormalization group equation

$$\left[\mu \frac{\partial}{\partial \mu} + \beta(g)\frac{\partial}{\partial g} - \gamma^{(N)i}(g)\right] C_{AS}^{(N)i}(q^2/\mu^2, g) = 0.$$

The solution is of the form

$$C^{(N)i}(q^2/\mu^2, g) \xrightarrow[-q^2 \to \infty]{} C^{(N)i}(-q^2)^{-\frac{1}{2}\gamma^{(N)i}(g_\infty)}$$

where g_∞ is a root of $\beta(g)$:

$$\beta(g_\infty) = 0 \text{ and } d\beta(g)/dg\Big|_{g = g_\infty} < 0$$

and $\gamma^{(N)i}$ is the anomalous dimension of $\mathscr{O}^{(N)i}$, $\gamma^{(N)i}(g = 0) = 0$; increases with N (positivity of νW_2) and is convex downwards ($\to 2\gamma_\psi$ as $n \to \infty$).

Consider $\mathscr{O}^{(N)i} = \Theta_{\mu\nu}$: conserved energy momentum tensor (D=4, J=2, twist 2). $\Theta_{\mu\nu}$ has no anomalous dimensions, $\gamma^\Theta = 0$. Now moment $x^N \longleftrightarrow$ spin N + 2 operators.

$$\int_0^1 dx\, \nu W_2 = \text{const} + \frac{C}{|q|^\nu}.$$

(const: contribution of $\Theta_{\mu\nu}$)

$\Theta_{\mu\nu}$ is isoscalar $\Rightarrow$

$$\int_0^1 dx\, \nu W_2^{\nu p}(\nu,q^2) = \int_0^1 \nu W_2^{\nu n}(\nu,q^2) \quad \text{as } -q^2 \to \infty$$

or, (with $\theta_c \to 0$)

$$\int_0^1 dx\, [\nu W_2^{\nu p} - \nu W_2^{\bar{\nu} p}] \to 0$$

We need consider two cases. [See, for example, S. Adler, NAL-Conf-74/39-THY] :

(1) $g_\infty \neq 0$. $\gamma^{(N)}(g_\infty) \neq 0$ in general $\Rightarrow$ moments go to zero as $(-q^2)^{-\frac{1}{2}\gamma^{(N)}}$.

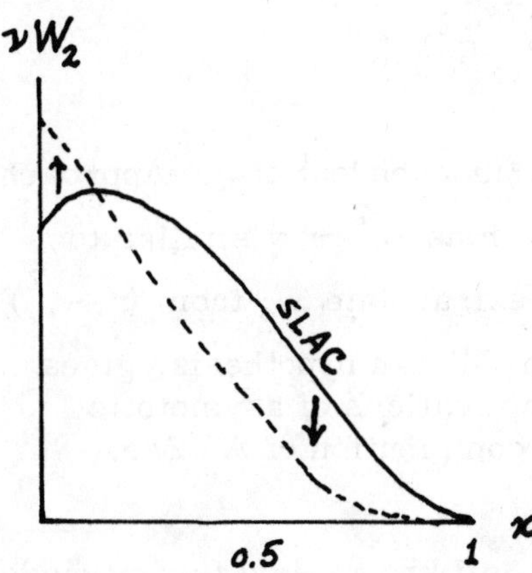

Fig. 9
Behavior of νW_2 as $q^2 \to \infty$ as predicted by field theories (schematic).

νW_2 will change as $-q^2 \to \infty$. (a) Near x = 1, decreases to make higher moments go to zero; (b) If $\gamma^{(2)i}$ all small, area remains $\simeq$ const; (c) near $x \simeq 0$ increases.

(2) $g_\infty = 0$ (<u>Asymptotic freedom</u>)

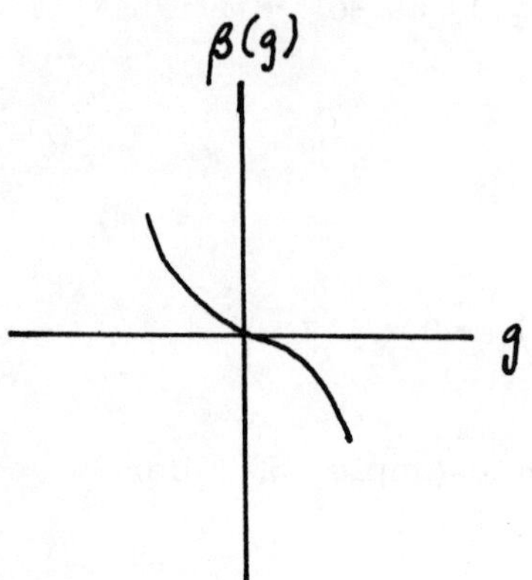

Fig. 10.
Behavior of β(g) in asymptotically free gauge theory of strong interactions.

This happens in nonabelian gauge theories.

$$C^{(N)i}(q^2) \underset{-q^2 \to \infty}{\to} (\ell n - q^2)^{-\frac{1}{2} a^{(N)i}}$$

where $a^{(N)i}$ are computable numbers, independent of g. Approach to asymptotic logarithmic; νW_2 behavior as $-q^2 \to \infty$ similar to above; since $a_{(n)i}$ known, possible to extrapolate W_2 from $(x, -q^2)$ to $(x, -q'^2)$. Combined with the Bloom-Gilman hypothesis, gives bounds on e.m. form factors. For implications of asymptotic freedom on neutrino physics, see the contribution of A. Zee.

3. Remarks

A. If exact scaling remains valid, all known field theory models of strong interactions in trouble.

B. Parton model works at SLAC and NAL ν energies, whether you believe in it or not--a case of Niels Bohr's horseshoe.

C. If scaling breaks down as field theory predicts, $\sigma_\nu - \sigma_{\bar\nu} \to 0$. Adler sum rule OK, but not Llewellyn Smith-Gross.

VI. GAUGE THEORIES

1. Gauge Theories of Weak Interactions.

General References:

B. W. Lee, Proceedings of Chicago-Batavia Conference, (1972), Vol. 4, p. 249.

S. Weinberg, Proceedings of Aix-en-Provence Conference (1973); Rev. Modern Physics, to be published.

i) Unification of electromagnetic and weak interactions in a gauge theory (nonabelian).

ii) Spontaneously Broken Gauge Symmetry: masses of weak bosons arise thru this mechanism. The symmetry breaking is especially "soft" → The renormalizability of the unbroken gauge theory is preserved.

In a renormalizable theory, bad high energy behavior of a tree graph has to be cancelled by another. Example:

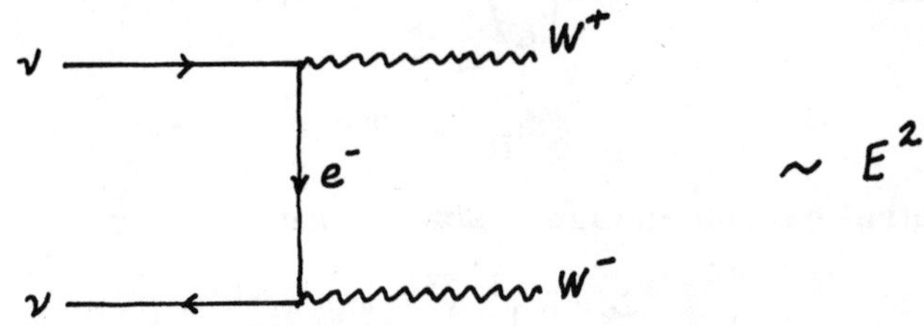

Fig. 11.
The process $\nu + \bar{\nu} \rightarrow W^{+} + W^{-}$ in the conventional theory.

The E^2-, E-terms must be cancelled by one or the other of the following graphs:

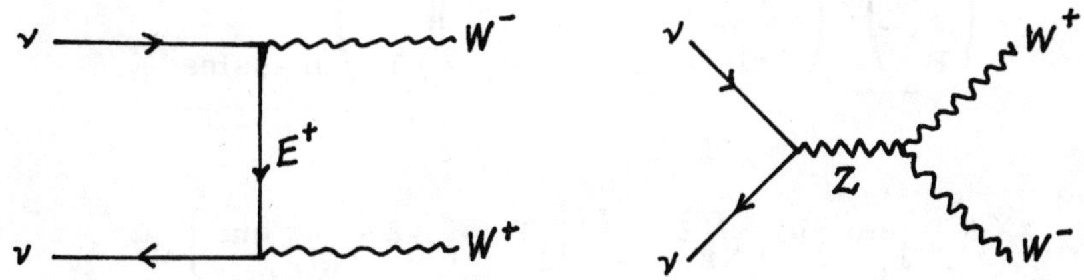

Fig. 12
Two tree diagrams for $\nu + \bar{\nu} \rightarrow W^{+} + W^{-}$ that appear in a renormalizable theory.

Because gauge theories are renormalizable, they must contain either heavy leptons, neutral currents or both.

iii) Important Issues:

Heavy leptons
Neutral currents
Charms
Nuclear physics

2. Neutral Current in the Weinberg-Salam Theory.

$$Q = T_3 + Y/2$$

(T_3 → weak isospin; Y → weak hypercharge)

Example: $\begin{pmatrix} \nu \\ e \end{pmatrix}_L$, $T = 1/2$, $Y = -1$

e_R , $T = 0$, $Y = -2$.

Neutral current-gauge boson coupling:

$$\mathscr{L} = g\, j^3_\mu A^{3\mu} + g' \underbrace{(j^{em}_\mu - j^3_\mu)}_{\frac{1}{2} \times \text{"hypercharge" current}} B^\mu .$$

Let

$$\begin{pmatrix} A^3_\mu \\ B_\mu \end{pmatrix} = \begin{pmatrix} \cos\theta_W, & \sin\theta_W \\ -\sin\theta_W, & \cos\theta_W \end{pmatrix} \begin{pmatrix} Z_\mu \\ A_\mu \end{pmatrix} \begin{matrix} \text{massive} \\ \text{massless } \gamma. \end{matrix}$$

Then

$$\mathscr{L} = e\, j^{em}_\mu A^\mu + \sqrt{g^2 + g'^2} \left[j^3_\mu - \sin^2\theta_W\, j^{em}_\mu \right] Z^\mu$$

with

$$e = g g' (g^2 + g'^2)^{-1/2}$$

$$\cos\theta_W = \frac{g}{\sqrt{g^2 + g'^2}} .$$

$$\mathscr{L}_Z = \sqrt{g^2 + g'^2}\, Z^\mu \left[J^3_\mu - \sin\theta_W\, J^{em}_\mu + \frac{1}{2} \bar{\nu} \gamma_\mu (1-\gamma_5)\nu + \ldots \right]$$

Effective Local Coupling:

$$\frac{g^2 + g'^2}{8 m_Z^2} \bar{\nu} \gamma^\mu (1-\gamma_5) \nu \Big[\underbrace{V^3_\mu - A^3_\mu}_{\substack{\text{isospin rotation} \\ \to \text{ charged current}}} - 2 \sin^2\theta_W \underbrace{J^{em}_\mu}_{\text{e.m. current}} \Big].$$

In the original construction of Weinberg and Salam, a particularly simple symmetry breaking mechanism was used, leading to the relation

$$\frac{m_Z^2}{m_W^2} = \frac{g^2 + g'^2}{g^2}, \text{ or } \frac{g^2 + g'^2}{m_Z^2} = \frac{g^2}{m_W^2}$$

In general, however, there need not be any relation between m_W and m_Z. Parameterize the strength of Z exchange by one parameter x:

$$\frac{g^2 + g'^2}{m_Z^2} = x \frac{g^2}{m_W^2}.$$

$$\Rightarrow \quad \frac{G_F}{\sqrt{2}} x \, \nu\gamma^\mu (1-\gamma_5)\nu \left[V_\mu^3 - A_\mu^3 - 2\sin^2\theta_W J_\mu^{em}\right].$$

3. Tests of the WS Theory in Hadronic Reactions.

While the properties of neutral current ought to be studied in the most general context, we will consider parameterizing neutral current effects a'la Weinberg-Salam theory.

i) Inclusive Reactions: Define

$$R_{\substack{\nu \\ \bar{\nu}}} = \sigma\begin{pmatrix} \nu_\mu \\ \bar{\nu}_\mu \end{pmatrix} + N \rightarrow \begin{pmatrix} \nu_\mu \\ \bar{\nu}_\mu \end{pmatrix} + B \Bigg) \Big/ \sigma\Bigg(\begin{matrix} \nu_\mu \\ \bar{\nu}_\mu \end{matrix} + N \rightarrow \begin{matrix} \mu^- \\ \mu^+ \end{matrix} + B \Bigg).$$

(A) The hadronic neutral current is in the quark language (with $\theta_c = 0$):

$$\bar{p}\,\gamma_\mu [a(1-\gamma_5) + b(1+\gamma_5)]\,p$$
$$+\,\bar{n}\,\gamma_\mu [c(1-\gamma_5) + d(1+\gamma_5)]\,n,$$

where

$$a = 1/2 - (2/3)\sin^2\theta_W, \quad b = -(2/3)\sin^2\theta_W$$

$$c = -1/2 + (1/3)\sin^2\theta_W, \quad b = (1/3)\sin^2\theta_W.$$

$$R_\nu = \frac{1}{2} - \sin^2\theta_W + \frac{20}{27}\sin^4\theta_W,$$

$$R_{\bar{\nu}} = \frac{1}{2} - \sin^2\theta_W + \frac{20}{9}\sin^4\theta_W .$$

[A. De Rujula et al., Revs. Mod. Phys., to be published.]

(B) Assuming scaling in electroproduction,

$$R_\nu \geq \frac{1}{2}\left\{1 - 2\sin^2\theta_W \sqrt{t}\right\}^2;$$

Assuming scaling both in eN and νN ($\bar{\nu}$N),

$$R_\nu \geq \frac{1}{2}\left\{\frac{2}{3} + \frac{1}{3}x - (1-x^2)t\right\}$$

where

$$t = \frac{G^2}{\pi}\frac{4}{3}m_N E_1 \int_0^1 dx\, F_2^{eN}(x)/\sigma(\nu_\mu + N \to \bar{\mu} + B).$$

Using

$$\int_0^1 dx\, F_2^{eN}(x) \simeq 0.14, \quad \sigma(\nu_\mu + N \to \bar{\mu} + B)$$

$$\frac{G^2}{\pi} m_N E_1 \cdot \ 0.52,$$

one gets $t \cong 0.36$.

Take $t = 1/3$. One gets

$$R_\nu = \frac{1}{2}\cdot\frac{1}{3}(1 + x + x^2),$$

$$R_\nu = \frac{1}{2}(1 - x + x^2).$$

[Pais and Treiman, Phys. Rev. D6, 2700 (1972); Paschos and Wofenstein, Phys. Rev. D7, 91 (1973)] .

ii) Exclusive Reactions

(A) Elastic: Theory predicts

$$0.15 \leq \sigma(\nu + p \to \nu + p)/\sigma(\nu + n \to \mu + p)$$

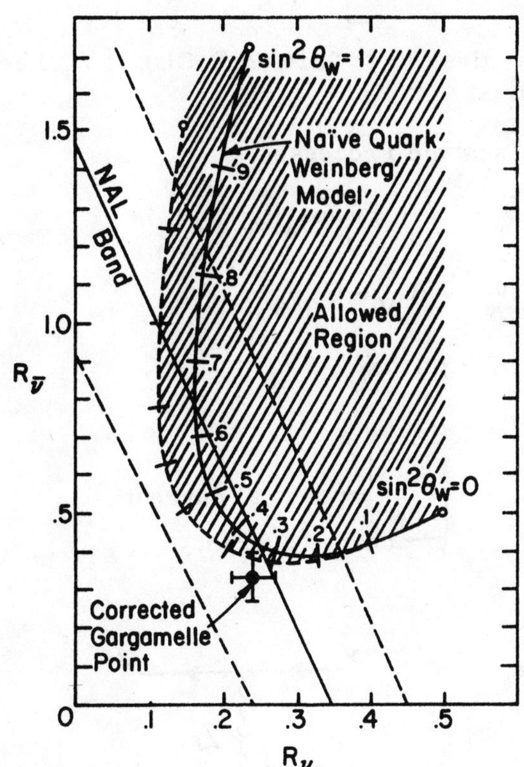

Fig. 13.
Comparison of theory and experiment for processes $(\nu, \bar{\nu}) + N \rightarrow (\nu + \bar{\nu}) +$ anything. From A. De Rujula et al., Revs. Mod. Phys., to be published.

$$\leq 0.25 \text{ for } \sin^2\theta \leq 1/2$$

Experiment (Gargamelle) gives 0.12 ± 0.06

(B) Δ Production. Define

$$R = \frac{\sigma(\nu_\mu + p \rightarrow \nu_\mu + p + \pi^\circ) + \sigma(\nu_\mu + n \rightarrow \nu_\mu + n + \pi^\circ)}{2\sigma(\nu_\mu + n \rightarrow \mu^- + p + \pi^\circ)}.$$

Δ dominance gives $R \geq 0.4 \sim 0.5$ for $\sin^2\theta_W < 1/3$.

Refined theory (Adler, Nussinov, and Paschos, Phys. Rev. D, to be published) gives

$\sin^2\theta_W$	Δ only	$+ I = 1/2$	+ charge exch. in Al
0.3	0.56	0.40	0.23
0.4	0.46	0.33	0.18

Experiment (W. Lee-Columbia) gives $R < 0.14$ (90% CL). [W. Lee, Phys. Letters 40B, 423 (1972)].

4. Tests of the WS Theory in Leptonic Processes.

$$H_W(\nu e \to e\nu) = \frac{G}{\sqrt{2}} \quad \bar{\nu}_\mu \gamma^\alpha (1-\gamma_5)\nu_\mu \bar{e}\gamma_\alpha (g_V - g_A\gamma_5)e$$

$$+ \bar{\nu}_e \gamma^\alpha (1-\gamma_5)\nu_e \bar{e}\gamma_\alpha (G_V - G_A\gamma_5)e$$

where

$$G_A \quad 1 + \quad g_A$$

$$G_V = 1 + g_V$$

g_V → neutral current

1 → charge current

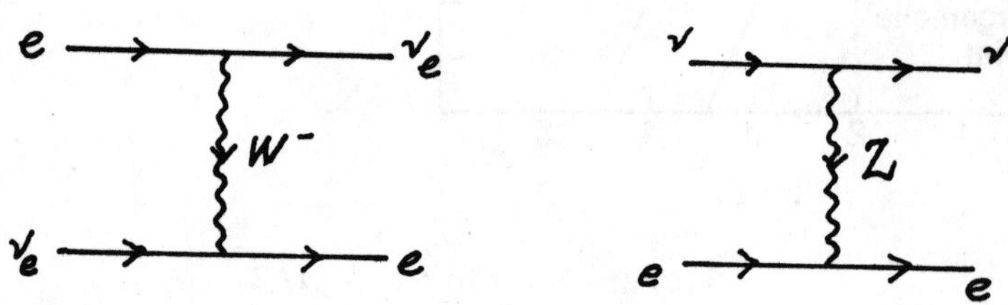

Fig. 14
Tree diagrams for $\nu + e \to \nu + e$.

$$g_V = (-\frac{1}{2} + 2\sin^2\theta_W)x$$

$$g_A = (-\frac{1}{2})\, x.$$

(A) $\sigma(\bar{\nu}_e + e \to \bar{\nu}_e + e) = C\, 10^{-41}\, cm^2\, (E_\nu / GeV)$.

$C = 0.54$ in standard V-A

$= 0.136 \sim 2.86$ in WS

Gurr, Reines and Sobel, Phys. Rev. Letters 28, 1406 (1972) find

$$\sigma \leq 3\,\sigma_{V-A} \quad (90\%\ cL).$$

(B) Gargamelle had 370,000 pix each for ν_μ and $\bar{\nu}_\mu$.

	WS	background (est)	observed
ν_μ	$0.6 \sim 6.0$	0.3 ± 0.2	0
$\bar{\nu}_\mu$	$0.4 \sim 8.0$	0.03 ± 0.02	1

[F. J. Hassert et al., Phys. Letters 46B, 121 (1973)].

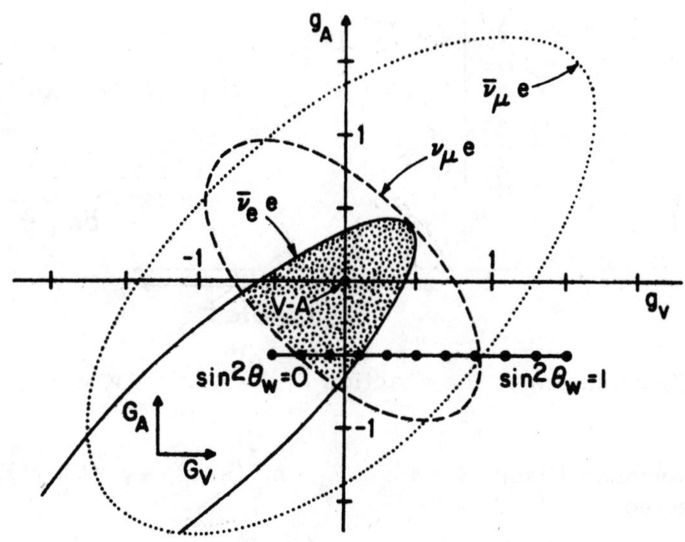

Fig. 15.
Various limits on g_A and g_V imposed on by $(\nu, \bar{\nu}) + e$ elastic scattering. From A. De Rujula et al., Revs. Mod. Phys. to be published.

5. Neutral Current in Nuclear Physics.

i) Nuclear coherent scattering of neutrinos.
Measures the strength of vector current, i.e.,

$$\langle I_3 - 2\sin^2\theta_W Q\rangle \equiv a \text{ in W-S.}$$

$$\sigma(100 \text{ MeV/c} < q_{recoil} < 300 \text{ MeV/c}) \text{ on } {}^{12}C$$

$$= a^2 \cdot 13.6 \times 10^{-39} \text{ cm}^2, \quad E_\nu > 1 \text{ GeV}$$

$$= a^2 \cdot 11.2 \times 10^{-39} \text{ cm}^2, \quad E_\nu \simeq 200 \text{ MeV}$$

[D. Z. Freedman, Phys. Rev. D, to be published.] Also astrophysical implications.!

ii) G.T. transitions by $\vec{A}^3$. [T. W. Donnelly et al., SLAC-Stanford preprint (1973).]

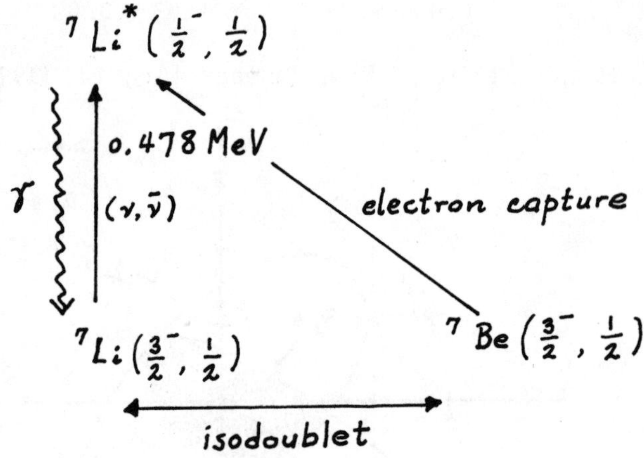

Fig. 16
Schematics of the reaction $(\nu, \bar{\nu}) + \text{Li} \rightarrow (\nu, \bar{\nu}) + \text{Li}^*$
$\rightarrow \text{Li} + \gamma.$

At Savannah River (2×10^{13} $\bar{\nu}_e/\text{cm}^2/\text{sec}$), 4γ's/day/Kg ^{7}Li are expected.

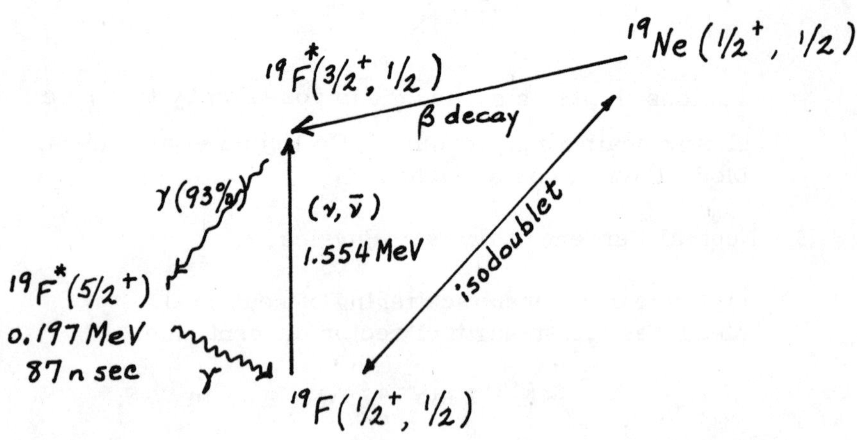

Fig. 17
Schematics of $(\nu, \bar{\nu}) + \text{F} \rightarrow (\nu, \bar{\nu}) + \text{F}^*$ and subsequent decays.

Delayed 2γ coincidence gives a unique signature!

iii) Giant dipole excitation. [Bilenky and Dadajan, Dubna

preprint (1973).] (V_0^3 and $\vec{A}^3$)

$$\sigma_{\nu+\bar{\nu}} \geq (1 - 2\sin^2\theta_W)^2\,\sigma_0 \quad (V_0^3 \text{ contribution}).$$

where σ_0 can be estimated from photoabsorption cross section.

σ_0 on a typical nucleus.

30 MeV	50	100
10^{-42}	$10^{-(41\sim 40)}$	10^{-39} cm^2

REMARKS ON NEUTRAL CURRENT INTERACTIONS*

J. J. Sakurai
University of California, Los Angeles, Ca. 90024

ABSTRACT

We discuss some experimental consequences of the hypothesis that the newly discovered neutral current may be identified with the SU(3) singlet baryon current.

The existence of a hadronic neutral current in high-energy neutrino interactions is, by this time, reasonably well established by four "independent" experiments.[1] Clearly our next major goal is to determine the symmetry properties of the newly discovered current. Even though most workers in this field are interested in analyzing the experimental data within the framework of unified gauge models à la Weinberg, Salam and Ward,[2] other models of neutral currents have not been ruled out. Last September I proposed a model based on the speculation that the hadronic neutral current may be identified with the SU(3) singlet baryon current,[3] which reads

$$j_\mu = i(\bar{u}\gamma_\mu u + \bar{d}\gamma_\mu d + \bar{s}\gamma_\mu s) \qquad (1)$$

in the quark model notation, accompanied possibly by its axial partner. Even though this model makes no attempt to unify the electromagnetic and weak interactions by introducing a mysterious angle, it has two features which may be considered attractive:

(i) The current is a unitary singlet invariant under a Cabibbo rotation

$$\begin{aligned} d &\to d\cos\theta + s\sin\theta, \\ s &\to -d\sin\theta + s\cos\theta, \end{aligned} \qquad (2)$$

hence no unnatural mechanism (e.g. a fourth quark with charm) is needed to eliminate unwanted $\Delta S = \pm 1$ currents.

(ii) The model may provide a framework for the simplest realization of "Universal Fermi Interactions."

I would like to discuss some experimental consequences of this model with special attention to the following two questions:[4]

(i) Is the interaction pure vector?

(ii) Is the current pure isoscalar?

*Supported in part by the National Science Foundation.

Let us start with the total cross section ratios. One of the things mentioned in my paper last September is the conjecture that the neutral current interaction may be of the pure vector type. As is well known, this can easily be tested; the absence of VA interference would imply

$$\sigma(\nu \to \nu) = \sigma(\bar{\nu} \to \bar{\nu}) \tag{3}$$

where $\sigma(\nu \to \nu)$ stands for the neutral-current inclusive cross section for ν on matter made up of equal numbers of protons and neutrons.[5] Notice that to the extent that the ratio of the $\bar{\nu}$ to ν cross section for <u>charged</u> current processes can be approximated by 1/3, Eq. (3) is equivalent to

$$R_{\bar{\nu}} = 3R_{\nu} \tag{4}$$

where

$$R_{\nu} \equiv \frac{\sigma(\nu \to \nu)}{\sigma(\nu \to \mu^-)}\ , \qquad R_{\bar{\nu}} \equiv \frac{\sigma(\bar{\nu} \to \bar{\nu})}{\sigma(\bar{\nu} \to \mu^+)}\ . \tag{5}$$

Shortly after preprints of my paper were circulated, I was notified by two members of the Gargamelle Collaboration (as well as by one of the referees of my paper) that the data of Hasert et al.[1] rule out the equality relation (3) by 1½ to 2 standard deviations. I understand, however, that the most recent results of the Harvard-Pennsylvania-Wisconsin Collaboration are quite consistent with (3). We'll hear more about this in the neutral current session tomorrow.

Turning now to specific final-state channels, let us first examine single-pion production. The soft pion theorem has important implications on

$$\nu + p \to \nu + \pi^+ + n, \tag{6a}$$

$$\nu + p \to \nu + \pi^o + p. \tag{6b}$$

Exploiting PCAC, one readily finds that the amplitudes for (6a) and (6b) are proportional to the nucleon matrix elements of $[Q_5^{1-i2}, j_\mu]$ and $[Q_5^3, j_\mu]$, respectively, where Q_5^α is the axial charge operator characterized by the isospin index α, and j_μ stands for the $\Delta S = 0$ neutral weak current. If j_μ is isoscalar as in my model, the above commutators vanish; so the soft pion theorem forbids single-pion production (6a) and (6b) near threshold. Furthermore it was already emphasized in my paper[3] that the baryon current model (or any pure isoscalar model) strictly forbids

$$\nu + p \rightarrow \nu + \Delta^{+} \, . \tag{7}$$

We therefore expect that <u>reactions (6a) and (6b) are highly suppressed at low energies</u>.

It was announced yesterday in the Washington Meeting by S. Barish[1] that the ANL Bubble Chamber Group now has several examples each of (6a) and (6b) with a cross section ratio compatible with pure $\Delta(1238)$ decay. Since the Argonne neutrino flux peaks at about 500 MeV, this does not look promising for my baryon current model. It is to be emphasized, however, that the model does not predict

$$\begin{aligned} &\nu + p \rightarrow \nu + \pi^{+} + \pi^{o} + n \, , \\ &\nu + p \rightarrow \nu + \pi^{o} + \pi^{o} + p \, , \\ &\nu + p \rightarrow \nu + \eta + p \, , \end{aligned} \tag{8}$$

to be suppressed. Since the π^{o} conversion efficiency is low in the ANL experiment, it is not inconceivable that these reactions may be confused with single-pion reactions (6a) and (6b). It is also worth examining if there are appreciable cross sections for the excitations of the nucleon into $P_{11}(1470)$, $S_{11}(1535)$ etc.

At energies above a few GeV diffractive ω production via Pomeron exchange

$$\nu + p(n) \rightarrow \nu + \omega + p(n) \tag{9}$$

may provide the best signature for the baryonic (or, more generally, isoscalar) vector current. In particular one may look for coherently produced ω mesons off complex nuclei in the Gargamelle data. Simple estimates based on comparison with ω electroproduction indicate that the cross section for (9) may be as large as ~10% of the total neutral current cross section for $E_{\nu} \gtrsim 5$ GeV, $q^2 \simeq 1$ GeV^2.

As first pointed out by Freedman,[6] coherence effects are expected also for purely elastic scattering off complex nuclei

$$\nu + A \rightarrow \nu + A \, . \tag{10}$$

Even though coherence is possible in the Weinberg model, the effect is predicted to be far more pronounced in the baryon current model. This may have some interesting applications in astrophysics.[7]

NOTE ADDED AFTER THE CONFERENCE

This note concerns a new absolute magnitude prediction for elastic νp scattering

$$\nu + p \rightarrow \nu + p \tag{11}$$

inferred from deep inelastic data.

Suppose we write the effective Lagrangian for the neutral current interaction as

$$L = (G\lambda/\sqrt{2})[\bar{\nu}\gamma_\mu(1+\gamma_5)\nu][\bar{u}\gamma_\mu u + \bar{d}\gamma_\mu d + \bar{s}\gamma_\mu s] \,. \tag{12}$$

At $q^2 = 0$ we have[8]

$$\left[(d\sigma/dq^2)_{\nu p \to \nu p} \Big/ (d\sigma/dq^2)_{\nu n \to \mu^- p} \right]\Bigg|_{q^2=0} = \frac{9\lambda^2}{1 + (g_A/g_V)^2} \tag{13}$$

where $(g_A/g_V)^2 \simeq 1.51$. To obtain the magnitude of λ we apply the standard quark-parton method (as described, for instance, in the summary talk by Feynman) to deep inelastic processes

$$\begin{aligned} \nu(\bar{\nu}) + N &\to \mu^-(\mu^+) + \text{any} \,, \\ \nu(\bar{\nu}) + N &\to \nu(\bar{\nu}) + \text{any} \end{aligned} \tag{14}$$

where N may stand for the proton or neutron. In the most naive version of quark parton models, the contributions from strange quarks and quark-antiquark pairs are ignored. We can then show that, apart from a common normalization factor, the differential cross-sections for (14) averaged over equal numbers of protons and neutrons are given by

$$\begin{aligned} \frac{d^2\sigma}{dxdy}(\nu \to \mu^-) &= [u(x) + d(x)]E \,, \\ \frac{d^2\sigma}{dxdy}(\bar{\nu} \to \mu^+) &= (1-y)^2\,[u(x) + d(x)]E \,, \\ \frac{d^2\sigma}{dxdy}(\nu \to \nu) &= \frac{d^2\sigma}{dxdy}(\bar{\nu} \to \bar{\nu}) \\ &= \lambda^2\left(1 - y + \frac{y^2}{2}\right)[u(x) + d(x)]E \end{aligned} \tag{15}$$

where u(x) and d(x) now stand for the distribution functions of the u-type and d-type quarks inside the proton. Notice that all four processes measure the same combination u(x) + d(x). Integrating (15), we readily obtain

$$\lambda^2 = \frac{3}{2}R_\nu = \frac{1}{2}R_{\bar{\nu}} = \frac{3}{4}(R_{\bar{\nu}} - R_\nu) \,, \tag{16}$$

a relation first derived by Rajasekaran and Sarma.[9] A more sophisticated analysis that does not ignore strange quarks and quark-antiquark pairs does not appreciably alter this result, as shown in a recent paper of McElhaney[10] who obtains

$$\lambda^2 = 1.25\ R_\nu\ ,$$
$$\lambda^2 = 0.535\ R_{\bar{\nu}}\ . \tag{17}$$

The Gargamelle and Harvard-Pennsylvania-Wisconsin data on R_ν and $R_{\bar{\nu}}$ discussed elsewhere in the Proceedings correspond to a value of λ^2 around 0.15 to 0.27. For definiteness we take

$$\lambda^2 = 0.2 \tag{18}$$

in the remaining part of this note.

When this value of λ^2 is inserted into (13) we obtain

$$\left[(d\sigma/dq^2)_{\nu p\to\nu p} \Big/ (d\sigma/dq^2)_{\nu n\to\mu^- p} \right]\Bigg|_{q^2=0} = 0.72\ . \tag{19}$$

At first sight it may appear that the predicted ratio is much too large. However, away from $q^2 = 0$, the VA interference term for the charged-current process $(\nu n \to \mu^- p)$ increases dramatically especially at low energies. As a result, the actual ratio of νp to $\mu^- p$ at moderate values of q^2 can be substantially smaller. Unfortunately the correction factor needed for finite values of q^2 depends on the shape of the various form factors as well as on q^2 and E. As a very crude estimate we may multiply the predicted ratio at $q^2 = 0$ by a suppression factor

$$\xi(E,\ q^2) = \frac{(d\sigma/dq^2)_{\nu n\to\mu^- p} + (d\sigma/dq^2)_{\bar{\nu} p\to\mu^+ n}}{2(d\sigma/dq^2)_{\nu n\to\mu^- p}}\ , \tag{20}$$

which should not be too bad if the various form factors fall off with q^2 in a similar manner. Typically, at E = 1 GeV, $q^2 = 0.5\ \text{GeV}^2$, we obtain from Fig. 11 of Llewellyn Smith's review article[11] $\xi \simeq 2/3$ which means

$$\left[(d\sigma/dq^2)_{\nu p\to\nu p} \Big/ (d\sigma/dq^2)_{\nu n\to\mu^- p} \right]\Bigg|_{q^2=0.5\ \text{GeV}^2,\ E=1\ \text{GeV}} = 0.48 \tag{21}$$

This still appears higher than the <u>upper limit</u> of Cundy et al.[12] (CERN propane data)

$$\frac{\sigma(\nu p \to \nu p)}{\sigma(\nu n \to \mu^- p)} \leq 0.12 \pm 0.06 \tag{22}$$

obtained under similar kinematical conditions. So we may have yet another reason for abandoning the simple baryon current model.

We often hear, however, that any neutral current limit quoted prior to 1973 is now suspect. We eagerly await new results from the ANL Bubble Chamber Group promised by M. Derrick for the London Conference.

REFERENCES

1. F. J. Hasert et al., Phys. Lett. 46B, 138 (1973);
   A. Benvunuti et al., Phys. Rev. Lett. 32, 800 (1974);
   B. Aubert et al. (Harvard-Pennsylvania-Wisconsin preprint);
   S. J. Barish, Bull. Am. Phys. Soc. 19, 566 (1974).
2. S. Weinberg, Phys. Rev. Lett. 19, 1264 (1967);
   A. Salam and J. C. Ward, Phys. Lett. 13, 168 (1964).
3. J. J. Sakurai, Phys. Rev. D9, 250 (1974).
4. Further consequences of this model are discussed in S. Pakvasa and S. F. Tuan, Phys. Rev. (to be published).
5. The equality relation (3) also holds if the interaction turns out to be of the scalar and/or pseudoscalar type. However, as pointed out during the Conference, if the neutral current cross section rises linearly with E (as seems to be the case experimentally), the differential cross section in the scalar and/or pseudoscalar case can be shown to be of the form

   $$\frac{d^2\sigma}{dxdy} = xy^2\ F(x)E$$

   where F(x) is the scalar structure function that satisfies Bjorken scaling. Notice that the y distribution in this case is completely different from what is predicted by any linear combination of V and A. The heretical possibility that the neutral current interaction may involve S, P and/or T is being examined in greater detail by S. P. Rosen, B. Kayser and collaborators.
6. D. Z. Freedman, Phys. Rev. (to be published).
   See also A. Pais and S. B. Treiman, Phys. Rev. (to be published).
7. J. R. Wilson, Phys. Rev. Lett. 32, 849 (1974).
8. I would like to take this opportunity to correct an error in Reference 3. The right-hand side of (16) should be multiplied by 9. This is because the baryonic charge of the proton is three times that of the quark. I am indebted to Dr. S. Pakvasa for pointing this out.

9. G. Rajasekaran and K. V. L. Sarma, Tata Institute preprint TIFR/TH/73-53.
10. R. McElhaney, University of Wisconsin preprint.
11. C. Llewellyn Smith, Phys. Reports 3C, 264 (1972). This particularly useful figure is taken from M. Derrick, Argonne preprint ANL/HEP 7017 (1970).
12. D. C. Cundy et al., Phys. Lett. 31B, 478 (1970).

NOTES ON CHARMED PARTICLE SEARCHES IN NEUTRINO EXPERIMENTS

Mary K. Gaillard *
National Accelerator Laboratory, Batavia, Il. 60510

WHY CHARM?

Renormalizability rests on gauge invariance which in turn requires a weak coupling of the form:

$$\mathscr{L}_W \sim g\left(J^+_\mu W^-_\mu + J^-_\mu W^+_\mu + J^0_\mu W^0_\mu\right) \tag{1}$$

where the neutral current is determined by the equal time commutator of the charged currents:

$$J^0_\mu = \frac{1}{\sqrt{2}}\left[J^+_\mu, J^-_0\right]_{e.t.} \quad . \tag{2}$$

From low energy phenomenology we know that the charged currents must contain, at least, the Cabibbo currents and the lepton currents:

$$J^+ = \bar{p}(n \cos\theta_c + \lambda \sin\theta_c) + \bar{\nu}_\mu \mu + \bar{\nu}_e e + (?) \quad . \tag{3}$$

If no other terms are added, the neutral current obtained from the commutator will contain a term:

$$\cos\theta_c \sin\theta_c (\bar{n}\lambda + \bar{\lambda}n) \tag{4}$$

which would induce the decay $K_L \to \mu\mu$ at a rate comparable to $K^+ \to \mu\nu$. Following Glashow, Iliopoulos and Maiani[1] (GIM), we rectify this unwanted feature by adding another piece to the charged current[2]:

$$(?) = \bar{p}'(\lambda \cos\theta_c - n \sin\theta_c) \tag{5}$$

where p' is a fourth quark carrying the charge of the proton, zero isotopic spin and strangeness, and a new quantum number called charm.

*On leave of absence from Laboratoire de Physique Théorique et Particules Elementaires, Orsay (Laboratoire associé au CNRS).

Many other constructions are possible (involving new leptons as well as quarks) which will eliminate the unwanted $|\Delta S| = 1$ piece of the neutral current. However, we shall limit our discussion to the above model which has several attractive features:

a) It is compatible with the observed hadron spectroscopy, as the Gell-Mann Zweig quark model[3] is unmodified. However charmed states, filling out SU_4 representations, are expected to appear in a higher mass region.

b) It is the most economical model. Once new particles have been introduced to obtain the desired structure of the neutral current, it is generally necessary to add still more particles in order to suppress higher order contributions to $|\Delta S| = 1$, $\Delta Q = 0$ transitions. In the above model, the p' does the job in every order.

c) This model is not yet ruled out by experiment. The leptonic couplings are those of the Weinberg-Salam[4] model which appears to be compatible with present data from neutrino experiments.

MASSES OF CHARMED PARTICLES

Consideration of the p, p' cancellation mechanism in higher order weak and/or electromagnetic processes places severe limits on the masses of charmed particles. The second order weak amplitudes for K decay into neutral lepton pairs are of the order

$$A(K_L \to \bar{\mu}\mu, K \to \pi \bar{\nu}\nu, \text{etc.}) \sim G_F \alpha \left(\Delta m^2 / m_W^2 \sin^2 \theta_W\right) \ln\left(m_W^2 / m^2\right) \qquad (6)$$

where $\Delta m^2 = m_{p'}^2 - m_p^2$, m is the largest quark mass, and θ_W is the Weinberg angle. The $K_L - K_S$ mass difference is of the order:

$$m_{K_L} - m_{K_S} \sim G_F \alpha \Delta m^2 / m_W^2 \sin^2\theta_W \quad , \qquad (7)$$

and the amplitude for $K_L \to \gamma\gamma$ is of the order

$$A(K_L \to \gamma\gamma) \sim G_F \alpha \Delta m^2 / m^2 \quad . \qquad (8)$$

The experimental observation that the first two amplitudes are highly suppressed with respect to $G_F\alpha$ while the third is not suppressed, tells us that the p, p' mass difference must be of the same order as the p' mass, but small compared to the mass scale of weak interactions:

$$\Delta m^2 \sim m_{p'}^2 << m_W^2 \sin^2\theta \approx (38 \text{ GeV})^2 \quad . \qquad (9)$$

Explicit evaluation of the above amplitudes using free quark diagrams gives the constraints:[5]

$$m_p \lesssim m_K << m_{p'} \lesssim 2\text{GeV} \quad . \tag{10}$$

Similar estimates based on pseudoscalar exchange rather than quark exchange indicate limits on the charmed pseudoscalar masses of the order

$$m_K << m_c \lesssim 5\text{GeV} \quad . \tag{11}$$

Since charmed particles have not been seen, and since too high a mass would imply the breakdown of the cancellation mechanism, we expect that charmed particle masses should lie in the range

$$2\text{GeV} \lesssim m_c \lesssim 10 \text{ GeV} \quad . \tag{12}$$

(Since quark masses are apparently small compared to hadronic masses, the PCAC mass formulae, relating squares of pseudoscalar masses to quark masses, may be relevant:

$$2m_p/m_\pi^2 \simeq m_\lambda/m_K^2 \simeq m_{p'}/m_c^2 \quad . \tag{13}$$

Then the p, p′ mass ratio must be extremely small:

$$m_p/m_{p'} \lesssim 1/200. \;) \tag{14}$$

The relevant conclusion of this discussion is that, if charm is to adequately account for the observed phenomenology of K-decay, charmed particles must be light enough to be produced at NAL energies. Indirect tests for the presence of charmed particles include:

a) threshold effects

and b) modification of the Adler sum rule.

However decisive tests must be direct, namely

c) the identification of a final state signature.

THRESHOLD EFFECTS

Near the threshold for the production of a charmed particle the cross section is expected to have the energy dependence[6]

$$\sigma_c \sim (E - E_{Th})^2 \tag{15}$$

where the threshold energy is related to the charmed particle mass by $(m_N << m_c)$:

$$E_{Th} \simeq m_c^2/2m_N \tag{16}$$

and m_N is the physical nucleon mass. Each channel will open slowly, but one expects many new channels to be appearing rapidly, with no startling effect on the energy dependence. Well above threshold the usual linear behavior, $\sigma_c \sim E$, is expected to set in again.

The allowed regions of phase space can be determined in terms of the usual kinematic variables $(m_N << E)$:

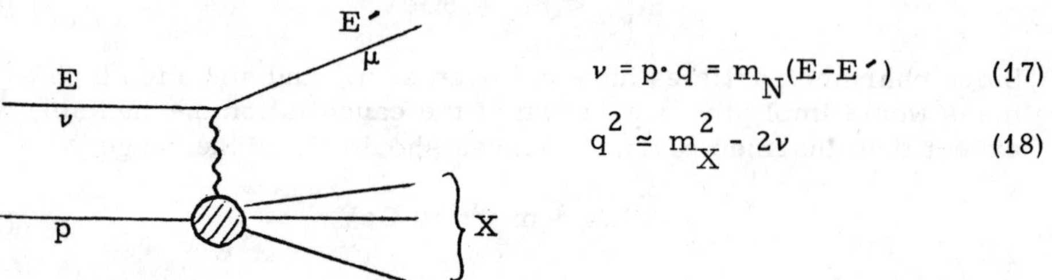

$$\nu = p \cdot q = m_N(E - E') \tag{17}$$

$$q^2 \simeq m_X^2 - 2\nu \tag{18}$$

or the "scaling" variables:

$$y = \nu / p \cdot p_\nu \quad , \qquad x = -q^2/2\nu \quad , \qquad 0 \leq x, y \leq 1. \tag{19}$$

The variable y is proportional to the total hadron energy in the final state; at threshold all the incident energy is used to produce the charmed particle:

$$E'_{Th} = 0 \quad ; \quad E_X^{Th} = E_{Th} \quad , \tag{20}$$

so above threshold:

$$y = (E - E')/E = E_X/E \geq E_{Th}/E \quad . \tag{21}$$

The variable x is related to the total invariant mass of the final state hadronic system:

$$x = -q^2/2\nu \simeq 1 - m_X^2/2\nu \leq 1 - m_c^2/2\nu < 1 - E_{Th}/E \quad . \tag{22}$$

Thus for energies near threshold charmed particle production occurs only in the small x, large y region.[7]

ADLER SUM RULE

The Adler sum rule[8] relates the difference of neutrino and anti-neutrino cross sections to the matrix element of the commutator of the weak charges. For the charge exchange process:

$$\nu(\bar{\nu}) + Z \to \mu^{\mp} + X \tag{23}$$

we have the sum rule:

$$\lim_{E\to\infty} \frac{\pi}{G_F^2}\left(\frac{m_W^2 - q^2}{m_W^2}\right)^2 \left[\frac{d\sigma^{\bar{\nu}}}{d|q^2|} - \frac{d\sigma^{\nu}}{d|q^2|}\right]_Z = \langle Z|[Q^+, Q^-]|Z\rangle \ . \tag{24}$$

For the usual Cabibbo theory the weak charges are:

$$Q^+ = p^{\dagger}(n \cos\theta_c + \lambda \sin\theta_c) = (Q^-)^{\dagger} \tag{25}$$

giving for the commutator:

$$\left[Q^+, Q^-\right] = 2I_3 + \sin^2\theta_c\left(\frac{3}{2}Y - I_3\right) \tag{26}$$

+ parity and/or strangeness changing terms.

With the GIM modification the commutator becomes:

$$\left[Q^+, Q^-\right] = 2I_3 - S + C + \text{parity odd terms.} \tag{27}$$

Since the target nucleus carries no charm (C) or strangeness (S), the effect of the modification is proportional to $\sin^2\theta_c \simeq 0.04$. However since for a heavy nucleus:

$$Y = B = N_p + N_n >> 2I_3 = N_p - N_n \tag{28}$$

the effect will be enhanced. Bég and Zee[9] found for iron (I_3=-2, Y=56):

$$<Z|\left[Q^+, Q^-\right]|Z> = \begin{cases} -0.56 & \text{Cabibbo} \\ -4 & \text{GIM} \end{cases} \ . \tag{29}$$

PRODUCTION OF CHARMED PARTICLES IN NEUTRINO EXPERIMENTS

Since the basic quark transitions are

$$p' \to \lambda \qquad \sim \cos\theta_c \tag{30}$$

$$p' \to \lambda \qquad \sim \sin\theta_c \tag{31}$$

charmed particles will decay predominantly into strange particles, providing a characteristic signature. However, the same amplitude ratio works against us in the production process. The elementary transitions are

$$\nu + n \rightarrow p' + \mu^- \qquad \sim \sin\theta_c \tag{32}$$

$$\nu + \lambda \rightarrow p' + \mu^- \qquad \sim \cos\theta_c \tag{33}$$

Since there are probably few λ's in the nucleon, production rates are expected to be damped.

In order to get some quantitative estimates, we use the parton model and assume that it is relevant at Gargamelle energies so that we may use existing data to determine the parton content of nucleons.[10, 11]

If $f_q(x) = xp_q(x)$ is the distribution function for quark q in the proton, weighted by its momentum fraction, we define the integral:

$$F_q \equiv \int dx \left[f_q(x) + f_{\bar{q}}(x)\right] \quad . \tag{34}$$

From the sum of electromagnetic cross sections on proton and neutron we can determine the quantity:

$$F_p + F_n + \frac{2}{5} F_\lambda + \frac{8}{5} F_{p'} \simeq 0.50 \pm 0.05 \quad . \tag{35}$$

On the other hand, the sum of neutrino and anti-neutrino cross sections for $\Delta S = 0$ transitions on heavy nuclei $\left[\sigma^{\nu Z}/A \simeq (\sigma^{\nu p} + \sigma^{\nu n})/2\right]$ determines the combination:

$$F_p + F_n + \tan^2\theta_c F_{p'} = 0.505 \pm 0.015 \tag{36}$$

Upon comparison of these two quantities, the positivity of the distribution functions, $F_q \gtrsim 0$, implies that the λ and p′ distributions in the nucleon are small. Specifically:

$$F_\lambda/(F_p+F_n) \lesssim 0.25 \ , \quad F_{p'}/(F_p+F_n) \lesssim 0.06 \quad . \tag{37}$$

A further indication that few p′'s are present in the nucleon is the fact that no surplus of events with $\Delta S = \pm 1$ has been observed[12], since a p′ converts most readily into a λ.

A further piece of information from the neutrino experiments is the relative number of anti-quark nucleons in the physical nucleon. Since particle-anti-particle cross sections for point particles are a third of particle-particle cross-sections, the total cross section ratio:

$$\frac{\sigma^{\bar{\nu}}}{\sigma^{\nu}} = \frac{F_{\bar{N}} + 1/3\ F_N}{F_N + 1/3\ F_{\bar{N}}} = 0.38 \pm 0.02 \tag{38}$$

determines the relative anti-quark content of the nucleon:

$$F_{\bar{N}}/F_N = 0.05 \pm 0.02 \tag{39}$$

where we have defined

$$F_N = \int dx \left[f_n(x) + f_p(x) \right] \tag{40}$$

$$F_{\bar{N}} = \int dx \left[f_{\bar{n}}(x) + f_{\bar{p}}(x) \right] \tag{41}$$

and have assumed $F_{p'} = 0$.

On the grounds that we expect the λ and $\bar{\lambda}$ content to be roughly equal $\left[\int \left(p_\lambda(x) - p_{\bar{\lambda}}(x) \right) dx = 0 \right]$, we shall assume the following:

$$F_{p'} \simeq 0 \ , \quad F_{\bar{N}} \simeq 0.05\ F_N \tag{42}$$

$$\int f_\lambda dx \simeq \int f_{\bar{\lambda}}\ dx \simeq F_\lambda/2 \lesssim 0.1\ F_N \quad .$$

Further, we assume an incident energy well above threshold so that the parton model is applicable:

$$E >> E_{Th} \tag{43}$$

and we assume that charmed states are not sufficiently long lived for their decay paths to be observable:

$$\gamma\beta c\tau << \text{cm} \quad . \tag{44}$$

On general dimensional grounds, since

$$\Gamma \sim \text{mass} \sim G_F^2 (\text{mass})^5 \quad , \tag{45}$$

we expect widths for charmed particle decays to scale with respect to widths for strange particle decays by a factor:

$$\Gamma_c \simeq \Gamma_S (m_c/m_s)^5 \cot^2\theta_c \quad . \tag{46}$$

There is an additional enhancement from the Cabibbo angle; since strange particle life times are typically of the order of 10^{-10} sec , if $m_c/m_s \gtrsim 2$, we have

$$\tau_c \lesssim 10^{-13} \text{ sec} , \quad c\tau_c \lesssim 3 \times 10^{-3} \text{ cm} . \tag{47}$$

In the following, we discuss possible signatures for charmed particle production, using the parton model results as input.

AN OPTIMISTIC ASSUMPTION: $F_\lambda \neq 0$

If there are as many as 10% of λ's and 10% of $\bar{\lambda}$'s in the nucleon, there will be an appreciable production of charmed particles with two characteristic signatures.

a) Enhancement of associated production. The elementary production and decay processes are (in terms of quarks):

$$\nu + \lambda \to \mu^- + p' \;\; \llcorner\!\!\to \lambda + \bar{n}p \tag{48}$$

and the charge conjugate process. Typical events will be of the form:

$$\nu + Z \to \mu^- + K^+ + (K^- \text{ or } Y) + X \qquad (\lesssim 17\% \text{ of total } \sigma^\nu) \tag{49}$$

$$\bar{\nu} + Z \to \mu^+ + K^+ + (K^- \text{ or } Y) + X \qquad (\lesssim 33\% \text{ of total } \sigma^{\bar{\nu}}) \tag{50}$$

where X represents non strange hadronic matter.

The increase in associated production could be observed as a threshold effect which would be about three times more prominent in anti-neutrino events. This is because the cross section for $\bar{\nu} + \bar{\lambda} \to \bar{p}' + \mu^+$ does not have the 1/3 suppression factor.

b) associated production accompanied by a di-lepton. In this case the p′ decays into leptons:

$$p' \to \lambda + \ell^+ \nu \quad . \tag{51}$$

Typical events are:

$$\nu + Z \to \mu^- \ell^+ + K\bar{K} + X \qquad (E^+ > E^-) \tag{52}$$

$$\bar{\nu} + Z \to \mu^+ \ell^- + K\bar{K} + X \qquad (E^- > E^+) \tag{53}$$

Near threshold most of the incident energy goes into producing the charmed particle; thus the decay lepton is expected to be faster than the production lepton.[13] This process provides a clearer signature, but it will be suppressed by the branching ratio for leptonic decay which may be small.

STRANGENESS CHANGING EVENTS

Independently of the λ content of the nucleon, one expects charmed particles to be produced at the level of $\sin^2\theta_c$. Again there are two characteristic signatures.

a) Apparent violation of the $\Delta S = \Delta Q$ rule. The elementary process for ν events is:

$$\nu + n \rightarrow p' + \mu^- \quad (p' \rightarrow \lambda + p\bar{n}) \tag{54}$$

which globally satisfies $\Delta S = -\Delta Q$. This process is enhanced relative to the usual $\Delta S = \Delta Q$ process which must occur by ν scattering on an anti-proton or a λ :

$$\nu + \begin{pmatrix}\lambda\\ \bar{p}\end{pmatrix} \rightarrow \begin{pmatrix}p\\ \bar{\lambda}\end{pmatrix} + \mu^- \tag{55}$$

The situation is reversed for $\bar{\nu}$ events, where charmed particle production:

$$\bar{\nu} + \bar{n} \rightarrow \bar{p}' + \mu^+ \quad (\bar{p}' \rightarrow \lambda + n\bar{p}) \tag{56}$$

is suppressed for want of anti-partons, but the $\Delta S = \Delta Q$ process is allowed:

$$\bar{\nu} + \begin{pmatrix}p\\ \bar{\lambda}\end{pmatrix} \rightarrow \begin{pmatrix}\lambda\\ \bar{p}\end{pmatrix} + \mu^+ \tag{57}$$

Using the parton content assumed above, we find the following cross sections relative to the total ν or $\bar{\nu}$ cross sections:

$$\sigma^{\nu}(\Delta S = \Delta Q) \lesssim 0.8\% \tag{58}$$

$$\sigma^{\nu}(\Delta S = -\Delta Q) \simeq 4\% \tag{59}$$

$$\sigma^{\bar{\nu}}(\Delta S = \Delta Q) \simeq 4\% \tag{60}$$

$$\sigma^{\bar{\nu}}(\Delta S = -\Delta Q) \simeq 0.6\% \tag{61}$$

b) Change of strangeness accompanied by a di-lepton. The leptonic decay of the charmed particles provides a unique signature:

$$\nu + Z \rightarrow \mu^-_{(slow)} + \ell^+_{(fast)} + (K^- \text{ or } Y) + X \tag{61}$$

$$\bar{\nu} + Z \rightarrow \mu^+_{(slow)} + \ell^-_{(fast)} + K^+ + X \quad . \tag{62}$$

However this process is again suppressed by the leptonic decay branching ratio:

$$\sigma^{\nu}/\sigma^{\nu}_{Total} \simeq 0.04\ B_{\ell} \tag{63}$$

$$\sigma^{\bar{\nu}}/\sigma^{\bar{\nu}}_{Total} \simeq 0.006\ B_{\ell} \tag{64}$$

WHAT IS THE LEPTONIC BRANCHING RATIO?

The strangeness changing leptonic and non-leptonic decays are a priori of the same order of magnitude:

$$p' \rightarrow \lambda + \ell^{+} \nu \qquad \sim \cos\theta_c \tag{65}$$

$$p' \rightarrow \lambda + p\bar{n} \qquad \sim \cos^2\theta_c \tag{66}$$

However the same is true for the decays of strange particles:

$$\lambda \rightarrow p + \ell^{-} \nu \qquad \sim \sin\theta_c \tag{67}$$

$$\lambda \rightarrow p + n\bar{p} \qquad \sim \sin\theta_c \cos\theta_c \tag{68}$$

while the experimentally determined amplitudes for non-leptonic decays are effectively of order 1 . Until the mechanism for the enhancement of these amplitudes is understood, one cannot predict whether it will also play a role in the decays of charmed particles.

If we consider quarks as very light and quasi-free within hadronic states, the fundamental decay mechanism is just a four fermion coupling; the partial width is proportional to the fifth power of the energy release:

$$\Gamma \sim G_F^2 Q^5 \tag{69}$$

If, as suggested by PCAC and the analysis of $\Delta S = 1$, $\Delta Q = 0$ amplitudes,

$$m_{\ell}\ ,\ m_p,\ m_n << m_{\lambda} << m_{p'} \tag{70}$$

we have $Q \simeq m_{\lambda}$ and $Q \simeq m_{p'} - m_{\lambda}$ for both leptonic and non-leptonic decays of strange and charmed particles, respectively. In any case, we expect both leptonic and non-leptonic decay rates to scale in the same way, roughly as the fifth power of the charmed to strange mass ratio.

For strange particle decays the leptonic branching ratio is very small; typically:

$$B_\ell^S \sim 10^{-3} \sim \sin^2\theta_c \, x \text{ (3-body phase space)} \quad . \tag{71}$$

All observed leptonic decay modes are suppressed either by three body phase space or by forbidden helicity states as in π or $K \to \ell\nu$. Phase space suppression should not be important in the decay of heavy charmed states. Therefore if the enhancement of non-leptonic amplitudes is operative for charmed particle decay, we expect:

$$B_\ell^c \sim \sin^2\theta_c \sim 4\% \tag{72}$$

In the case of no enhancement the leptonic branching ratio could be much higher:

$$B_\ell^c \sim 50\% \tag{73}$$

However we should not count on such a high leptonic decay rate.

SUMMARY OF SIGNATURES

We recapitulate the characteristic signatures for charmed particle production (specific to the Weinberg-Salam model as modified by GIM) in neutrino experiments.

a) Increased associated production appearing in the low x, high y region. Above threshold the effect could rise to, say, 10% of the total ν cross section and 20% of the total $\bar{\nu}$ cross section if the λ, $\bar{\lambda}$ sea comprises 10% of the parton distribution.

b) Associated production accompanied by a di-lepton: slow μ^-, fast ℓ^+ in ν events; slow μ^+, fast ℓ^- in $\bar{\nu}$ events. The effect is three times more prominent in $\bar{\nu}$ events, but is suppressed by the leptonic branching ratio.

c) Apparent violation of the $\Delta S = \Delta Q$ rule. Well above the charm threshold energy we would see 4% of $\Delta S = -1$ events and about half a percent of $\Delta S = +1$ events in both ν and $\bar{\nu}$ beams.

d) Change of strangeness accompanied by a di-lepton: slow μ^-, fast ℓ^+, $\Delta S = -1$ in ν events; slow μ^+, fast ℓ^-, $\Delta S = +1$ in $\bar{\nu}$ events. The effect should be about 10 times more prominent in the ν beam, but suppressed by the leptonic branching ratio.

The first two signatures depend on the presence of λ-partons in the nucleon and may be entirely absent. If the GIM mechanism is the correct one, signatures c) and d) must be present at the level of $\sin^2\theta_c \sim 4\%$ at NAL energies.[14]

These notes behefitted from lively discussions with B. W. Lee, E. Paschos and G. Snow.

REFERENCES

1. S.L. Glashow, J. Iliopoulos, L. Maiani, Phys. Rev. D2, 1285 (1970).
2. S. Weinberg, Phys. Rev. Letters 27, 1688 (1971).
3. M. Gell-Mann, Phys. Letters 8, 214 (1964); G. Zweig, CERN preprints TH 401, 412 (1964).
4. S. Weinberg, Phys. Rev. Letters 19, 1264 (1967); A. Salam, Proc. Eighth Nobel Symposium (Almquist and Wiksell, Stockholm, 1968).
5. E. Ma, Phys. Rev. (to be published), M.K. Gaillard and B.W. Lee, NAL preprint 74/21-THY (Submitted to Phys. Rev.).
6. For a two body final state and neglecting the lepton mass, the c.m. energy of the final state lepton is: $k'_0 \simeq \sqrt{m_p/E}(E-E_{th})/2$. There is one factor of $k'_0 = |\vec{k}'|$ from phase space and one from the leptonic structure function: $\ell_{\mu\nu} = T_r \not{k}\gamma_\mu \not{k}' \gamma_\nu(1-\gamma_5)$.
7. A. De Rujula, Howard Georgi, S.L. Glashow and H.R. Quinn, "Fact and Fancy in Neutrino Physics", to be published in Rev. Mod. Phys.
8. S.L. Adler, Phys. Rev. 143, 1144 (1966).
9. M.A.B. Bég and A. Zee, Phys. Rev. D8, 2334 (1973). See also A. DeRujula and S. Glashow, Phys. Letters 46B, 381 (1973).
10. The neutrino data used here is taken from P. Musset, Proc. 2nd Aix-en-Provence Conf. on Elementary Particles, Sept. 1973.
11. Similar estimates for charmed particle production in neutrino experiments have been made by G. Altarelli, N. Cabibbo and L. Maiani, Phys. Letters B48, 435 (1974).
12. D. Perkins, Proc. XVI International Conf. on High Energy Physics, Vol. 4, 189 (Chicago-Batavia, 1972).
13. G. Snow has pointed out that this effect will tend to be washed out due to the fact that production near threshold is suppressed [$\sigma_{th} \sim (\Delta E)^2$]. Furthermore, the heavier charmed states will lose much of their energy through a strong decay cascade before the lepton is emitted by the lowest energy state. He estimates that if charmed masses are as low as 2 or 3 GeV the lepton energy distributions, averaged over NAL neutrino energies, will be roughly the same.
14. More general searches for charmed particles have been discussed by G. Snow, Nucl. Phys. B55, 445 (1973) and Electromagnetic and Weak Interactions (8th Rencontre de Moriond), 379 (CNRS, 1973), and by J. Rosner (unpublished).

NEUTRINO CROSS-SECTIONS AT VERY HIGH ENERGIES†,††

A. Zee*
Princeton University, Princeton, N. J. 08540

ABSTRACT

The implications of asymptotic freedom for neutrino experiments at very high energies are discussed.

I would like to describe some results of work done in collaboration with S. Treiman and F. Wilczek. Since Professor B. Lee has just reviewed for you the kinematics of electro- and neutrino-production in the deep inelastic region, I need only remind you that $1 \leq \omega \equiv 2\nu/Q^2 < \infty$ and $0 < x = Q^2/2\nu \leq 1$. Our theoretical starting point is the discovery[1] made by Gross, Wilczek, Politzer and 't Hooft last year that in a Yang-Mills theory of strong interaction something can actually be said about the structure functions, namely that for large Q^2 the moments

$$F_n(Q^2)=\int_0^1 dx\, x^n F_2(x,Q^2) = C_n[\lambda(Q^2)]^{-a_n}\,[1 + 0(\frac{1}{\log Q^2})] \tag{1}$$

Here $\lambda(Q^2) \equiv \log(Q^2/\mu^2)/\log(Q_0^2/\mu^2)$. The important thing to note is that a_n is a theoretically computable and known function of n. On the other hand, C_n is theoretically unknown but may be considered as known experimentally to the extent that $F_2(x,Q_0^2)$ has been measured. μ^2 is a scale that appears in the renormalization group equation. Its precise value is not known but is expected to be small since scaling sets in so rapidly at SLAC. We will assume that a value for Q_0^2 may be chosen, once and for all, such that for $Q^2 > Q_0^2$ the correction term to Eq. (1), indicated to be an inverse power of $\log Q^2$, is negligible. This is purely an optimistic assumption and may well be wrong. We will return to it later.

In summary, Eq. (1) allows us to generate the moments of the structure function for $Q^2 > Q_0^2$ once they are known for $Q^2 = Q_0^2$.

The a_n's are simply related to the anomalous dimensions of operators of spin $J = n+2$. These operators may be classified as singlet or non-singlet under some internal symmetry group (SU(2) or SU(3)). The singlet operators govern the behavior of $F_2{}^{e(p+n)}$ and $F_2{}^{\nu+\bar{\nu}}$, for example, while the non-singlet operators govern

†This is an expanded version of a talk given at the IVth International Conference on Neutrino Physics and Astrophysics.

††Work supported in part by the National Science Foundation under Grant No. GP-3078X.

*Alfred P. Sloan Foundation Fellow

$F_2^{e(p-n)}$ and $F_2^{\nu-\bar{\nu}}$. Let us discuss the singlet a_n's. We are more interested in their functional dependence on n. Their precise numerical values depend on the gauge group chosen. The singlet a_n's have the following properties[2]:

(I) $a_o=0$. This follows from the theorem that the stress-energy tensor can have no anomalous dimension[3]

(II) $a_n > 0$ for positive even integers. This fact together with (I) implies that $F_2(x,Q^2)$ must fall for $x \sim 1$ and rise for $x \sim 0$ as Q^2 increases.

(III) $a_n \to$ const log n as $n \to \infty$. This is characteristic of gauge theories, abelian as well as non-abelian.

(IV) Since $F_2 \geq 0$ we may analytically continue $F_n(Q^2)$ into the complex plane. Standard Regge lore suggests that $F_2(\omega,Q_o^2) \to$ constant as $\omega \to \infty$ for fixed Q^2, so that c_n will have some singularity, perhaps a simple pole or a fixed cut, at n = -1. We find by consulting Ref. 2, that

$$a_n \to \frac{-a}{n+1} \quad \text{as} \quad n \to -1 \text{ with } a > 0 \tag{2}$$

We now recall Eq. (1) and see that this simple pole gets transmuted into an essential singularity of the moments $F_n(Q^2)$ by the renormalization group! This essential singularity controls the large ω or small x behavior of $F_2(x,Q^2)$. This behavior ought to be determinable by approximating Eq. (1) for $n \approx -1$ and inverting. Thus we set $C_n = \frac{c}{n+1}$ and $a_n = \frac{-a}{n+1}$ in Eq. (1) and obtain

$$F_2(\omega,Q^2) \xrightarrow[\omega \to \infty]{} \frac{c}{2\sqrt{\pi}} \frac{e^{2\sqrt{a \log\lambda(Q^2)\log \omega}}}{[a \log \lambda(Q^2)\log\omega]^{\frac{1}{4}}} \tag{3}$$

(The choice for C_n is arbitrary. Different choices only affect the denominator in Eq. (3). See Ref. 4 for details.) The exponential in Eq. (3), which is a direct consequence of the essential singularity we discussed, says that $F_2(\omega,Q^2)$ grows faster than any power of log ω (but slower than any power of ω) as $\omega \to \infty$. It would be interesting to see whether or not forthcoming experiments in electroproduction confirms this peculiar behavior. This phenomenon is characteristic of gauge theories and may be attributed to the presence of spin 1 gauge fields, which produce in a_n a pole at n = -1 (i.e. J = 1).

The non-singlet a_n's, in contrast, do not have a pole at n=-1. This leads to a suggested behavior perhaps like $F_2^{e(p-n)}(\omega,Q^2) \to \omega^{-\frac{1}{2}}(\log Q^2/\mu^2)^c$ with c > 0. At large ω the singlet operators dominate and thus

$$F_2^{\nu t}(\omega,Q^2) = F_2^{\bar{\nu}t}(\omega,Q^2) = 3F_2^{et}(\omega,Q^2) \quad \text{(large } \omega) \tag{4}$$

where t = any target. These matters, as well as further details, are discussed in a recent paper[4].

I will now discuss the implications of these considerations for neutrino experiments at high energies. Let us recall a standard kinematical formula

$$\frac{d\sigma^{(\nu/\bar{\nu})}}{dx\, dy} = \frac{G^2 ME}{\pi} \left[(1-y+\frac{y^2}{2}) F_2(x,y) \mp y \left(1-\frac{y}{2}\right) xF_3(x,y) \right] \quad (5)$$

The variable $y \equiv \nu/mE = Q^2/2mEx$ varies between 0 and 1. In writing down Eq. (5) we have used the Callan-Gross relation and charge symmetry. If Bjorken scaling holds, then F_2 and F_3 are independent of y and $d\sigma/dy$ and σ are controlled by the moments $\int_0^1 dx\, F_2(x)$ and $\int_0^1 dx\, x\, F_3(x)$. In particular

$$\left(\frac{\pi}{G^2M}\right) \frac{1}{E} \sigma^{\nu,\bar{\nu}} = \frac{2}{3} \int_0^1 dx\, F_2(x) \mp \frac{1}{3} \int_0^1 dx\, x\, F_3(x) \quad (6)$$

In the CERN neutrino experiments[5] the cross-section indeed rises linearly with E. Why this should happen already at such low energies $E \lesssim 10$ GeV is a mystery since, as we shall see, even at very high energy the behavior of the structure functions at small Q^2 influences the cross-section substantially. The CERN experiment also finds that $\sigma^{\nu}/\sigma^{\bar{\nu}} \simeq 3$ which implies that $\int_0^1 dx\, xF_3(x) = -\int_0^1 dx\, F_2(x)$ and thus $xF_3 = -F_2$ since $|xF_3| \leq F_2$ by positivity. This leads to the prediction

$$\frac{1}{E} \frac{d\sigma^{\nu,\bar{\nu}}}{dy} \propto \begin{cases} 1 & \text{for } \nu \\ (1-y)^2 & \text{for } \bar{\nu} \end{cases} \quad (7)$$

The present experimental evidence is consistent with a remarkably simple picture based on the parton model. I refer you to Professor Feynman's talk at this conference for a review. I would now like to discuss deviations from this lovely picture.

To begin with, the cross-sections are no longer given simply in terms of moments. Instead, one has

$$\frac{\pi}{G^2M} \frac{1}{E} \frac{d\sigma^{\nu/\bar{\nu}}}{dy} = \left(1 - y + \frac{y^2}{2}\right) \int_0^1 dx\, F_2(x, Q^2{=}2mEy\, x) \mp y\left(1 - \frac{y}{2}\right) \int_0^1 dx\, x\, F_3(x, Q^2{=}\, 2mEy\, x) \quad (8)$$

It is difficult to reach any conclusion without doing a detailed calculation since the behavior of the functions

$$H(E) \equiv \int_0^1 dx\, F_2(x, Q^2{=}2mEx) \quad (9)$$

and

$$K(E) \equiv -\int_0^1 dx\, xF_3(x, Q^2{=}2mEx) \quad (10)$$

hinges on delicate cancellation between the $x \simeq 1$ region and the $x \simeq 0$ region. If we are only allowed to use the general properties of the moments $F_n(Q^2)$ we can only obtain two bounds:

$$\text{const.} \frac{E}{(\log E)^{\gamma}} < \sigma^{\nu+\bar{\nu}}(E) < \text{const.} E(\log E)^{\beta} \tag{11}$$

Here β is a positive number arbitrarily close to zero while γ is a positive number which depends on the gauge group. For the derivation of these bounds, see Ref. 6.

To proceed further, we have to make the assumption mentioned in the beginning of this talk that a value of Q_o^2 may be chosen once and for all.[7] This is necessary since we now wish to invert Eq. (1). Note that the large n behavior of $F_n(Q^2)$ controls the threshold ($x \simeq 1$) region[8].

By changing variables to $t = \log \omega$ and $s = n + 1$, Eq. (1) may be re-written as

$$\int_0^\infty dt\, e^{-st} F_2(t,\lambda) = C_n \lambda^{-a_n} \tag{12}$$

We take $F_{input}(t) \equiv F_2(t,\lambda=1)$ to have the form measured at SLAC; then

$$C_n = \int_0^\infty dt\, e^{-st} F_{input}(t) \quad \text{is known.}$$

By the convolution theorem

$$F_2(t,\lambda) = \int_0^t d\tau\, K(\tau,\lambda)\, F_{input}(t-\tau) \tag{13}$$

where the kernel K is the solution of the equation

$$\int_0^\infty dt\, e^{-st} K(t,\lambda) = \lambda^{-a_n} \tag{14}$$

Eq. (14) cannot be solved analytically if the exact form for a_n (as computed in Ref. 2) is used. On the other hand, it is possible to find an approximate form for a_n which is accurate to within 1%, respects the general properties of a_n, and allows $K(t,\lambda)$ to be written down analytically in terms of modified Bessel functions. For details, see Ref. 6. We then numerically compute $F_2(t,\lambda)$ and H(E), and similarly $F_3(t,\lambda)$ and K(E). It turns out that the low Q^2 region contribute substantially to H(E) and K(E) until $E \gtrsim 50$ GeV. For this and other reasons it is not possible to give a detailed quantitative description. Only qualitative trends may be discerned.

We find that for $E \gtrsim 50$ GeV H(E) rises very slowly and K(E) falls, also very slowly. Thus at large E

(A) $(1/E)\ d\sigma^{\nu}/dy$ shows deviation from $(1-y)^2$ by going to a non-zero constant as $y \to 0$.

(B) $(1/E)\ \sigma^{\bar{\nu}}$ rises slowly.

(C) $(1/E)\ \sigma^{\nu}$ is essentially constant to within 1% for E up to ~400 GeV but probably rises slowly.

These effects are all quite small because the variations of $F_2(x,Q^2)$ in the x~1 region and the x~0 region tend to cancel each other and these are further washed out after the integration over y to obtain σ. The quantity one should examine in detail experimentally is probably $(1/E)\ d\sigma/dx$. We find that

(D) $(1/E)\ d\sigma/dx$ rises for small x and falls for large x as E increases.

For further details see Ref. 6.

REFERENCES

1. D. Gross and F. Wilczek, Phys. Rev. Letters 30 1343 (1973).
   H. D. Politzer, Phys. Rev. Letters 30 1346 (1973).
   G. 't Hooft, (unpublished).
2. D. Gross and F. Wilczek, Phys. Rev. D9, 980 (1974).
3. K. Wilson, Phys. Rev. 179 1499 (1969).
4. A. de Rujula, S. L. Glashow, H. D. Politzer, S. B. Treiman, F. Wilczek and A. Zee, Princeton preprint (1974).
5. D. H. Perkins, Proceedings of the XVI International Conference on High Energy Physics, V. 4, 189.
6. S. Treiman, F. Wilczek and A. Zee, Princeton preprint (1974).
7. We thank Professor R. P. Feynman for a discussion on this point.
8. D. Gross, Rockefeller preprint COO-2233B-46.

NEUTRINO EXOTICS

George Marx
Department of Atomic Physics
Roland Eötvös University
Budapest

ABSTRACT

This report gives a review of the theoretical ideas and observational evidences concerning the possible neutrino instability, neutrino rest mass and exotic neutrino coupling.

EXOTIC PROPERTIES OF NEUTRINOS

A neutrino has been considered to have normal properties in the orthodox sense, if it is one of the four neutrino states:

ν_e	$\bar{\nu}_e$	ν_μ	$\bar{\nu}_\mu$
left	right	left	right

if it has vanishing rest mass and vanishing magnetic moment, if it has a charged V-A current weak coupling via

$$\ell_\alpha^{(-)} = \bar{e}\gamma_\alpha \frac{1+\gamma_5}{2}\nu_e + \bar{\mu}\gamma_\alpha \frac{1+\gamma_5}{2}\nu_\mu$$

and if it has a gravitational coupling via

$$t_{\alpha\beta} = \sum_{\nu_e,\nu_\mu} \bar{\nu}\left(\gamma_\alpha \overrightarrow{\partial}_\beta + \gamma_\beta \overrightarrow{\partial}_\alpha - \gamma_\alpha \overleftarrow{\partial}_\beta - \gamma_\beta \partial_\alpha\right)\frac{1+\gamma_5}{2}\nu$$

This means, that the neutrinos have been described by two /two-component/ Weyl spinors, and the two additive lepton numbers

$$L_e = N(e^-) - N(e^+) + N(\nu_e) - N(\bar{\nu}_e),$$

$$L_\mu = N(\mu^-) - N(\mu^+) + N(\nu_\mu) - N(\bar{\nu}_\mu)$$

are conserved separately. Up to the seventies these assumptions were in complete agreement with all the observational evidences and they offered the simplest theoretical explanations for each reaction involving neutrinos.

The iniciative to start the speculations about more exotic possibilities was given recently by the following questions:

1. The orthodox charged V-A current theory of weak interactions is not renormalizable. Salam-Ward and Weinberg have proposed a field theory of weak interactions, which contains also a weak neutral current. Its leptonic terms are the following:

$$l_\alpha^{(0)} = \frac{1}{4\cos\Theta}\Big[\bar{\nu}_e\gamma_\alpha(1+\gamma_5)e - \bar{e}\gamma_\alpha(1-4\sin^2\Theta+\gamma_5)e + \\ + \bar{\nu}_\mu\gamma_\alpha(1+\gamma_5)\mu - \bar{\mu}\gamma_\alpha(1-4\sin^2\Theta+\gamma_5)\mu\Big].$$

   This theory is already renormalizable. It is a remarkable fact, that the $\nu_\mu + e \to e + \nu_\mu$ and ν_μ + hadron $\to$ hadrons $+ \nu_\mu$ scatterings seem to have been observed at CERN /the second possibly also at NAL/. The CERN evidence indicates

$$\sin^2\Theta = 0.35 \pm 0.05$$

   This problem will be discussed in a separate report.
2. The neutrinos, which should be produced in the β^+ decays, associated with the thermonuclear reactions in the Sun, do not seem to reach their target in South-Dakota, according to the results reported by Prof. Davis.
3. The largest mass concentrations of the universe seem to produce stronger gravity, than expected from their visible mass. This focuses the astrophysical interest onto the gravitational properties of the cosmological neutrinos. This question will be discussen in details by Prof. S. Bludman.
4. There is a convincing evidence for the existence of a very faint interaction with a coupling strength at least thousand times smaller, than that of the weak interactions. This interaction violates the CP and T symmetry. One may start to speculate about the involvement of neutrinos in this faint interaction. /These speculations were supported by some discrepancies in $\beta\beta$ decay life times./

The lesson, given to the speaker by Prof. Primakoff had been to review about the possible exotic couplings of the neutrinos.

OSCILLATING NEUTRINOS ?

The question, whether $\bar{\nu}_e$ is different from ν_e or not, led to the investigation of $\beta\beta$ decays, pioneered by Prof. Primakoff. Evidently, a nuclear decay

$$\mathcal{N}(A,Z) \rightarrow \mathcal{N}(A,Z+2) + 2e^- + 2\bar{\nu}_e$$

is in complete agreement with the normal properties of neutrinos, but

$$\mathcal{N}(A,Z) \rightarrow \mathcal{N}(A,Z+2) + 2e^-$$

would produce a change in the lepton charge, $\Delta L = 2$. The phase space is more favourable in the second case, consequently the corresponding half life times are shorter, than in the first one. The limits on ^{48}Ca and other isotopes and the geological determination of the ^{130}Te half life excluded the possibility of a ν-less $\beta\beta$ decay with a strength comparable to the weak interactions and they were in agreement with the first /conventional/ decay mode[1].

The trouble was produced by the geological determination of ^{128}Te half life, published by Takaoka and Ogata[2]. The ration of the ^{128}Te, ^{130}Te half lives were compatible only with the ν-less $\beta\beta$ decay. This evidence encouraged Pontecorvo to suggest the idea, that the CP-violating faint interaction violates also the lepton charge conservation[3]. Primakoff developed this idea into an attractive unified theory of T and L violation[4]. In this theory the fundamental asymmetry was built into the weak leptonic current:

$$l_\alpha^{(-)} = \bar{e}\,\gamma_\alpha\left[(1+\gamma_5)+\eta(1-\gamma_5)\right]\nu_e + \text{similar for } \mu \text{ and } \nu_\mu .$$

Here ν_e is a Majorana neutrino, being charge conjugated to itself:

$$\nu_e^c = \nu_e .$$

The small parameter $|\eta| \simeq 10^{-3}$ was able to explain the milliweak CP violation in the $K_L \rightarrow \pi\pi$ decay and the milliweak L violation in the ^{128}Te $\rightarrow$ ^{128}Xe + $2e^-$ decay at the same time. The introduction of the Majorana neutrino allowed a finite rest mass and a finite magnetic moment for the neutrino.

One of the most interesting consequences of the

faint L violation could be the phenomenon of neutrino oscillation, advocated by Pontecorvo[5] :

$$\underset{\text{left}}{\nu_e} \rightleftarrows \underset{\text{left}}{\bar{\nu}_e} \quad \text{or /may be/} \quad \underset{\text{left}}{\nu_e} \rightleftarrows \underset{\text{left}}{\nu_\mu}$$

The lefthanded antineutrinos on the low energy muon neutrinos are sterile, they do not interact with the detectors. This decrease in the counting rate offered an explanation of the Davis experiment. The decrease is 50 % in average, but under very convenient circumstances, in the case of monoenergetic ν_e detector it may be as big as 100 %.

The oscillation time turns out to be finite only in the case of a nonvanishing ν mass. If $m_\nu \to 0$, then $\tau \to \infty$ as a consequence of time dilatation. The solar neutrinos have a broad spectrum, so the neutrino oscillation hardly can give practical decrease in the counting rate, which is equal to 90 % [6] . Taking the recent sharp limits of Prof. Davis into account, it is inprobable, that the negative result can be explained by a two-neutrino oscillation. /A milliweak oscillation produces an oscillation wave length, which is much smaller, than the solar distance, which results in a 50 % decrease. A superweak oscillation produces an oscillation wave length, which is much larger, than the solar distance, in this case no decrease may be expected. But these estimations concerning ν masses and splittings are of rather dubious value./

If one wants to decrease the counting rate of the Davis experiment stronger, than 90 %, demanded by the theoretical expectation and actual expectation limit, one has to postulate several neutrinos:

$$\nu_e \rightleftarrows \nu_1 \rightleftarrows \cdots \rightleftarrows \nu_m .$$

The assumption of the four components and of the new types of neutrinos makes the oscillation theory less attractive, than in its original form.

The last development in this field was the re-measurement of the ^{128}Te life time by the Minnesota University[8] . It turned out, that the transition probability $^{128}Te \to {}^{128}Xe$ ought to be 100 times smaller, that reported by Takaoka and Ogata, so the discrepancy in the $\beta\beta$ decay disappeared. There is no laboratory indication for any violation of the lepton charge conservation. /To be more specific, the experiments show, that $L_e + L_\mu$ and $(-1)^{L_e - L_\mu}$ are good quantum numbers up to milliweak order of magnitude, but the evidence for the independent conservation of the two additive quantum numbers L_e and L_μ are very poor, they need a more careful experimental investigation[9] ./

DECAYING NEUTRINOS ?

If neutrinos are produced continuously at the center of the Sun, but they don't reach the Earth, the most simple-minded explanation may be, that they decay on their way. This idea has been exploited by Bahcall, Cabibbo and Yahil[10]. In order to prevent the infinite time dilatation, they assumed a finite rest mass for the neutrino. So they arrived at a decay probability

$$\frac{1}{\tau} = \frac{g^2}{16\pi} \cdot \frac{m_\nu c^2}{\hbar} \cdot \frac{m_\nu c^2}{E_\nu}$$

Here g is the coupling constant of the new /unknown/ $\nu_e \Rightarrow \nu' + S$ or $\nu_e \Rightarrow \nu' + V$ coupling. /Here S or V denotes a new very light scalar or vector particle./

The $\nu_e \Rightarrow \nu_1 + \nu_2 + \nu_3$ decay is excluded, because it would modifiy the μ life time $/\mu^+ \Rightarrow e^+ + 4\nu$ via virtual ν_e/. A $\nu_e \Rightarrow \nu' + S$ or $\nu_e \Rightarrow \nu' + V$ decay does not contradict the μ life time, if

$$\tau\,(E_\nu \sim \mathrm{MeV}) \simeq 0.1\,\mathrm{sec}, \quad \text{i.e.}\ \frac{g^2}{4\pi} < 2.10^{-9}\, m_\nu^2,$$

if m_ν has been expressed in eV units. Evidently, the negative result of the Davis experiment can be explained, if

$$c\tau \leq d,$$

d being the radius of the orbit of the Earth. The restrictions give the following limits on the strength of the new coupling:

$$4.10^{-12} \leq \frac{g^2}{4\pi} \cdot \left(\frac{m_\nu c^2}{1\mathrm{eV}}\right)^2 \leq 2.10^{-9}.$$

The new coupling must be apparently very weak. Evidently, it is rather improbable, that τ be just of the order d/c, consequently the decay hypothesis predict, that the counting rate of the Davis experiment is not only very small, but practically zero.

Reines, Sobel and Gurr[11] reinvestigated the reactor neutrino experiments from the point of view of the ν stability. They concluded, that $c\tau >> d$ for the $\nu_e \Rightarrow \nu' + \gamma$ decay, otherwise photons would be counted in the 0.1–0.5 MeV energy range. So the possibility $\nu_e \Rightarrow \nu_\mu + \gamma$ has been ruled out: the decay products cannot be known particles.

Up to this point it was assumed, that an unstable neutrino is necessarily massive. The photon decay was investigated already long ago[12]. The decay possibility of a massless particle was reinvestigated by Parkinson and Vasholz quite recently[13]. From phase space considerations they concluded, that the many particle decay of a massless particle is definitely excluded, but a two-body decay is still possible. The two decay products may share the momentum and energy of the original neutrino, they both are necessarily massless. A decay

$$\nu_e \rightarrow \nu_e + \gamma$$

is excluded. The only possibilities are $\nu_e \rightarrow \nu' + S$ or $\nu_e \rightarrow \nu' + V$ as in the massive case.

There is, however, a fundamental difference between the decays of massless and massive particles. In the case of the vanishing mass the decay products move with the same speed in the same direction, their separation does not increase. There is a possibility for the inverse reaction $\nu_1 + S \rightarrow \nu_e$ and the net result will be an oscillation $\nu_e \rightleftarrows \nu_1 + S$, which is equivalent to a dressed stable neutrino, but its dress is on the mass shell. One can express himself also on another way: $\nu_1 + S$ hit the same target at the same time. They carry all the quantum numbers of the original ν_e, so very probably they are able to produce the same reaction, as the original ν_e. The quantum theory of the decay of a massless particle is not yet clarified enough, but there is only little hope, that this can help in explaining the Davis experiment. At the end of this point let us conclude in the following:

A. If all the light particles are known and if lepton charge is conserved and if the neutrinos are massless, then the neutrinos must be stable particles. There is no direct or indirect laboratory indication for any ν instability.

B. The negative result of the Davis experiment can be explained by ν decay or ν oscillation only, if the neutrino has a finite rest mass, if there are new -unobserved- light particles and if there exists a new very weak interaction among ν and the unobserved light particles. If one is unwilling to accept any of these assumptions, one must look for an astrophysical solution of the Davis puzzle.

STRONGLY INTERACTING NEUTRINOS ?

Untill recently, the only well-established interaction

of neutrinos was its participation in the weak charged "current" $l_\alpha^{(\circ)}$. Now there is an experimental indication for the existence of a neutral neutrino "current" $\bar{\nu}\Gamma\nu$, which interacts weakly with other currents /elastic scattering of ν_μ on charged leptons and on nuclei/, and possibly with itself /elastic scattering of ν_e and ν_μ on ν_e and ν_μ /. Bialynicka-Birula[14] and Bardin-Bilenki-Pontecorvo[15] called the attention to the possibility, that the ν-ν force may be quite strong. As a matter of fact, this is the strongest possible coupling of neutrinos, which is not ruled out by the available -very indirect- experimental evidence.

The best reactions to observe the ν-ν coupling are the following:

$$\pi^+ \rightarrow e^+ + \nu_e + \nu_e + \bar{\nu}_e$$

$$K^+ \rightarrow e^+ + \nu_e + \nu_e + \bar{\nu}_e$$

$$\nu_\mu + p \rightarrow \mu^+ + n + \nu_\mu + \nu_\mu$$

$$\nu_\mu + p \rightarrow e^+ + n + \nu_e + \nu_\mu$$

All these reactions are transmitted by a virtual neutrino. From the analysis of the available experimental limits the authors concluded, that the four-neutrino coupling constant F is related to the conventional coupling constant G by the following inequality: $F \lesssim 10^6 G$, i.e.

$$F. m_p^2 \leq 20.$$

This means, that the possibility of a strong inter-neutrino force is not excluded at all. Obviously there is plenty of room for improvement of the experimental accuracy.

It is true, that there is no convincing theoretical indication for such a strong inter-neutrino force, but its principal possibility puts the question of neutrino and neutretto rest mass into new light.

MASSIVE NEUTRINOS ?

It is a common belief, that the neutrinos are massless and they are described by two-component Weyl-spinors. The foundation of this belief is mainly of esthetical character. The laboratory upper limits on the neutrino rest masses are rather crude[16]:

$$m(\nu_e) < 60 eV$$

from tritium decay spectrum and

$$m(\nu_\mu) < 0.8\ MeV$$

from K decay spectrum. Untill the $e-\mu$ mass splitting is not completely understood, one cannot exclude the possibility of a considerable ν_μ rest mass theoretically. The struggle for understanding the negative outcome of the Davis experiment made the discussion about the neutrino rest mass interesting again[17].

The weak interactions are really weak at low energies, where a rest mass could manifest itself, so there is no hope for a fast improvement under laboratory conditions. Zel'dovich was the first, who directed his attention to the other well accepted interaction of the neutrinos: to their gravity.[18] He pointed to the fact, that the neutrinos were necessarily produced in the early hot area of the expanding Universe. Due to their large degree of freedom($\nu_e, \bar{\nu}_e, \nu_\mu, \bar{\nu}_\mu$), they may be the most aboundant particles of the Universe. Their number is a bit larger, than the number of the photons, observed in the 2.7 °K background radiation. The photons are 10^8 times more aboundant, than the nucleons. This has the consequence, that if $m_\nu > 10$ eV, the relict neutrinos may give a larger contribution to the gravitational mass of the Universe, than the atoms, and this may influence the history of the Universe on an important way.[19] /As a matter of fact, there are some observational indications into this direction. This point will be discussed by Prof. S. Bludman in details/ Here I mention only, that cosmology has supplied us with the information

$$m(\nu_e) \text{ and } m(\nu_\mu) < 10\ eV,$$

based on the gravitational coupling of neutrinos. This is a much more accurate information about ν masses /and about the absence of strong ν interactions/, than anything from laboratories.

NORMAL NEUTRINOS??

At the beginning of this talk we raised some questions about the exotic properties of the neutrinos. We know, that they participate in charged and neutral weak currents and probably also in the gravity-producing energy-momentum tensor. But we did not succeed to find any convincing experimental of theoretical indication for a new degree of freedom, for a new exotic interaction or for a disturbing instability. It may be, finally, that they are well-behaved particles and we understand them quite well. What makes neutrino science interesting nowadays, is not an unexpected face of this particle, but just its well-understood behaviour. They are not objects of research any longer, but much more tools of research for explaining the really exotic bodies, like the proton or the Sun or the Universe.

REFERENCES

1. T. Kirsten et al., Phys. Rev. Letters 20,1300 /1968/
2. N. Takaoka - K. Ogata, Z. Naturforsch. 21A,84 /1966/
3. B. Pontecorvo, Phys. Letters 26B,630 /1968/; Journal Exptl. and Theor. Physics USSR 53,1717 /1967/ ; "Old and New Problems in Elementary Particles" 251 /1970/ ; "Interrogatives about Neutrinos" Dubna Preprint El - 5439 /1970/ ; Uspehi Fiz. Nauk USSR 104 3 /1971/
4. H. Primakoff - D. H. Sharp, Phys. Rev. Letters 23, 501 /1969/
   H. Primakoff - S. P. Rosen, Phys. Rev. 184,1925 /1969/
5. B. Pontecorvo, Chalk River Report /1946/
   V. Gribov - B. Pontecorvo, Phys. Letters 28B,493 /1969/
6. J. N. Bahcall - S. C. Frauchi, Phys Letters 29B, 623 /1969/
7. T. de Graaf, Nuovo Comento Letters 2 /1971/
8. E. Fiorini, private communication /1974/
9. G. Marx, "Lepton Charge Conservation" Proceedings of the Neutrino'72 Conference /Balaton, Hungary/ Technoinform, Budapest, 123 /1972/
10. J. N. Bahcall - N. Cabibbo - A. Yahil, Phys. Rev. Letters 28,316 /1972/
    S. Pakvasa - K. Tennakoe, Phys. Rev. Letters,28 1415 /1972/
11. F. Reines - H. W. Sobel - H. S. Gurr, Phys. Rev. Letters 32,18o /1974/
12. F. Rohrlich, private communication
    P. Havas, Amer. J. of Physics 34,/1966/ 753; 35, 545,/1967/
13. M. T. Parkinson - D. P. Vasholz, Phys. Letters 45B,376 /1973/
14. Z. Bialynicka - Birula, Nuovo Cimento 33,1484 /1964/
15. D. Yu. Barden - S. M. Bielenky - B. Pontecorvo, Phys. Letters 32B,121 /1970/
16. Particle Data Group: Review of Particle Properties, Lawrence - CERN
17. N. Cabibbo - G. Marx, Internt'l Conf. on High Energy Physics, Batavia Vol. 2. /1972/
18. Ya. B. Zeldovich - Novikov: "Relativistic Astrophysics" Vol. 2.
19. G. Marx - A. S. Szalay, Proc. "Neutrino'72"Conf. Technoinform, Budapest - Hungary 191 /1972/
    - Internt'l Centre for Theor. Physics IC-73-71 preprint, Trieste /1973/- Acta Phys. Hung. 35 113 /1974/

ON THE WEAK ENERGY DIFFERENCE BETWEEN MIRROR MOLECULES

Eva Gajzago - George Marx
Department of Atomic Physics
Eötvös University
Budapest

ABSTRACT

This report gives a review of the possibilities, how weak electron-proton force /neutral current coupling/ could manifest itself by producing parity impurities in the atomic and molecular spectroscopy. The experimental implications have been also discussed in some details.

INTRODUCTION

Several gauge theories predict a weak neutral current coupling between leptons and hadrons /e.g. the Weinberg model [1] /or at least between charged leptons and hadrons /e.g. the Lee-Prentki-Zumino model [2] /. This charge retaining parity-asymmetric weak force is transmitted by the massive Z^o boson. The experimental indication of the neutrino-nucleon and neutrino-electron forces support this idea [3] . If the Hamiltonian of the atomic and molecular systems contains also a term which anticommutes with parity, one may expect parity impurities in atomic spectroscopy [4-10] and in quantum chemistry [11-12].

As a consequence of the large mass of the Z^o boson, the parity-asymmetric correction to the Coulomb force has a very short range. In electron-electron interactions this contact potential is mostly negligible because of the Pauli "repulsion" [6] . The weak odd-parity contribution to the electron-nucleus interaction turns out to be [6]

$$H' = -\frac{G}{4\sqrt{2}\, m_p c}\left\{\underline{\sigma}\,\underline{p}\,\delta(\underline{x}-\underline{x}_n)+\delta(\underline{x}-\underline{x}_n)\,\underline{\sigma}\,\underline{p}\right\}Q_n, \qquad (1)$$

where in the Weinberg gauge model [1] the weak charge of the nucleus is given by the formula

$$Q_n(Z,N) = (4\sin^2\theta - 1)Z + N. \qquad (2)$$

/In the numerical estimations we shall use the value $\sin^2\Theta = 0.35$ for the Weinberg angle Θ, compatible with the Gargamelle evidence [3], giving $Q_M \simeq A - 0.6Z$./

PARITY IMPURITY IN ATOMIC STATES

H' is of the first order in weak coupling constant G and so are the parity impurities included by this weak force. For simplicity, let us consider the $2s_{1/2}$ state of the H atom. As a rule, the nearest level produces the largest perturbation. In this case the separation of the $2s_{1/2}$ and the $2p_{1/2}$ state is given by the Lamb shift:

$$\Delta E_L = E(2s_{1/2}) - E(2p_{1/2}) = \frac{1}{2}\alpha^5 m_e c^2 = 5.10^{-6}\,eV. \quad (3)$$

So the perturbed $2s_{1/2}$ state is described by the state vector

$$\psi(2s_{1/2}) = \varphi(2s_{1/2}) + \frac{\langle H' \rangle}{\Delta E_L}\varphi(2p_{1/2}). \quad (4)$$

The perturbing matrix element $\langle H' \rangle$ is very small compared to the atomic energies:

$$\langle H' \rangle = \langle 2p_{1/2} | H' | 2s_{1/2} \rangle = -\frac{i}{32\pi\sqrt{3}}\frac{G\hbar}{m_e c}a^{-4} =$$

$$= -\frac{i}{32\pi\sqrt{3}}(G m_e^2)\alpha^4(m_e c^2) = -i(1.6\times 10^{-16}\,eV). \quad (5)$$

Here a is the innermost radius, its minus fourth power comes from the expression $\psi_s(0)^*\,\mathrm{grad}\,\psi_p(0)$. The combination $(G m_e^2)\alpha^4(m_e c^2) = 3\times 10^{-14}$ eV is the characteristic value of the weak contributions to the atomic and molecular energies. Due to the smallness of the energy denominator, the parity impurity $2p_{1/2}$ in the $2s_{1/2}$ state turns out to be rather significant: $\langle H' \rangle : \Delta E_L = 10^{-10}$. To observe this, one must look for first order effect, e.g. for interference phenomena. In the one- photon decay of this physical state a forbidden magnetic dipole transition $2s \to 1s\,(M)$ is mixed with an allowed electric dipole transition $2p \to 1s\,(E)$:

$$\frac{|E|^2}{|M|^2} = \frac{\Gamma(2p\to 1s)}{\Gamma(2s\to 1s)} = \frac{9.10^8\,s^{-1}}{10^{-7}\,s^{-1}} \simeq 10^{16}. \quad (6)$$

The interference of the electric and magnetic radiation produces a partial circular polarization:

$$\mathcal{C} = 2\frac{\langle H' \rangle}{\Delta E_L}\cdot\frac{|E|}{|M|} = 2\frac{\langle H' \rangle}{\Delta E_L}\frac{\Gamma(2p \to 1s)^{1/2}}{\Gamma(2s \to 1s)^{1/2}} \sim 10^{-3} (7).$$

This is a surprisingly high value, smaller effects have been observed already in nuclear gamma transitions. Unluckily, the $2s \to 1s$ transition for the H spectrum is inconvenient for practical observation, because of the disturbing two-photon decays. M.A. Bouchiat prepares an experiment with the $7s \to 6s$ transition of the Cs atom at the Ecole Normale Superiore [7]. Here the larger value of Q_n /and the sharper increase of the Dirac wave function at the nucleus/ compensates the larger energy denominator, so the expected circular polarization is $\sim 10^{-4}$.

A more favourable situation may be found in the case of the muonic atoms, where the larger mass produces a smaller Bohr radius a, consequently the muonic wave function concentrates stronger onto the nucleus. The weak matrix element H' will be larger by a factor $(m_\mu/m_e)^3$, than that for the electron, i.e. $\langle H' \rangle \sim$ $\sim 10^{-10}$ eV. Unluckily, the energy denominator increases, too, proportional to m_μ/m_e, or even stronger, due to vacum polarization effects, so the expected parity impurity increases less, than $(m_\mu/m_e)^2$, compared to the normal H atom. The experimental possibilities are still remarkable, as stressed by J. Bernabeu, T.E.O. Ericson and C. Jarskog [10]. The energy splitting due to vacuum polarization and that due to nuclear size have opposite effects and they make the energy denominator to be the smallest around the ^{6}Li muonic atom $(\sim 1$ eV$)$. If we look at the correlation between photon momentum and muon spin $(\underline{p}_\gamma \underline{\sigma}_\mu)$, a few percent effect is predicted. The muon polarization is made observable by the momentum of the decay electron. In the most favourable case of the one photon decay in the muonic ^{6}Li atom the $(\underline{p}_\gamma \underline{p}_e)$ asymmetry may reach 9% ! /The real experiment is made more difficult by the disturbing two-photon decays, which do not show this large parity violation./

These examples show, that the weak parity impurities may result in a detectable effect in atomic transitions.

PARITY IMPURITIES IN MOLECULAR STATES

Another interesting domain to look for the consequences of the non-commutativity of the molecular Hamiltonian and parity is the chemistry of reflection-asymmetric molecules. The ground state energies of mirror molecules will not be equal necessarily [11, 12]. The essential aim of this report is to start the exploration of this phenomenon, which may be interesting also from biochemical point of view.

It is evident from the form of the weak perturbation /1/, that the contribution of an electron pair with compensated spin will cancel. Let us denote the orbital wave function of an odd electron /possessing an uncompensated spin within the molecule/ with $\psi_e(\underline{x})$. In first approximation the perturbation /1/ will produce an energy shift

$$\int \psi_e^*(\underline{x}) H' \psi_e(\underline{x}) d^3x . \qquad (7)$$

After a simple manipulation this gives the following energy difference between mirror molecules:

$$\Delta E = \frac{G}{\sqrt{2}} \sum_n \sum_e \underline{\sigma}_e \underline{j}_e(\underline{x}_n) Q_n(Z,N) . \qquad (8)$$

The summations are extended over all electrons /e/ with uncompensated spin and over all nuclei /n/ of the molecule. Here $\underline{j}_e(\underline{x})$ is the conventional Schrödinger current,

$$\underline{j}_e(\underline{x}) = \frac{i\hbar}{2m_e} \left\{ \underline{\nabla} \psi_e^*(\underline{x}) \cdot \psi_e(\underline{x}) - \psi_e^*(\underline{x}) \cdot \underline{\nabla} \psi_e(\underline{x}) \right\}, \qquad (9)$$

to be taken at the nuclei. $\underline{\sigma}_e$ is an axial vector, $\underline{j}_e$ is a polar vector, so the weak energy shift will have opposite sign in mirror nuclei, if it is non-vanishing at all.

The simplest examples of asymmetric molecules are the two-atomic molecules, made of different atoms and having an additional magnetic field, e.g. the MgH. Here the three valence electrons feel the electric field of Mg^{++} and H^+, they will have bottle shaped orbital wave functions /superpositions of atomic s wave functions into bonding and non-bonding molecular orbitals/. The electron spin will be influenced by the proton magnetic moment. The molecular axis $\underline{x}$ (Mg → H) and the electron spin $\underline{\sigma}_e$ may give a pseudoscalar term $\underline{x} \cdot \underline{\sigma}_e$. Such a term appears, however, only in a Hamiltonian, which is asymmetric also from the point of view of time reversal. Our weak Hamiltonian /1,8/ has a symmetric form from the point of view of time reversal, so it does not give any mass splitting between the two mirror MgH molecules.

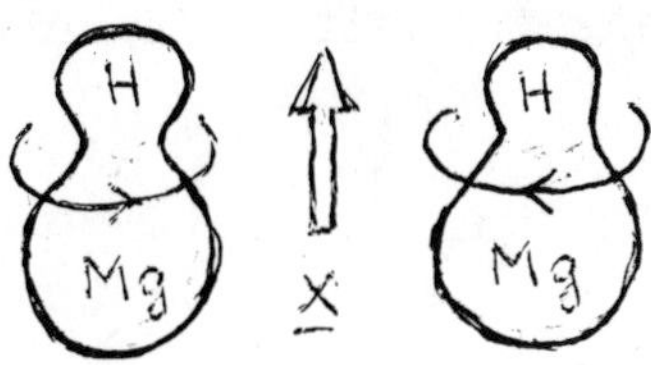

This negative conclusion has a wider validity. Let us assume, that the molecular Hamiltonian contains only kinetic energy and Coulomb energy terms. This Hamiltonian is completely real. Its ground state eigen function will be a real expression. This statement is not modified by Pauli principle. /Even the degenerate excited states may be described by real eigenfunctions./ In the case of real functions the expression /9/ vanishes. There is no steady electron current at the place of the nucleus, so the weak contribution /8/ to the molecular energy vanishes exactly. If we take only the Coulomb interactions into account in shaping the orbital wave functions, the P -odd, T -oven weak perturbation /1,8/ will produce no energy difference between mirror molecules in the first order of the perturbation theory.

In the case of molecules with paramagnetic properties the situation will be different. If the electrons with uncompensated spin build up a state with maximum spin multiplicity, this spin magnetism will be coupled to the orbital magnetic momentum of the electrons. In this case not the standing wave, but the running wave solutions will be favourable even for the ground state: one may expect a nonvanishing value of $j_z(\underline{x})$ in certain regions. So the appearence of a weak energy shift /8/ is not excluded. Formally speaking: the Pauli spin-energy-term and the spin-orbit coupling destroy the real character of the Hamiltonian: the complex eigenfunctions may give a nonvanishing value for the current /9/.

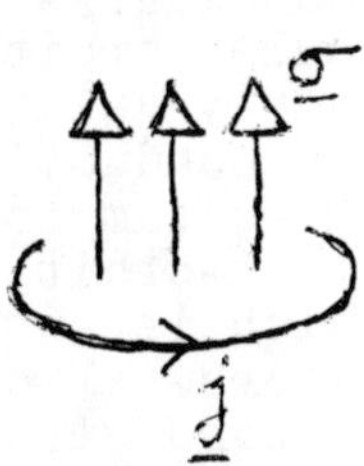

To be more specific, the second approximation of perturbation theory will give an energy shift even in the case of the real ground state eigenfunction, as a consequence of an interplay between the weak Hamiltonian(1) and of the spin-orbit coupling. If one starts e.g. from an sp_x hybrid state with spin pointing in z direction, one will get a weak matrix element, similar to the formula /5/, a spin-orbit matrix element of the order

$\alpha^4 m_e c^2$ and an energy denominator of the order $\alpha^2 m_e c^2$. This means, that the energy shift is in general by a factor α^2 smaller, than the weak transition matrix element /5/:

$$\Delta E \sim (G m_e^2) \alpha_Z^6 Q_m z (m_e c^2).$$

Here

$$\alpha_Z = a : \frac{\hbar}{m_e c} = \frac{Z_{eff} e^2}{\hbar c \cdot n_{eff}^2} = \frac{1}{137} \frac{Z_{eff}}{n_{eff}^2}$$

is the effective fine structure constant of the unpaired electrons. z is the number of these electrons, taking part in producing this weak energy shift. One can win something, if there is a nearby perturbing molecular energy level: $E_n - E_m \sim 1$ eV. On the other hand, the π bonding states /being the most sensitive for the molecular shape/ give a small electron density, but large gradient at the nucleus. This may produce an extra factor R/a, if R is the nuclear radius. Putting everything together, an estimate $\Delta E \sim 10^{-14}$ eV is rather optimistic one for the energy splitting between the mirror molecules.

A realization of this situation is offered by the example, which has been investigated by P. Hraskó from a different point of view[13]. Let us consider a helix-shaped molecule with delocalized molecular orbitals. /A closed helix may eliminate the boundary value problems./ If an electron moves along a conducting right-handed screw it has an orbital momentum component parallel to its linear momentum. The spin-orbit coupling will direct its spin into position, which is antiparallel to orbital momentum.

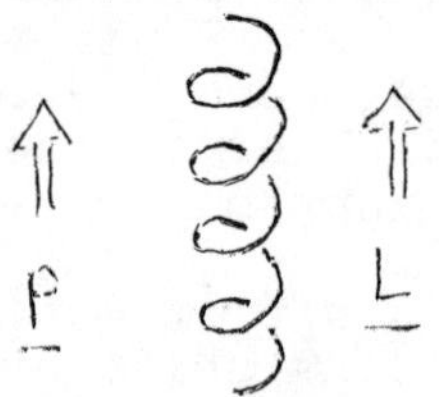

The electrons with opposite spin and momentum will form a "helical electron gas". The negative helicity may give a weak energy term, due to the Hamiltonian /1/. The effect will be made smaller by the π-bonding character of the delocalized molecular orbitals. The important point is, however, that we expect a difference in the ground state energy of right-handed screw-shaped and left-handed screw-shaped molecules[12]. <u>The mirror molecules, showing weak energy</u>

differences, are expected to be rather complicated ones.

There is a very tiny hope, that this small energy difference may be observable directly. If we drive a tunable laser with the light of a right-handed molecule, it will radiate a frequency ν_R. We can beat this frequency with the optical frequency ν_L of the left-handed molecule, a beating may be produced. By measuring the beating frequency ω one finds the frequency difference $\nu_R - \nu_L$ [14]. Principally, one can go down to $\Delta\nu/\nu = 10^{-12}$ or even beyond this ratio by this metod, but a few orders of magnitude are still missing. Practical difficulties /e.g. line width/ probably do not allow the realization of this experiment.

W. Thiemann and K. Wagener reported about observing a small difference in the crystallization speed of enantiomorphic compounds [15]. They interpreted this effect as an indication for a difference in the lattice energies:

$$\Delta E/E \simeq 10^{-5}$$

This value is definitely larger, than anything in our theoretical estimations. So we propose, that the observed effect - if real - should be explained by energy differences. It is still possible, however, that the transition probabilities are more sensitive to the weak correction /1/. We have seen on the example of the hydrogen atom, that H' can lead to considerable value of a transition matrix element $\langle b|H'|a\rangle$, but it gives only small energy shift $\langle a|H'|a\rangle$ in molecules /if any at all/. The barrier penetration probability may be very sensitive for weak shape-dependent corrections. It is still worth to think about the possibility, that the amazing asymmetries of life are amplified manifestations of the weak asymmetry of fundamental forces, but these weak asymmetries may work through a chain of transitions, not in steady states [16-17].

The authors are indebted to Dr. A. Garay, Ya. B. Zel'dovich and L. Wolfenstein for valuable discussions and suggestions.

REFERENCES

1. S. Weinberg, Phys. Rev. D5 1412 /1972/
2. B. W. Lee, Phys. Rev. D6 1188 /1972/
   J. Prentki - B. Zumino, Nucl. Phys. B47 /1972/
3. Gargamelle Collaboration, Phys. Letters B46 121 /1973/ and more recent results, to be published.
4. Ya. B. Zel'dovich, J. Exptl. Theoret. Phys. USSR 33 1531 /1957/ 36 964 /1959/. Talk given at the Triangle Seminar on the Neutral Weak Currents at Low Energies /Debrecen, Hungary, 1974/.
5. B. Sakitt - G.Feinberg, Phys. Rev. 151,1341 /1966/
6. F. Curtis - Michel, Phys. Rev. B138,4o8 /1965/
7. M. A. Bouchiat - C. Bouchiat, Phys. Letters B48, 111 /1974/
8. C. A. Piketty, "Parity Violation in Atoms" Orsay, LPTHE 73/2o preprint
9. A. N. Moskalev, Exptl. Theoret. Phys. USSR Letters 19,229 /1974/
1o. J. Bernabeu - T. E. O. Ericson - C. Jarskog CERN preprint /1974/
11. D. Rein, talks given at the International Symposium on Generation and Amplification of Asymmetry in Chemical Systems /Jülich, September 1973/ and at the Triangle Seminar on Neutral Weak Currents at Low Energies /Debrecen, Hungary, March 1974/
12. G. Marx, "Weak Interactions outside the Laboratory". Proceedings of the XIII. Internationale Universitätswochen für Kernphysik /Schladming, Austria, February 1974/ J. Springer, Vienna
13. P. Hraskó, KFKI Preprint 73/4o /Budapest, Hungary/
14. Z. Bay, private communication
15. W. Thiemann - K. Wagener, Angewandte Chemie, English edition, 9 74o /197o/ - See also the report presented at the Triangle Seminar on Neutral Weak Currents at Low Energies /Debrecen, Hungary, March 1974/.
16. A. S. Garay, Nature 219,338 /1968/
    A. S. Garay - L. Keszthelyi - I. Demeter - P. Hraskó, Chem. Phys. Letters 23,549 /1973/
17. H. Yamagata, J. Theoret. Biol. 11,495-498 /1966/

TWO-BODY NEUTRINO REACTIONS: FORM FACTORS*

P. A. Schreiner
Argonne National Laboratory, Argonne, Illinois 60439

ABSTRACT

A review of experiments to measure hadronic form factors in two-body and quasi-two-body neutrino reactions is presented.

INTRODUCTION

I will review experiments to measure hadronic form factors in two-body and quasi-two-body charged-current neutrino reactions. There is data on seven reactions:

$$\nu\, n \rightarrow \mu^- \, p \tag{1}$$

$$\bar{\nu}\, p \rightarrow \mu^+ \, n \tag{2}$$

$$\bar{\nu}\, p \rightarrow \mu^+ \, \Lambda^o \tag{3}$$

$$\nu\, p \rightarrow \mu^- \, \Delta^{++} \tag{4}$$

$$\nu\, n \rightarrow \mu^- \, \Delta^+ \tag{5}$$

$$\bar{\nu}\, p \rightarrow \mu^+ \, \Delta^o \tag{6}$$

$$\bar{\nu}\, n \rightarrow \mu^+ \, \Delta^- . \tag{7}$$

The data[1] on reactions (2), (5), (6), and (7) are so limited that little physics analysis is possible. The only results are that the cross sections are consistent with what one would expect based on form factors measured in the charge symmetric neutrino reactions; I will not discuss them further.

The data on processes (1), (3), and (4) come from spark chamber experiments at CERN, Brookhaven, and Argonne, and from bubble chamber exposures at CERN and Argonne. Fig. 1 shows the energy distribution of the various ν beams, which all employ one or two magnetic focusing devices for the parent pions and kaons. The uncertainty in the normalization was about ± 30% in the old (< 1968) experiments. The second and third generation experiments

*Work performed under the auspices of the U. S. Atomic Energy Commission.

have less than ± 15% flux errors, although the errors are often larger at the low and high energy extremes of the spectrums. The neutrino energy in these experiments is essentially below 6 GeV. Within the next few months, we can expect new low energy results from the 7-ft bubble chamber at BNL and very high energy data from the 15-ft chamber at the National Accelerator Laboratory.

THE REACTIONS

The simplest charged-current neutrino reaction is $\nu n \rightarrow \mu^- p$. The hadronic current matrix element for the reaction is[2]

$$J_\lambda = \frac{G}{2} \overline{U}_p \{\gamma_\lambda F_1^V + \frac{i\sigma_{\lambda\nu} q^\nu F_2^V}{2M} (\mu_p - \mu_n) + \frac{q_\lambda F_3^V}{M} \tag{8}$$

$$+\gamma_\lambda \gamma_5 F_1^A + \frac{q_\lambda \gamma_5 F_p^A}{M}$$

$$+ \frac{\gamma_5 (P_n + P_p)_\lambda}{M} F_3^A\} U_n ,$$

where q is the four-momentum transfer between the leptons, M is the nucleon mass, μ_p and μ_n are the proton and neutron magnetic moments, and the F's are the hadronic form factors. The form factors are complex functions of q^2. Present day event samples consist of only a few hundred events, so one must make some assumptions to simplify the above expression if any progress is to be made.

(1) The Conserved Vector Current Hypothesis - This hypothesis states that the divergence of the vector current is zero, and therefore the F_3^V form factor must be zero.

(2) Time Reversal Invariance - This forces all the form factors to be relatively real.

(3) Charge Symmetry - This symmetry requires all the form factors to be real except F_3^A which is purely imaginary. Thus assumptions (2) and (3) force F_3^A to be zero. At this point, it should be noted that F_3^A is also required to be zero if there are no Second Class axial currents. Remember that First Class currents transform under G parity like $GVG^{-1} = V$ and $GAG^{-1} = -A$; Second Class axial currents transform like $GAG^{-1} = A$.

If Second Class currents do exist, they will be quite difficult to detect in $\nu n \rightarrow \mu^- p$; one must measure the final state baryon polarization[3] or measure precisely the differential cross section at very large momentum transfers.

(4) Isotriplet Current Hypothesis - This well-known hypothesis allows one to identify F_1^V and F_2^V with the Dirac and Pauli isovector form factors measured in electron-nucleon scattering experiments. F_1^V and F_2^V are usually written in terms of the magnetic and electric form factors of the proton and neutron. To an accuracy of 7%, these functions are well represented by dipoles

$$\frac{G_m^p}{\mu_p} = \frac{G_m^n}{\mu_n} = G_e^p = \frac{1}{(1 - q^2/m_v^2)^2} \tag{9}$$

with the vector mass m_V equal to 0.84 ± 0.03 GeV. Figure 2 shows some recent data on G_m^p and a comparison of G_m^p with eq. (9).

(5) Negligible F_p^A - Induced Pseudoscalar Form Factor - Since the axial current is believed to be only partially conserved, $F_p^A \neq 0$. Assuming F_p^A to be dominated by the pion pole, it can be expressed in terms of known coupling constants and a pion propagator. Near $q^2 = 0$, F_p^A is quite large, but since it enters the cross section multiplied by m_μ^2, it is at most a 1% effect.

(6) The last assumption is that the q^2 dependence of F_1^A is that of a dipole:

$$F_1^A(q^2) = 1.23/(1 - q^2/m_A^2)^2 \ . \tag{10}$$

This shape is used only because a dipole fits the vector form factors so well.

With these six assumptions, there is only one number, m_A, to measure. Present experiments have used the two obvious methods to measure m_A -- the shape of the differential cross section and the value of the total cross section. Theoretically, the shape method is not nearly as sensitive as the rate method, but because of the ν flux errors, the two methods are nearly equal in worth in current experiments.

Since there are no bare neutron targets available, one must use some nucleus as a target and then correct for Fermi motion and the Pauli exclusion principle. The ratio R of the nuclear cross section to that off a free neutron has been calculated by many authors.[4,5] Figure 3 shows results for Fe^{56} and d. The theoretical error in heavy nuclei calculations such as for Fe is about $\pm 7\%$; and in d, the error is $\pm 4\%$. While the ratio R is of course strongly dependent upon q^2, it is only weakly dependent upon the ν beam energy.

Nearly all attempts to measure the axial form factor suffer from common problems. They must have low contamination of photons and neutrons in the beam to prevent large backgrounds of $\gamma n \to \pi^- p$ and $nn \to n\pi^- p$. The two serious neutrino-induced backgrounds are $\nu n \to \mu^- p\pi^0$ and $\nu p \to \mu^- p\pi^+$, since both of these cross sections are comparable with that of $\nu n \to \mu^- p$; if the photons from the π^0 decay go undetected or the π^+ scatters and is absorbed in a heavy nucleus target, one-pion production can result in a quite large contamination. Also in heavy nucleus experiments, the final state proton may scatter in the nucleus so that its momentum is not well determined.

Table I summarizes the five experiments to measure $\nu n \to \mu^- p$.[6-10] The first three suffer from large π^0 and π^+ contamination and large ν flux errors; they measured m_A by using the shape of the dN/dq^2 distribution. The last two experiments used dN/dq^2 and the magnitude of the cross section. Let me discuss these two experiments in some detail.

The CERN experiment used a C_3H_8 fill of 0.5 m^3 bubble chamber. The candidates for $\nu n \to \mu^- p$ had one muon track and one or two protons; the second proton can come from rescattering effects in the nucleus. In order to remove the neutron, photon, and charged hadron background, two cuts were made: the visible momentum along the known neutrino beam direction had to be greater than 0.3 GeV/c and the visible energy had to be greater than 0.3 GeV. After these cuts 130 events remain. In an attempt to remove $\nu p \to \mu^- p\pi^+$ background, one considers the quantity $m^2 = (P_\nu + P_T - P_\mu)^2$ the missing-mass squared recoiling from the muon; Fig. 4a, b shows the distribution of this variable for events with one and two protons, respectively. The one-proton distribution looks reasonable compared with the solid curve -- a Monte Carlo calculation based on the Fermi gas model of the nucleus. However, the two-proton data is shifted to high mass values. These events are believed to be strongly contaminated by inelastic events and were, therefore, removed from the physics sample; corrections were made to account for their removal. In the one-prong m^2 distribution, the dotted line indicates the estimated π^o contamination. In order to reduce this background, a cut was made on the m^2 distribution at 1 GeV^2. Sixty-six events remain. Figures 5 and 6 show the differential cross section and energy variation of the production cross section after corrections were made for all of the cuts. A fit to the shape and rate gives m_A = (0.7 ± 0.2) GeV; the curves in Figs. 5 and 6 show the results of the fit.

A second analysis of this data was also performed. In order to remove the effects of the Pauli exclusion principle, events with q^2 less than 0.3 $(GeV/c)^2$ were removed; and to reduce some of the uncertainties in the ν flux, an $E_\nu \geqslant 1$ GeV cut was made. Only 26 events remain. These events represent a $\nu n \rightarrow \mu^- p$ rate that also corresponds to m_A = (0.7±0.2) GeV.

Next consider the continuing Argonne experiment using a deuterium fill of the 12-foot bubble chamber (11.1 m^3 fiducial volume). By using deuterium as a target, the experiment avoids many of the problems of these early studies. The candidates appear in the 12-foot chamber as 1-, 2-, or 3-prong events depending upon the spectator proton momentum and q^2. The 1-prong events are not used in this analysis. About 30% of the proton spectators are seen (3-prongs) in the 12-foot chamber, as compared to about 40% usually seen in small deuterium bubble chambers.

In order to separate the $\mu^- p$ events and to investigate background contamination, consider the 0-constraint "fit" to $X^o + d \rightarrow \mu^- p p_s$, where X^o has unknown mass, energy, and direction. The mass-squared of X^o is plotted in Fig. 7 versus the angle of X^o relative to the known neutrino beam direction for the events with a seen proton spectator. A clear enhancement is seen at zero mass and angle. Figure 8 shows the same distribution for the unseen spectator events. Again, a clear signal is observed, although it is somewhat washed out compared to that in Fig. 7. Most of the other events in these plots are charged hadron scatters (eq. $\pi^+ n \rightarrow \pi^o p$). By carefully examining the distributions, the non-neutrino induced background is determined to be less than 2%. The contamination from the $\nu d \rightarrow \mu^- p p_s \pi^o$ process is determined to be (1 ± 1)%. This value was measured by selecting $\nu d \rightarrow \mu^- p \pi^+ n_s$ events in the same film, deleting the π^+ track, and treating the events as $\nu d \rightarrow \mu^- p n_s$. (The $\Delta I = 1$ rule was used to estimate the $\mu^- p p_s \pi^o$ cross section relative to that of $\nu d \rightarrow \mu^- p \pi^+ n_s$.)

Now, for all of the 2- and 3-prong events, 3-constraint kinematic fits were attempted to $\nu d \rightarrow \mu^- p p_s$. For the unseen spectator events, the standard bubble chamber method was used of assigning a momentum of 0 ± 50 MeV/c to the spectator proton. Figure 9 shows the spectator proton momentum distribution. (The spectator proton is defined to be the one with lower momentum.) There is a small excess of events at high momentum, which can be attributed to rescattering effects. Figures 10 and 11 display the dN/dq^2 and $\sigma(E)$ distributions, respectively. A fit just to the shape of the differential cross section gives $m_A = 0.79^{+0.14}_{-0.15}$ GeV. A fit just to the rate gives $m_A = 1.04^{+0.20}_{-0.17}$ GeV. Since these results are consistent, we do an overall shape plus rate fit and obtain $m_A = (0.90 \pm 0.10)$ GeV. This result is stable against subdivisions of the data into low/high energy, low/high q^2, seen/unseen spectator, etc. If one allows vector mass parameter m_V to be variable as well as m_A and makes a two-parameter fit (using dipole form factors for both), we find $m_V = 0.70^{+0.19}_{-0.14}$ GeV and $m_A = 1.14^{+0.18}_{-0.32}$ GeV. So the data is consistent with the conserved vector current hypothesis. One can also try fitting the data with a monopole axial vector form factor, with the vector form factors fixed by the electron-scattering results. The monopole form gives an equally acceptable fit with $m_A = (0.54 \pm 0.08)$ GeV.

The weighted average value of m_A from the five experiments shown in Table I is (0.87 ± 0.08) GeV. Thus, the vector and axial mass parameters are equal to each other within errors.

Theoretically, Riazuddin[11] has derived a relation between F_1^A and the magnetic form factor of the proton. He assumes scaling to hold and the Llewellyn-Smith sum rule[11], which relates the structure functions of inclusive neutrino-proton and neutron scattering to those of electron nucleon scattering. At $q^2 \to \infty$, one obtains the simple relation $F_1^A/G_m^p = 1/2\beta(1 + \mu_n/\mu_p)$, where $\beta = 6$ in the quark model and 2 in the Sakata model. If one makes the strong assumption that both form factors are dipoles at large q^2 values, then one obtains the result that $m_A = 1.01$ GeV and 0.76 in the quark and Sakata models, respectively. So in this simple calculation, the data lies between the two predictions.

The measurement of m_A in neutrino reactions is also of interest because one can use it to test current-algebra and PCAC ideas. This comes about because m_A can be determined in the reaction $ep \to en\pi^+$ at threshold. The differential cross section for the latter can be written as

$$\frac{d\sigma}{dE'd\Omega} = s\,\sigma_{Mott}\, q_\pi^* + \left(\frac{d^2\sigma}{dE'd\Omega}\right)_{\Delta^+_{1236}}, \qquad (11)$$

where the first term is the s-wave contribution and the second term is the p-wave Δ^{+}_{1236} contribution (q^{*}_{π} is the π^{+} momentum in the $\pi^{+}n$ rest frame). Now a measurement of the slope parameter s can be used along with current-algebra and PCAC to determine F^{A}_{1} in the soft pion limit. In fact, s is given simply by

$$s = \frac{g^2}{16\pi^2 (M+m_\pi)^2} \{2 (\frac{F^A_1}{F^A_1(0)} - \frac{q^2}{2M^2 - q^2} G^n_m)^2$$

$$- 8M^2 q^2 (\frac{G^n_e}{2M^2 - q^2})^2 \tag{12}$$

$$+ 4 \tan^2 \theta/2 \,(1 - \frac{q^2}{4M^2}) (\frac{F^A_1}{F^A_1(0)} - \frac{q^2}{2M^2 - q^2} G^n_m)^2\}$$

The problem is that one needs to correct this expression for the finite pion mass, and there seems to be no unique way to do this at present. I have chosen to use the recent analysis of Brian Read.[12] Figure 12 shows measurements of F^{A}_{1} by four groups[13-16] studying $ep \to en\pi^{+}$. The curve is the average value of m_A obtained from $\nu n \to \mu^{-} p$. There appears to be some disagreement and the disagreement persists no matter what currently popular formulation of the finite mass corrections is used.

The next two-body reaction I want to review is $\bar{\nu} p \to \mu^{+} \Lambda^{o}$. Only one experiment has observed this process -- the current CERN experiment[17] using a freon fill of the Gargamelle bubble chamber. They observe 13 events which corresponds to a cross section off freon of $(1.3^{+0.7}_{-0.5})\, 10^{-40}$ cm^2/proton. The important question is, what is the cross section off free protons? Detailed Monte Carlo studies show that the probability for Λ absorption is ~15%, and the probability that a Σ will convert into a Λ in the freon nucleus is ~20%. Now the cross section for Σ production divided by that for Λ^{o} production is about 0.75, independent of the value of m_A, assuming exact SU(3) to hold. So the cross section scale factor to transform the freon result to a free proton result is determined to be 1.0 ± 0.5. The $\bar{\nu} p \to \mu^{+} \Lambda$ cross section is thus $(1.3^{+1.0}_{-0.8})\, 10^{-40}$ cm^2. To do better, one clearly needs data in a hydrogen or propane bubble chamber.

Figure 13 shows the energy distribution of the events as a function of q^2. Note that below 3 GeV, essentially all of the antineutrinos come from π^- decays; only above ~4 GeV do K^- decays contribute strongly to the $\bar{\nu}$ flux. So presumably $\bar{\nu}_{\pi^-}$ can produce strange particles.

While the data in Fig. 13 looks reasonable, there is clearly not enough events to determine the form factors from the dN/dq^2 shape. To determine m_A from the cross section, one can write the matrix element for the reaction as

$$J_\lambda = -\frac{G}{2}\sqrt{\frac{3}{2}} \sin\theta_c \, \overline{U}_\Lambda \left[\gamma_\lambda F_1^p + \frac{i\sigma_{\lambda\nu} q^\nu}{2m} \mu_p F_2^p + \frac{1+2x}{3} \gamma_\lambda \gamma_5 F_1^A\right] U_p \qquad (13)$$

where I have made the usual assumptions to simplify the expression. Furthermore, if one assumes $m_A = 0.84$ GeV, exact SU(3) symmetry, and an F/D ratio of 0.45/0.70 independent of q^2, then the cross section averaged over the CERN $\bar{\nu}$ flux distribution is $2.4 \times 10^{-40}\ cm^2$. So the data is consistent with this simple theory and an axial vector form factor equal to that measured in $\Delta S = 0$ reactions.

The final reaction I want to review is $\nu p \rightarrow \mu^- \Delta^{++}_{1236}$. This process was first studied using a "free proton" target in the old CERN propane bubble chamber.[18] In this experiment, there was 85 3-prong events with track ionization and delta rays consistent with the $\mu^-\pi^+p$ final state. To separate out the $\mu^-\pi^+p$ events off free protons, the group used longitudinal and transverse momentum unbalance. Figure 14 shows a scatter plot of these two variables for the $\mu^-\pi^+p$ candidates. An enhancement is certainly present at low momentum unbalance. Cuts were made at $|\Delta P_x| \leq 120$ MeV/c and $P_\perp < 240$ MeV/c. Fifty-two events remain. To estimate the contamination from $\nu p \rightarrow \mu^-\pi^+ p\pi^o$, one uses the 12 events where one or both photons from the π^o decay convert into e^+e^- pairs; after applying the longitudinal and transverse momentum cuts, the background from π^o production is 4.3 ± 2.5 events. The contamination from carbon interactions is measured by comparing the propane events with data from a freon fill of the chamber; the carbon interaction background is thus estimated as 4.6 ± 1.5 events. So the $\mu^-\pi^+p$ data sample contains 15% contamination.

Figure 15 shows a scatter plot of the π^+p invariant mass as a function of q^2 for the 52 events. A strong Δ^{++}_{1236} signal is seen above a low background; the events are mainly produced at low q^2 values.

Before discussing the CERN data further, let's introduce the Argonne data[19] obtained from hydrogen and deuterium fills of the 12-foot chamber. To separate the $\mu^-\pi^+p$ events, consider all 3-prong events whose tracks reconstruct as $\mu^-\pi^+p$. For the 410 such events in H_2, we perform the 0-constraint "fit" to $X^o + p \rightarrow \mu^-\pi^+p$ where, as before, X^o has unknown mass, energy, and direction. Figure 16 shows a scatter plot of $(MM)^2_{X_o}$ versus $(\phi)_{X^o}$; a clear signal at zero mass and angle is seen. The background is mainly from π^o production and is estimated to be 2%. In D_2, one has the corresponding reaction $X^o + d \rightarrow \mu^-\pi^+pn_s$, where the neutron spectator is undetected. Figure 17, which shows a histogram of $(MM)^2_{Xo}$ when (ϕ) is less than 20°. The cross-hatched events give a 3-constraint fit to $\mu^-\pi^+p$; the background is seen to be quite low, and by carefully examining the events in the wings, is determined to be $\leq 5\%$. In order to perform the 3-constraint fit in D_2, the starting value for the momentum of the neutron spectator was set at 0 ± 50 MeV/c. As expected, there is then a loss of high momentum spectators. This loss is estimated to be $(17 \pm 5)\%$ based on the seen proton spectators in the $\nu d \rightarrow \mu^-pp_s$ reaction.

Figure 18 shows the three invariant mass combinations for the combined H_2 and D_2 data. As in the CERN data, the final state is dominated by Δ^{++}_{1236} production. A fit to the mass spectrum gives the resonant fraction as $(95 \pm 5)\%$. There is no evidence for μ^-p or $\mu^-\pi^+$ mass enhancements. Figure 19(a) shows the q^2 distribution for the events; the mean value is 0.3 $(GeV/c)^2$. However, q^2 is not a very interesting variable to look at because the kinematic limits of q^2 depend upon the neutrino energy, and this is a variable energy experiment. Figure 19(b) shows instead the $\cos\theta^*_{\nu\mu}$ distribution; the data is forward peaked, but not as strongly as suggested by the q^2 plot.

Figure 20 shows the energy variation of the cross section for the CERN and ANL data. All flux uncertainties are included in the error bars. The Argonne results are somewhat lower than that of CERN, but perhaps consistent within the error bars.

More information about the production process, and therefore the form factors, can be learned by studying the decay angular distribution of the Δ^{++}. Figure 21 shows the coordinate system defined by S. Adler; it has the y-axis along the production plane normal and the z-axis along the three-momentum transfer direction between the two leptons. In this system, the decay distribution can be represented by the sum of four terms, assuming time reversal invariance to hold in the production process and the absence of background:

$$D(\theta,\phi) = \frac{1}{\sqrt{4\pi}} \{ Y_o^o - \frac{2}{\sqrt{5}} (\tilde{P}_{33} - \frac{1}{2}) Y_2^o - \frac{4}{\sqrt{10}} \tilde{P}_{3-1} \mathrm{Re} Y_2^2 + \frac{4}{\sqrt{10}} \tilde{P}_{31} \mathrm{Re} Y_2^1 \} \tag{14}$$

where the $\tilde{P}_{mn}$ are the density matrix elements and the Y_L^M are the usual spherical harmonics. Figure 22 shows histograms of cos θ and ϕ in the CERN and ANL experiments. The distributions are relatively featureless, although there are some weak indications of "illegal" moments in two of the plots.

Now the question is, what information about form factors can be learned from all of these distributions? To answer that we must first examine the hadronic current matrix element. There are several sets of variables used in the literature to study Δ1236 production: multi-pole moments, helicity amplitudes, and Rarita-Schwinger form factors. Since the title of this talk is "Form Factors," let me use the form factor notation, and to be precise, I will use that given by Llewellyn-Smith in his review paper.[2] The matrix element is thus

$$M = \sqrt{\frac{3}{2}}\, G \bar{\psi}_\alpha \{ [\frac{C_3^V}{m} \gamma_\lambda + \frac{C_4^V}{m^2} (P_\Delta)_\lambda + \frac{C_5^V}{m^2} (P_p)_\lambda] \gamma_5 F^{\lambda\alpha} + C_6^V j^\alpha \gamma_5 + [\frac{C_3^A}{m} \gamma_\lambda + \frac{C_4^A}{m^2} (P_\Delta)_\lambda] F^{\lambda\alpha} + C_5^A j^\alpha + \frac{C_6^A}{m^2} (q)^\alpha q \cdot j \} U_p \tag{15}$$

where j_α is the lepton current, $F^{\lambda\alpha} = q^\lambda j^\alpha - q^\alpha j^\lambda$, and the $C_i^{V,A}$ functions are the four vector and four axial form factors. One would, of course, like to simplify the expression. First, assume time reversal invariance; this forces all the form factors to be relatively real. Second, assume the isotriplet current hypothesis so that the vector form factors can be determined from electroproduction data on the Δ1236. One finds[20] that $C_5^V = C_6^V = 0$, $C_4^V \approx -\frac{M}{W} C_3^V$ and C_3^V is well represented by the dipole expression:

$$C_3^V = \frac{2.05}{(1 - q^2/0.54)^2} \quad . \tag{16}$$

However, note that the mass parameter is 0.73 GeV instead of 0.84 GeV as in elastic scattering. Figure 23 compares C_3^V with G_m^p.

Consider next the axial form factors. By measuring the differential cross section and three density matrix elements at different values of neutrino energy, the four axial form factors are far over determined.[21] But present day experiments just do not have enough events to perform the measurements. So one must consider model predictions for the form factors. C_3^A and C_4^A are completely model dependent. C_6^A, the induced pseudoscalar form factor, is always taken to contain a pion pole and is given by:

$$C_6^A = \frac{m g_{\Delta P\pi +} F_\pi}{2\sqrt{3}\,(m_\pi^2 - q^2)} \quad . \tag{17}$$

Note that the cross section is not sensitive to C_6^A since it enters multiplied by m_μ^2. To determine C_5^A, one may use the PCAC hypothesis. Letting $m_\pi^2 \to 0$ and $q^2 \to 0$ and forcing the divergence of the axial current to equal 0, one has

$$C_5^A(0) = 1.2 \ .$$

Table II summarizes the predictions for the dominant axial form factors at $q^2 = 0$ in several different models.[21-25] The q^2 dependence of the form factors in the various models are somewhat different, but they all contain a common dipole factor of $(1 - q^2/m_A^2)^{-2}$, where m_A is the same mass parameter measured in elastic scattering.

To compare these models with data, we have performed maximum likelihood fits to the experimental production and decay angular distributions and the energy dependence and magnitude of the cross section. m_A was allowed to be a free parameter in these fits. Table III gives our fitted values of m_A and the corresponding density matrix elements, averaged over all production angles (only a subsample of 121 events were available at the time these fits were done). Salin's and Bijtebier's models disagree badly with the data. However, Adler and Zucker do give reasonable density matrix elements and cross sections. For these two models, the production and differential cross sections are compared in Figs. 20 and 24 respectively. As can be seen, the data is not yet good enough to separate the two models. The revised Zucker model[26] (which has $C_5^A = 1.2$ and thus satisfies PCAC) also gives a good fit to the data.

PROGRESS IN THE NEAR FUTURE

Considering the possible progress one can expect on form factor measurements in the next few years, the major limitation will still be neutrino flux. With more data from Argonne and new data from Brookhaven and NAL, one can hope for the following:

$\nu N \to \mu N$
- test if the dipole axial form factor is the correct shape.
- resolve the possible discrepany with $ep \to en\pi^+$ measurements of m_A.
- test the isotriplet current hypotheses that $m_V = 0.84 \pm 0.03$ GeV.

$\bar{\nu} N \to \mu^+ Y$
- test the SU_3 and $\Delta I = 1$ predications for $\bar{\nu} n \to \mu^+ \Sigma^-$, $\bar{\nu} p \to \mu^+ \Sigma^o$, and $\bar{\nu} p \to \mu^+ \Lambda^o$.

$\nu N \to \mu\Delta_{1236}$
- actual measurements of the axial form factors, not just tests of model predictions.
- test of $\Delta I = 1$ rule for the $(\nu n \to \mu^- \Delta^+)/(\nu p \to \mu^- \Delta^{++})$

REFERENCES

1. E. Young, CERN Yellow Report 67-12 (1967).
2. C. H. Llewellyn-Smith, Phys. Reports 3C, 261 (1973).
3. R. Tarrach, Nucl. Phys. B70, (1974).
4. J. S. Bell and C. H. Llewellyn Smith, Nucl. Phys. B28, 317 (1971).
5. S. K. Singh, Nucl. Phys. B36, 419 (1972).
6. A. Orkin-Lecourtois and C. A. Piketty, Nuovo Cimento 50A, 927 (1967).
7. M. Holder, et al., Nuovo Cimento 57, 338 (1968).
8. R. L. Kustom et al., Phys. Rev. Letters 22 1014 (1969).
9. I. Budagov et al., Lett. Nuovo Cimento 2, 689 (1969).
10. W. A. Mann et al., Phys. Rev. Letters 31, 844 (1973).
11. Riazuddin, Phys. Letters 40B, 583 (1972).

12. B. Read, "Threshold Electroproduction Experiments and the Axial Vector Form Factors," Paper 38 at 1973 International Symposium on Photon and Electron Physics at Bonn.
13. E. Amaldi et al., Phys. Letters 41B, 216 (1972).
14. P. Brauel et al., DESY Preprint 73/24 (1973).
15. D. R. Bolterill et al., Daresbury Preprint DNPL/ p 158 (1973).
16. E. D. Bloom et al., Phys. Rev. Letters 30, 1186 (1973)
17. T. Eichten et al., Phys. Letters 40B, 593 (1972).
18. I. Budagov et al., Phys. Letters 29B, 524 (1969).
19. J. Campbell et al., Phys. Rev. Letters 30, 335 (1973).
20. A. Dufner and Y. Trai, Phys. Rev. 168, 1801 (1968).
21. P. Schreiner and F. von Hippel, Nucl. Phys. B58, 333 (1973).
22. P. Salin, Nuovo Cimento 48A, 506 (1967).
23. S. Adler, Annals of Physics 50, 189 (1968).
24. J. Bijtebier, Nucl. Phys. B21, 158 (1970).
25. P. Zucker, Phys. Rev. 4D, 3350 (1971).
26. P. Zucker, Private Communication.

TABLE I - Measurements of the Axial Vector Mass

Experiment	No. Events	Contamination	Flux Error	Method	M_A (GeV)
CERN BC CF_3Br	88	15%	± 30%	$\frac{dN}{dq^2}$	0.75 ± 0.24
CERN Sp C $A\ell$	74	45%	± 30%	$\frac{dN}{dq^2}$	0.65 ± 0.42
ANL Sp C Fe	~200	25%	± 30%	$\frac{dN}{dq^2}$	1.05 ± 0.20
CERN BC C_3H_8	26 66	10%	± 15%	$\frac{dN}{dq^2}$, σ	0.70 ± 0.20
ANL BC D_2	288	3%	± 15%	$\frac{dN}{dq^2}$, σ	0.90 ± 0.10

TABLE II - Predications for the Dominant Axial Vector Form Factor at $q^2 = 0$

Model	$C_3^A(0)$	$C_4^A(0)$	$C_5^A(0)$
Salin	0.0	2.7	0.0
Static Quark Adler	0.0	-0.3	1.2
Bijtebier	0.0	~3.0	1.2
Zucker	1.8	-1.8	1.9

TABLE III - Experimental density-matrix elements and, for each model: fitted values of m_A, integrated density-matrix elements and chi-squared (over the number of data points) for density matrix elements.

	m_A(GeV)	$\tilde{\rho}_{33}$	$\tilde{\rho}_{3-1}$	$\tilde{\rho}_{31}$	$\chi^2(\rho_{mn})$
Data		0.58 ± 0.09	-0.24 ± 0.11	-0.18 ± 0.11	
Salin	0.73 ± 0.14	0.46	0.05	0.22	$\frac{22}{3}$
Adler	0.96 ± 0.16	0.67	-0.01	-0.13	$\frac{6}{3}$
Bijtebier	0.69 ± 0.12	0.66	-0.05	0.13	$\frac{12}{3}$
Zucker	0.74 ± 0.13	0.65	-0.03	-0.19	$\frac{4}{3}$

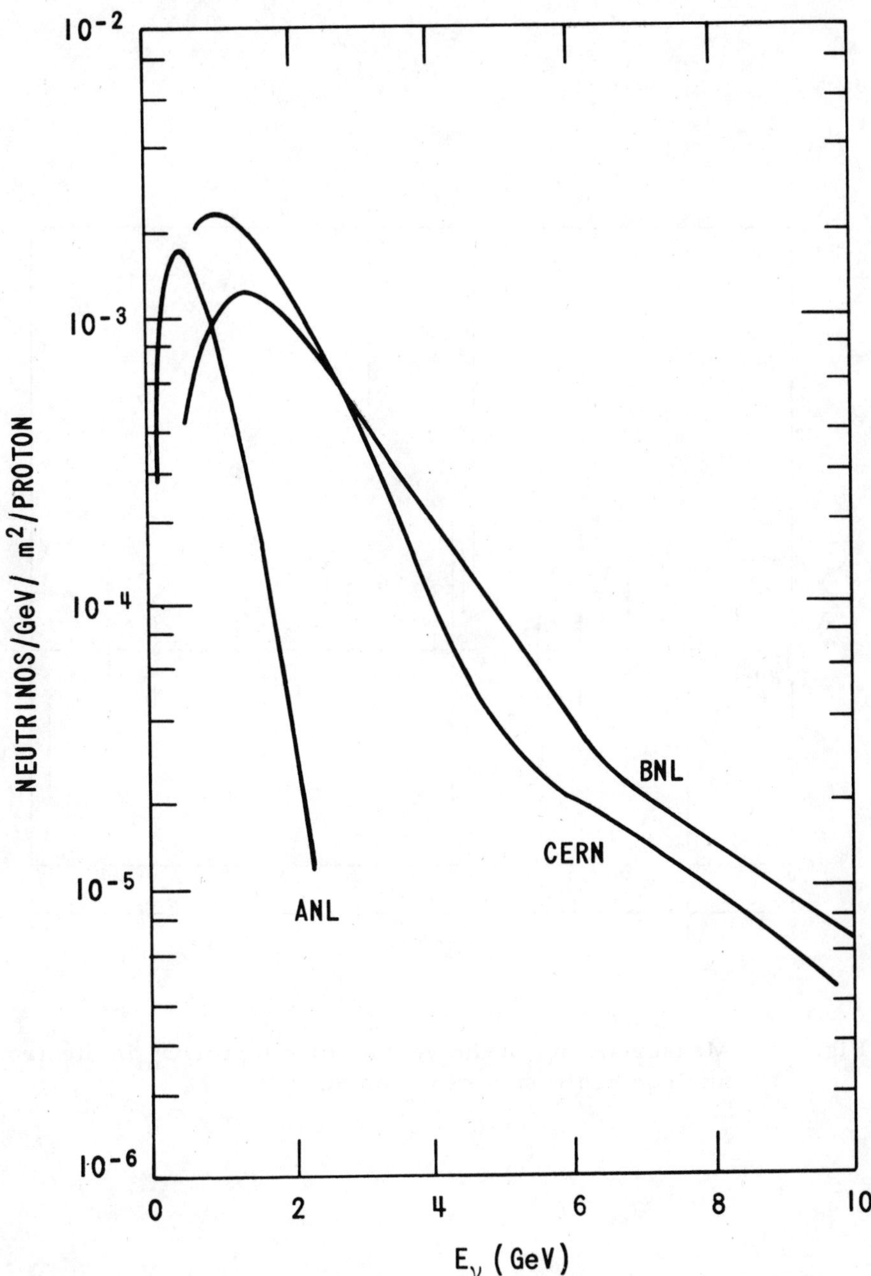

Fig. 1 Neutrino Flux distributions at Argonne, Brookhaven, and CERN.

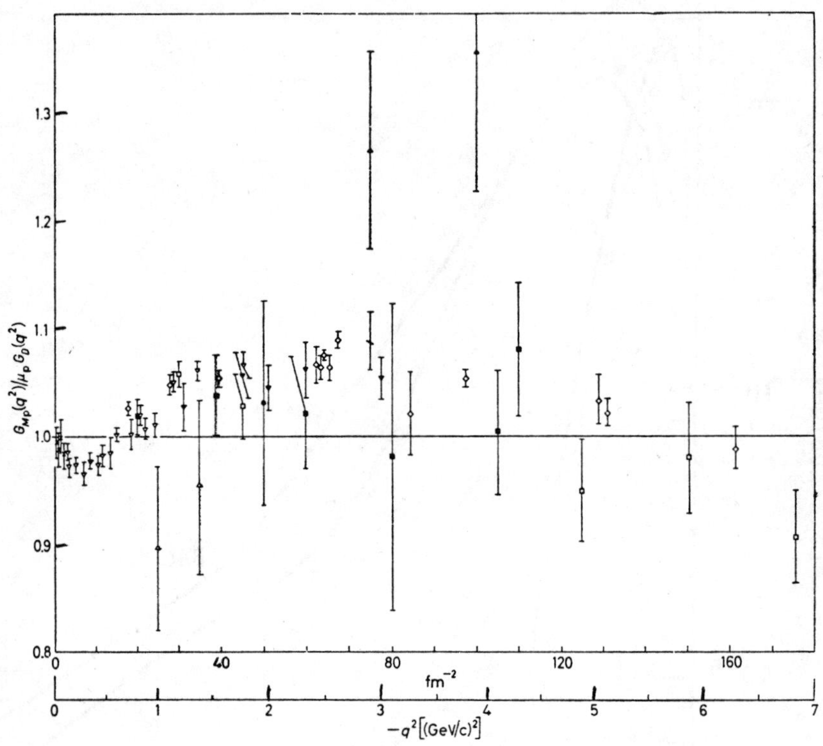

Fig. 2 Measurements of the vector form factor G_m^ρ in electron-nucleon scattering experiments.

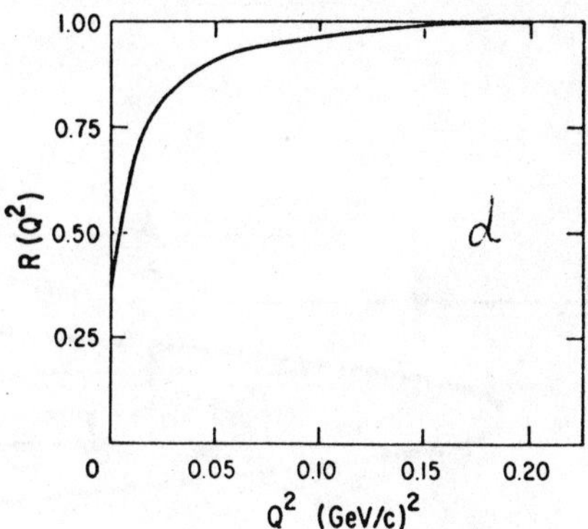

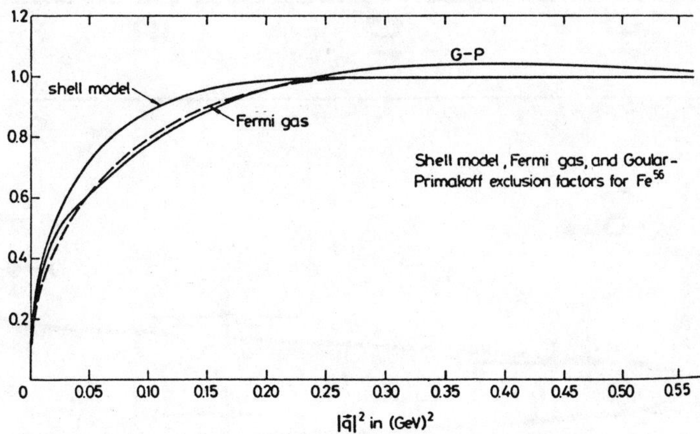

Fig. 3 Pauli exclusion ratios for iron and deuterium - see Ref. 4.

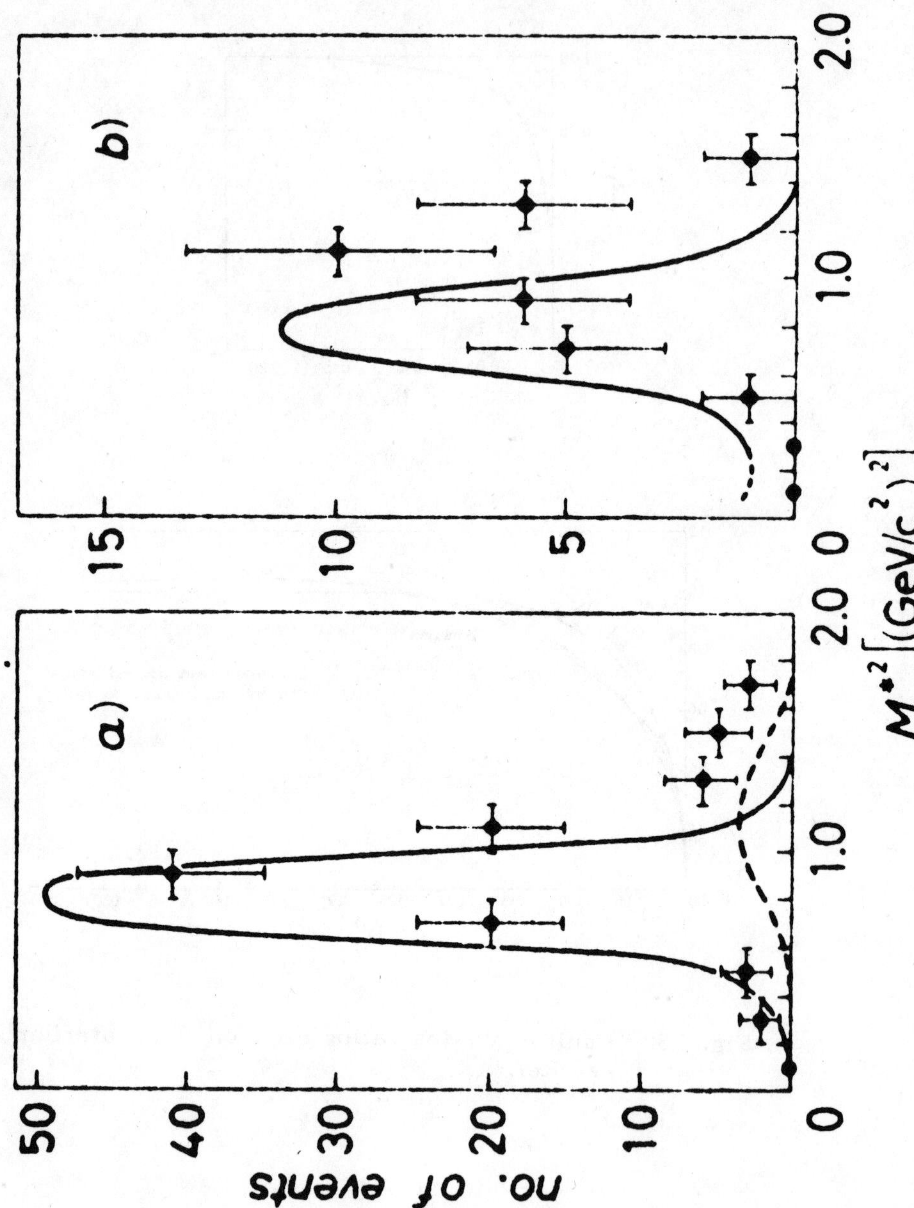

Fig. 4 Missing-mass squared distribution for $\nu\, n \rightarrow \mu^- \, MM$ in CERN propane experiment.

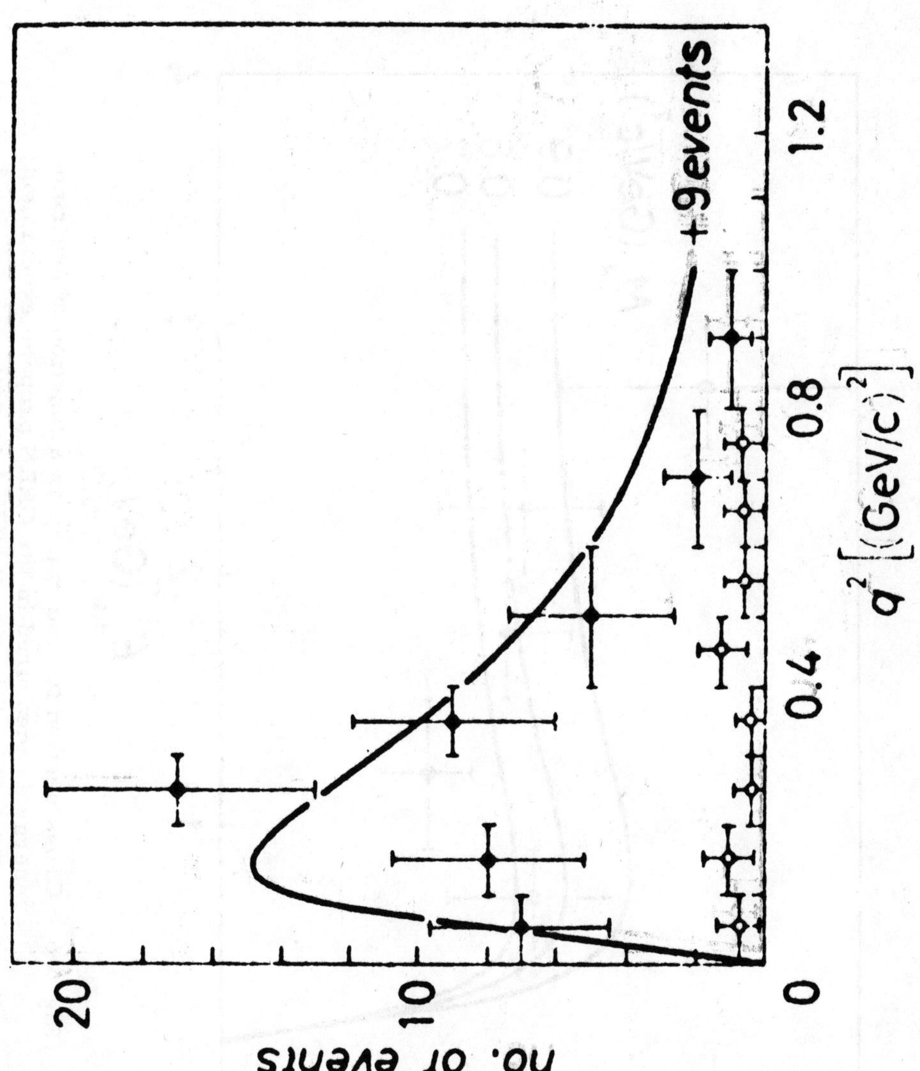

Fig. 5 Differential cross section for $\nu\, n \rightarrow \mu^- p$ in CERN propane experiment; solid curve is V-A theory with m_A = 0.7 GeV.

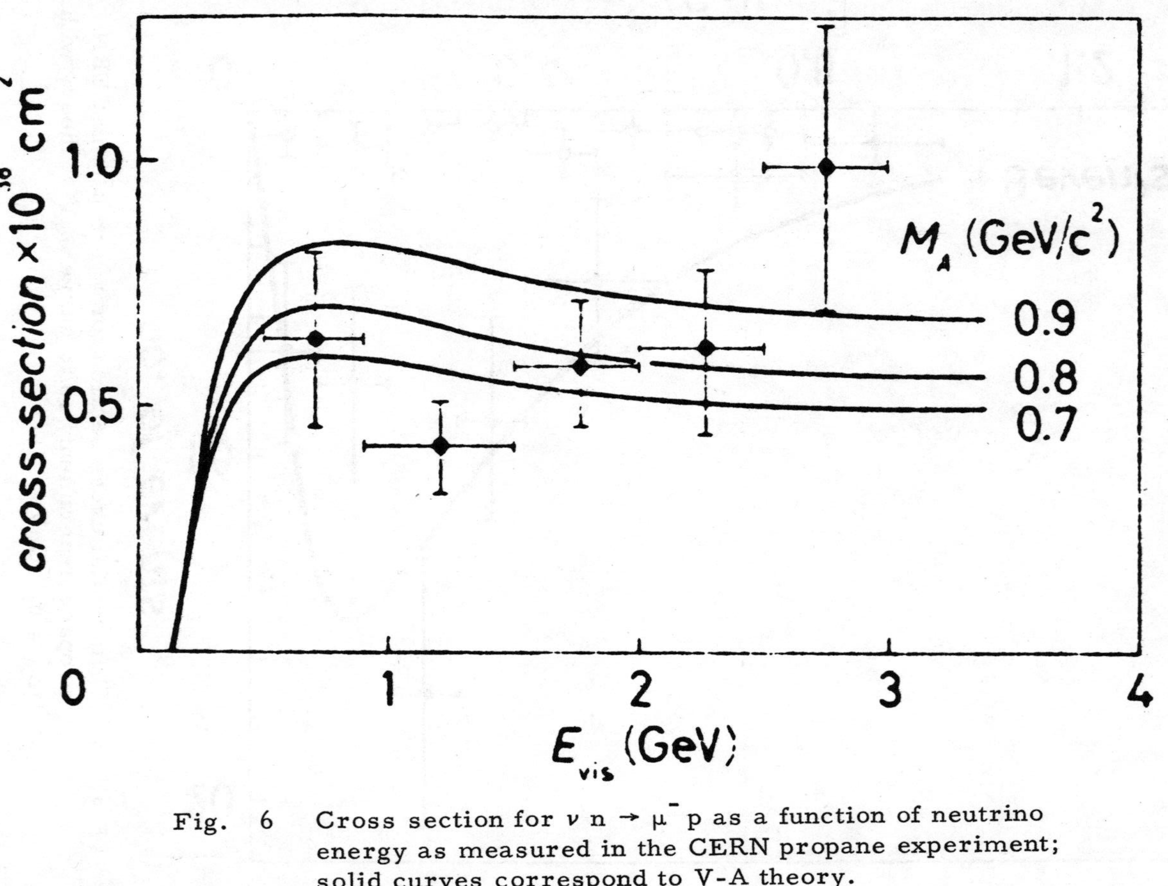

Fig. 6 Cross section for $\nu\, n \rightarrow \mu^- p$ as a function of neutrino energy as measured in the CERN propane experiment; solid curves correspond to V-A theory.

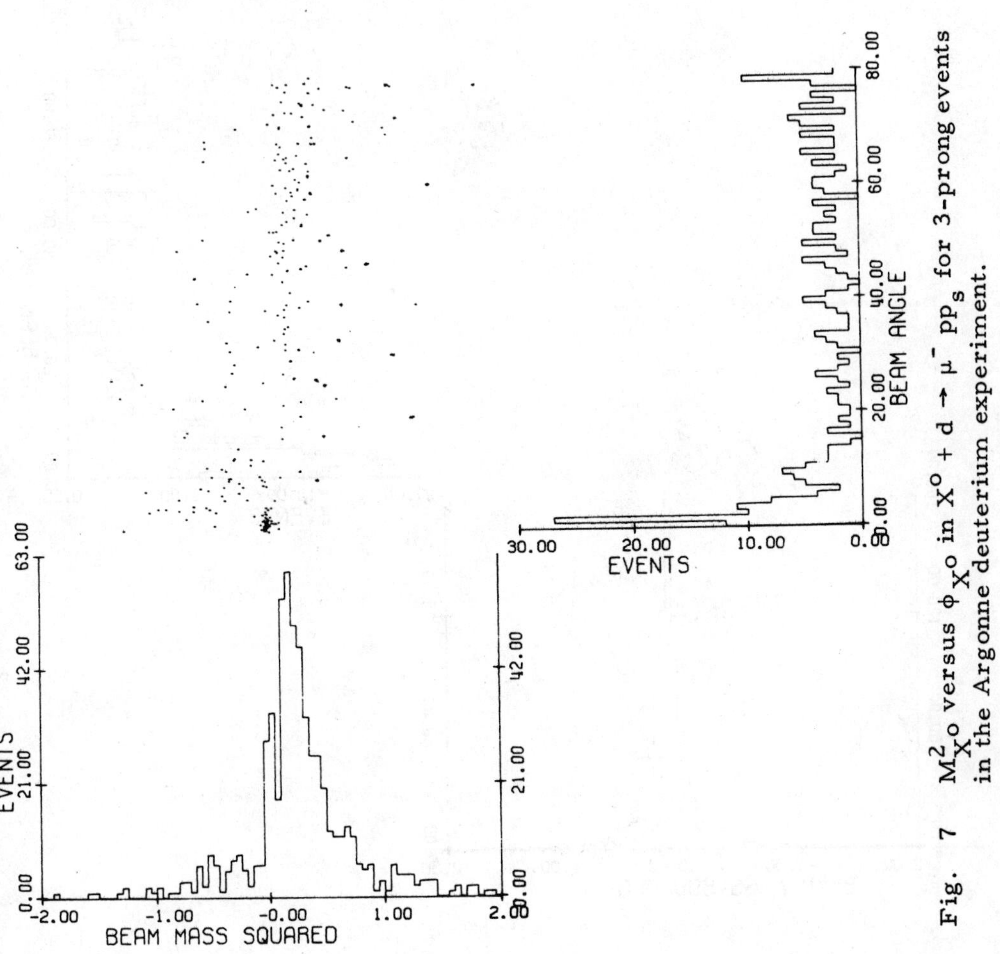

Fig. 7 $M^2_{X^o}$ versus ϕ_{X^o} in $X^o + d \to \mu^- pp_s$ for 3-prong events in the Argonne deuterium experiment.

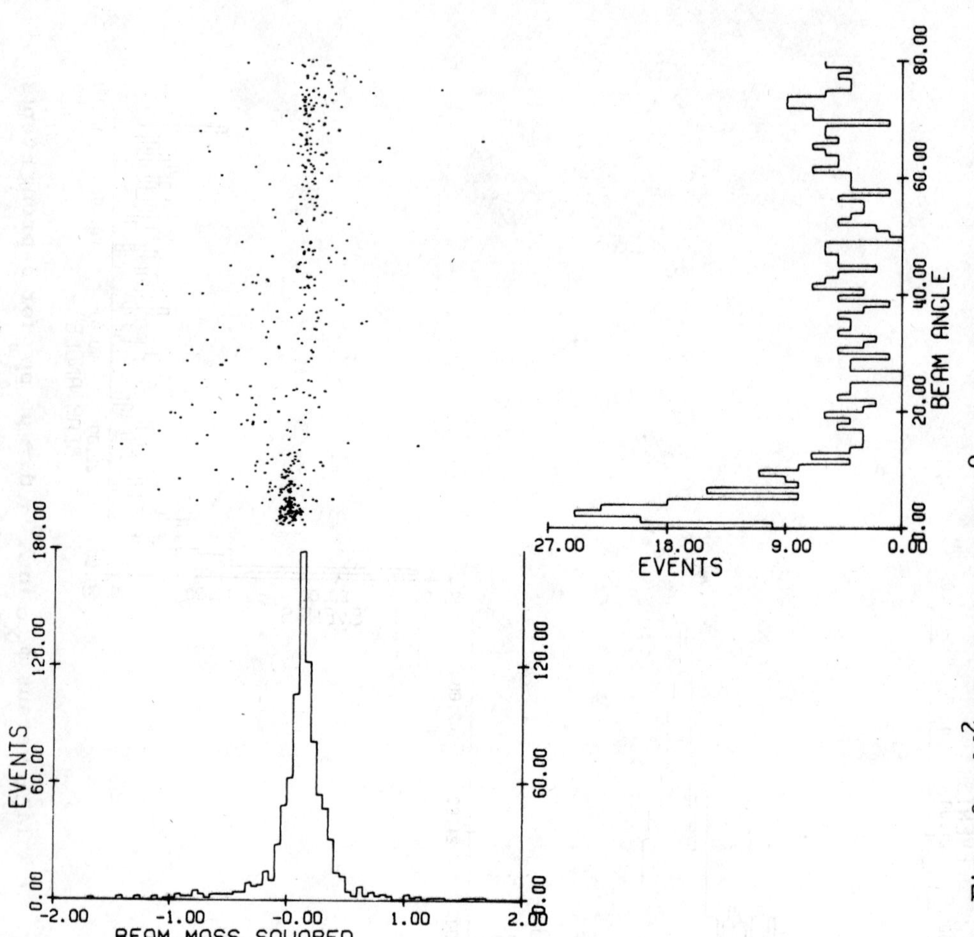

Fig. 8 M^2 versus

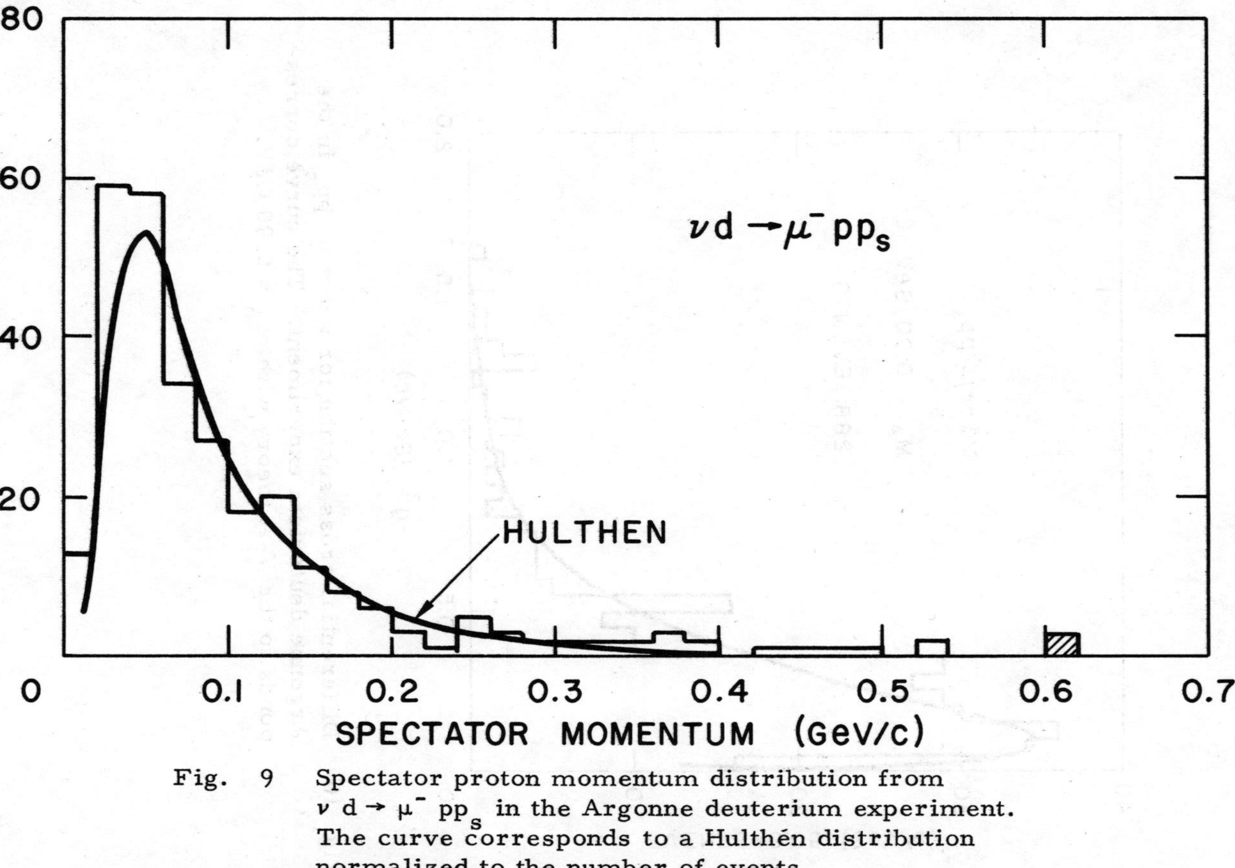

Fig. 9 Spectator proton momentum distribution from $\nu\, d \rightarrow \mu^-\, pp_s$ in the Argonne deuterium experiment. The curve corresponds to a Hulthén distribution normalized to the number of events.

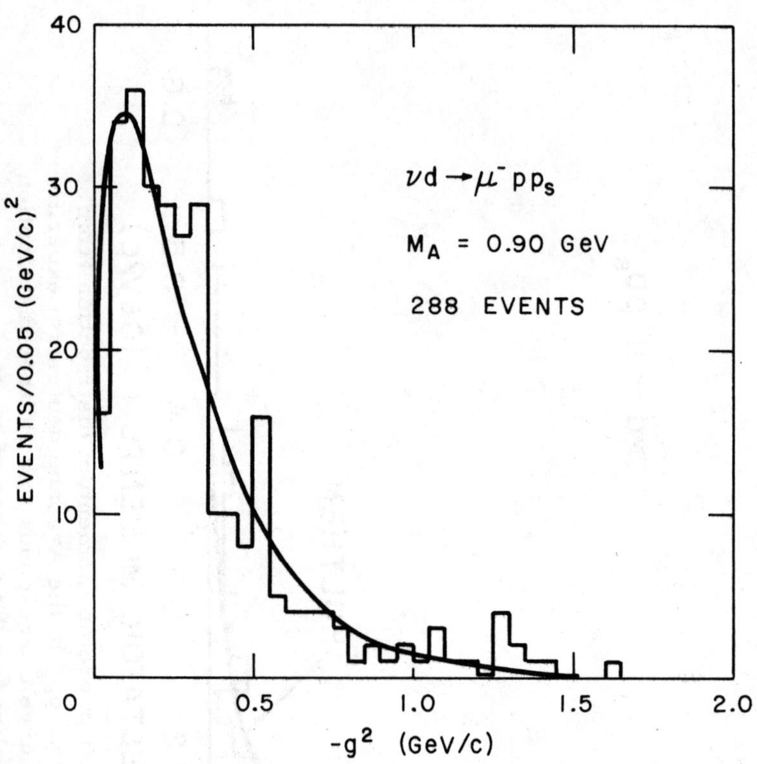

Fig. 10 Differential cross section for $\nu\, d \to \mu^- pp_s$ in the Argonne deuterium experiment. The curve corresponds to the V-A theory with m_A = 0.90 GeV.

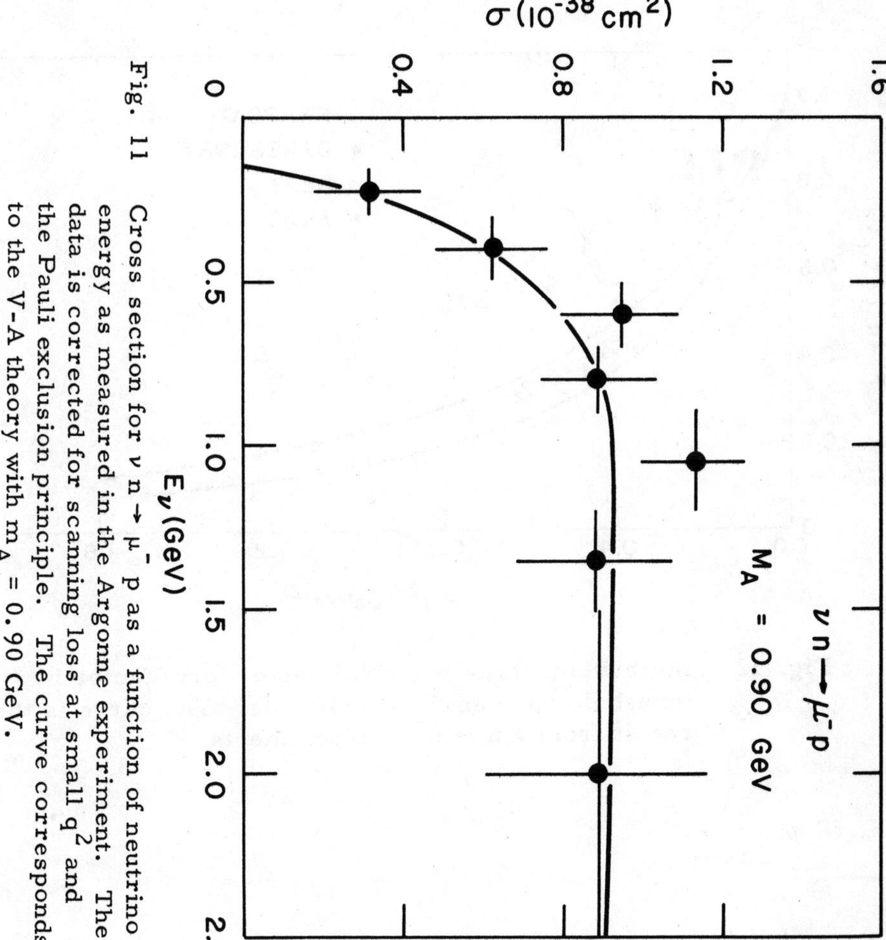

Fig. 11 Cross section for $\nu\, n \rightarrow \mu^- p$ as a function of neutrino energy as measured in the Argonne experiment. The data is corrected for scanning loss at small q^2 and the Pauli exclusion principle. The curve corresponds to the V-A theory with $m_A = 0.90$ GeV.

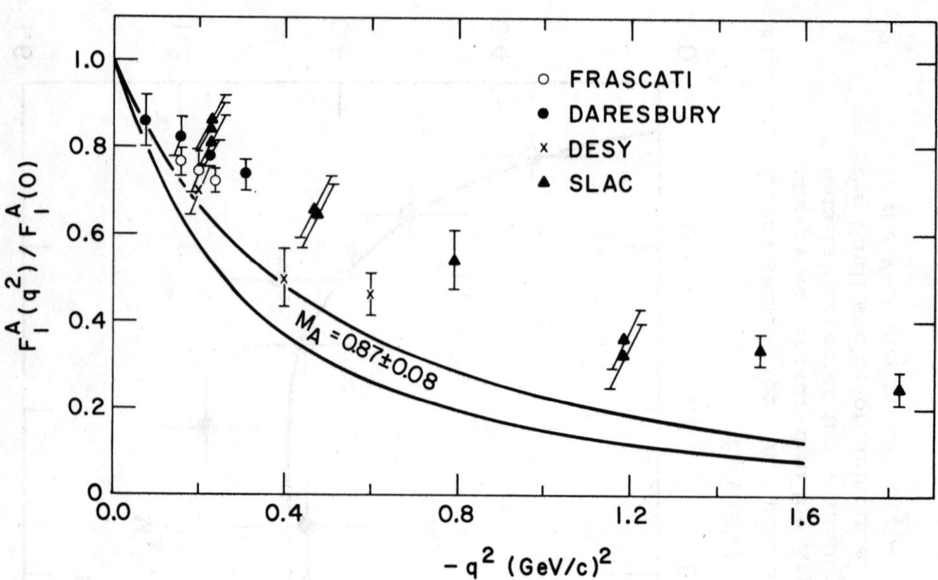

Fig. 12 Measurements of the axial vector form factor in threshold $ep \rightarrow en\pi^+$ experiments. The curve is the result from $\nu\, n \rightarrow \mu^- p$ experiments.

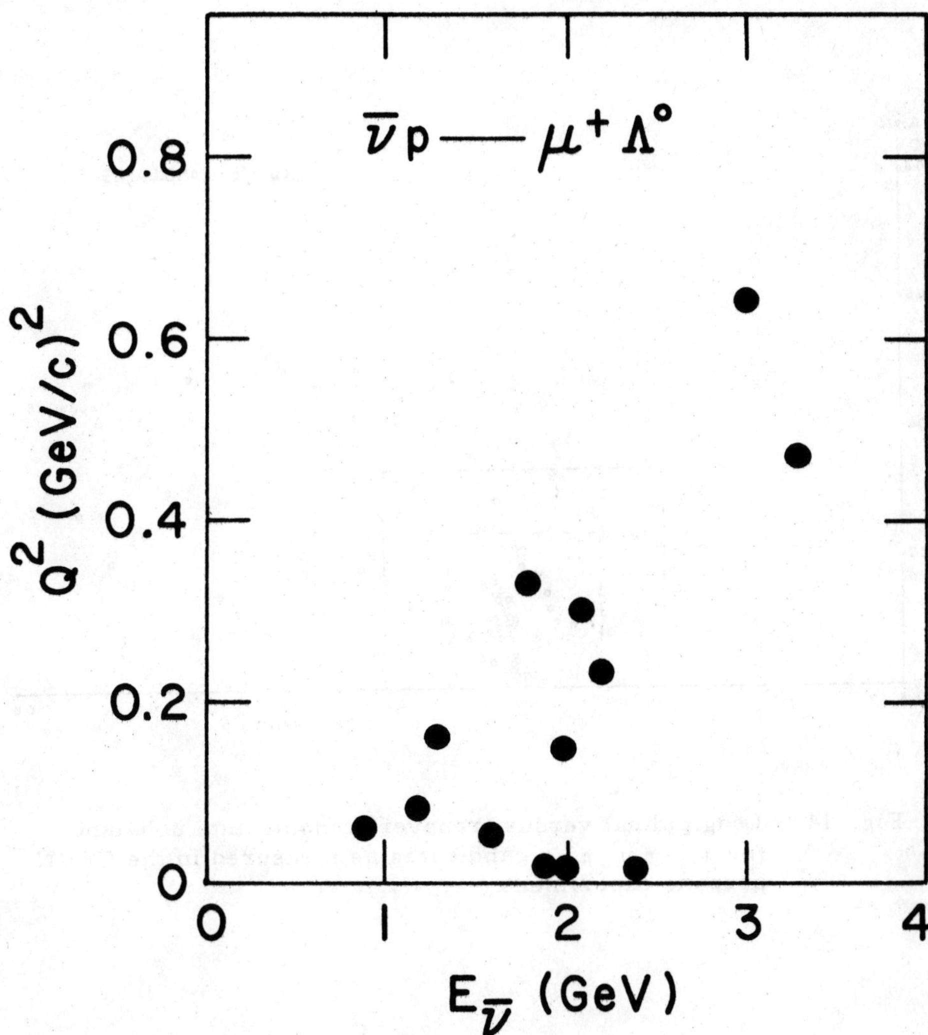

Fig. 13 Anti-neutrino energy versus q^2 for $\bar{\nu}\, p \rightarrow \mu^+ \Lambda^o$ events observed in the Gargamelle experiment.

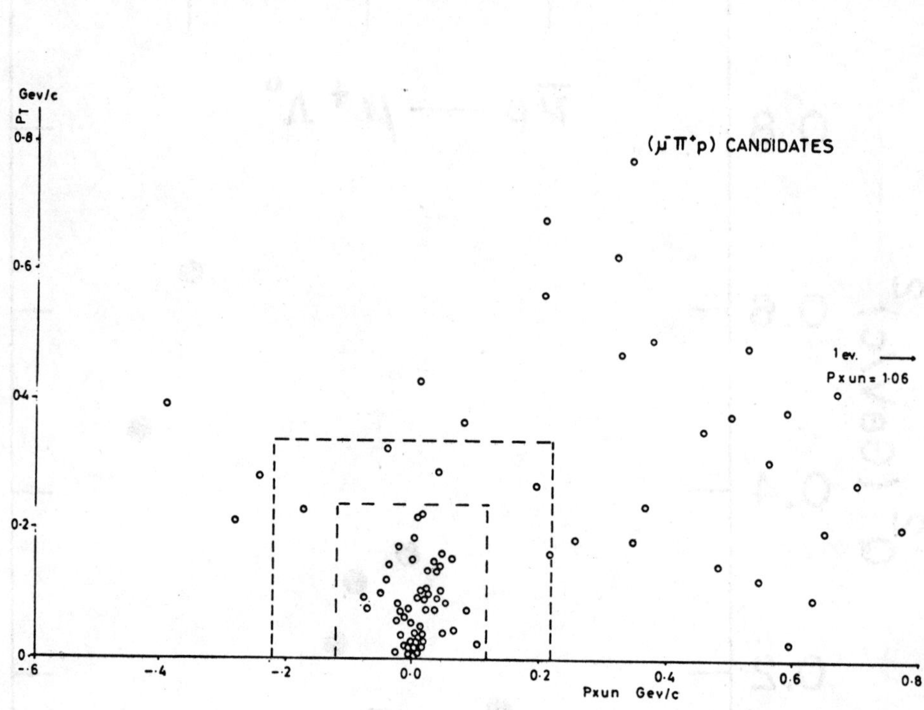

Fig. 14 Longitudinal versus transverse momentum unbalance for $\nu p \to \mu^- \pi^+ p$ candidates as measured in the CERN propane experiment.

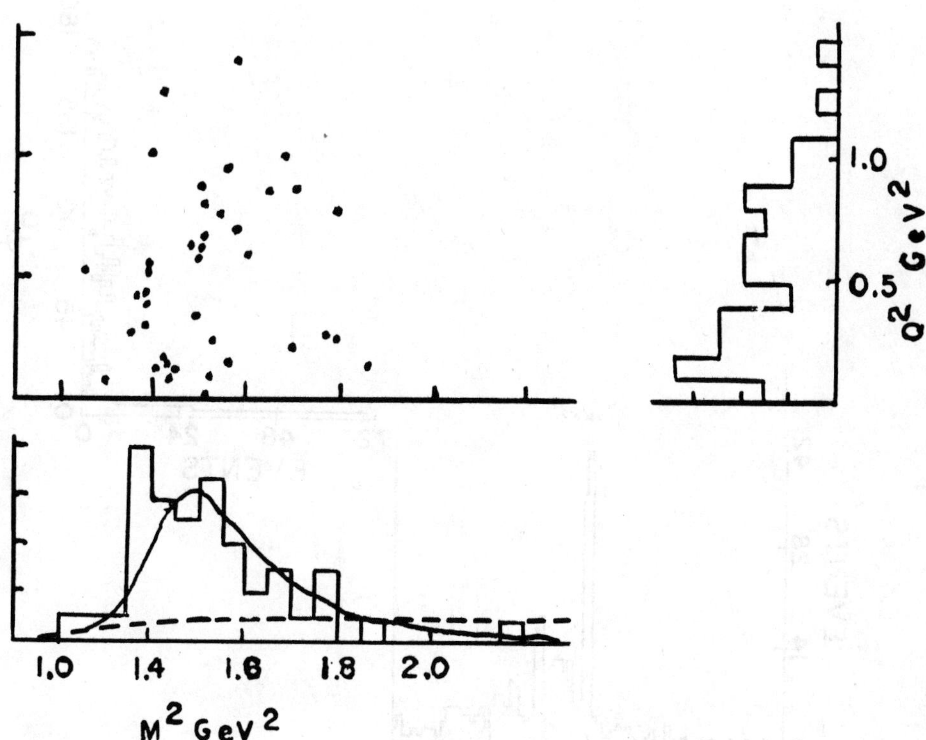

Fig. 15 Chew-Low plot $\nu\, p \rightarrow \mu^- \pi^+ p$ in the CERN propane experiment.

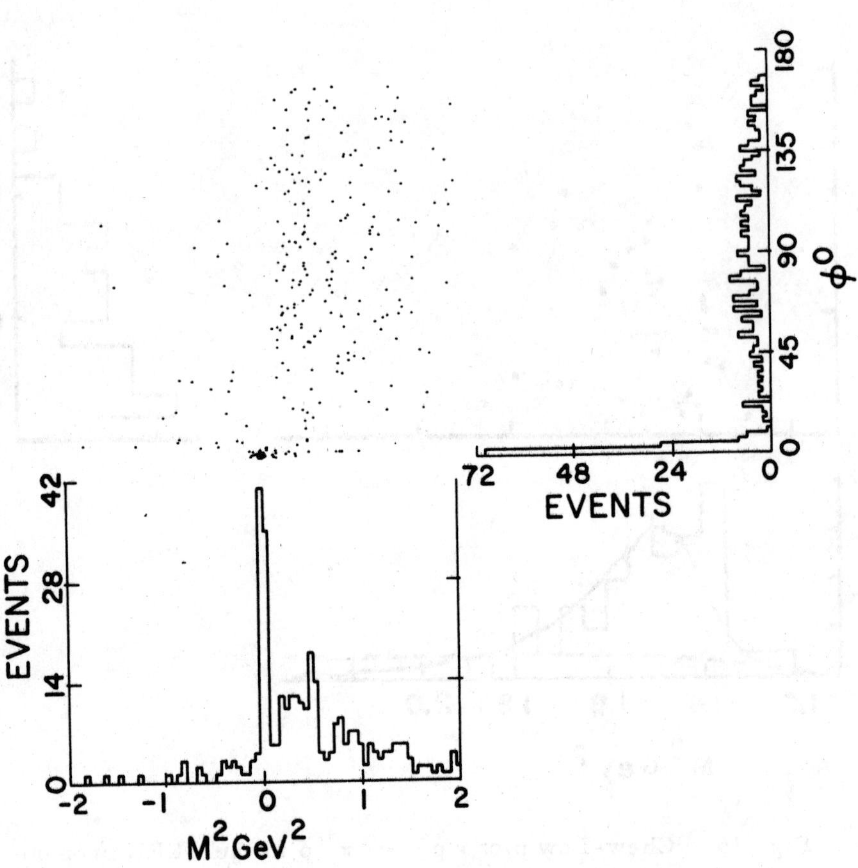

Fig. 16 $M^2_{X^o}$ versus ϕ_{X^o} in $X^o + p \to \mu^- \pi^+ p$ in the Argonne hydrogen experiment.

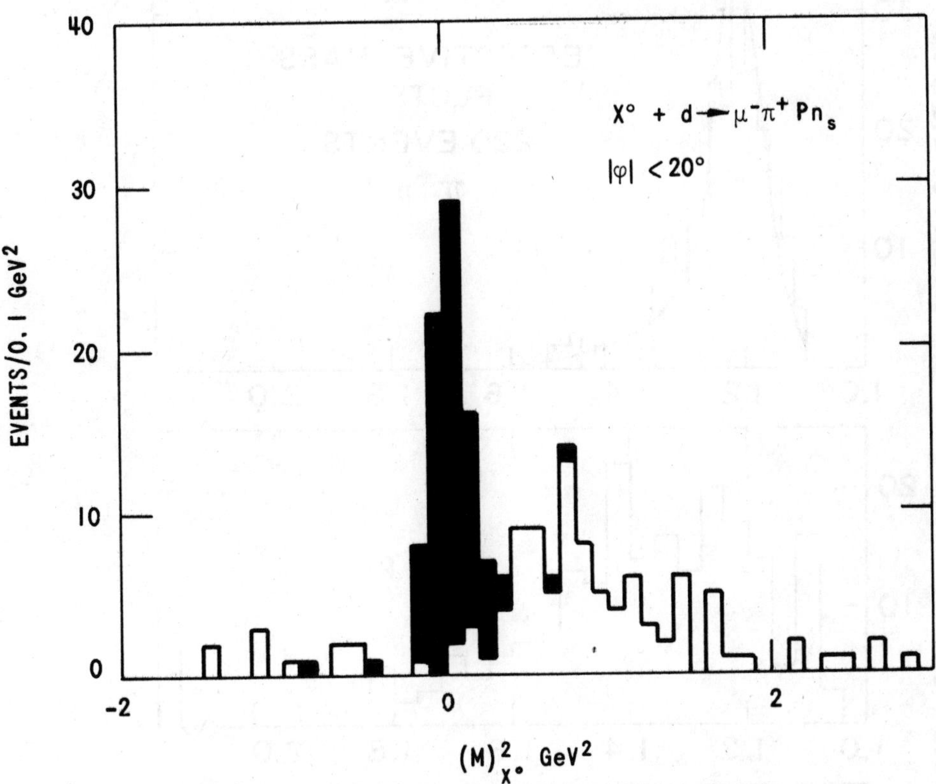

Fig. 17 Histogram of $(MM)^2_{X^o}$ when $\phi_{X^o} \leq 20^o$ in the process $X^o + d \rightarrow \mu^- \pi^+ p n_s$ in the Argonne deuterium experiment.

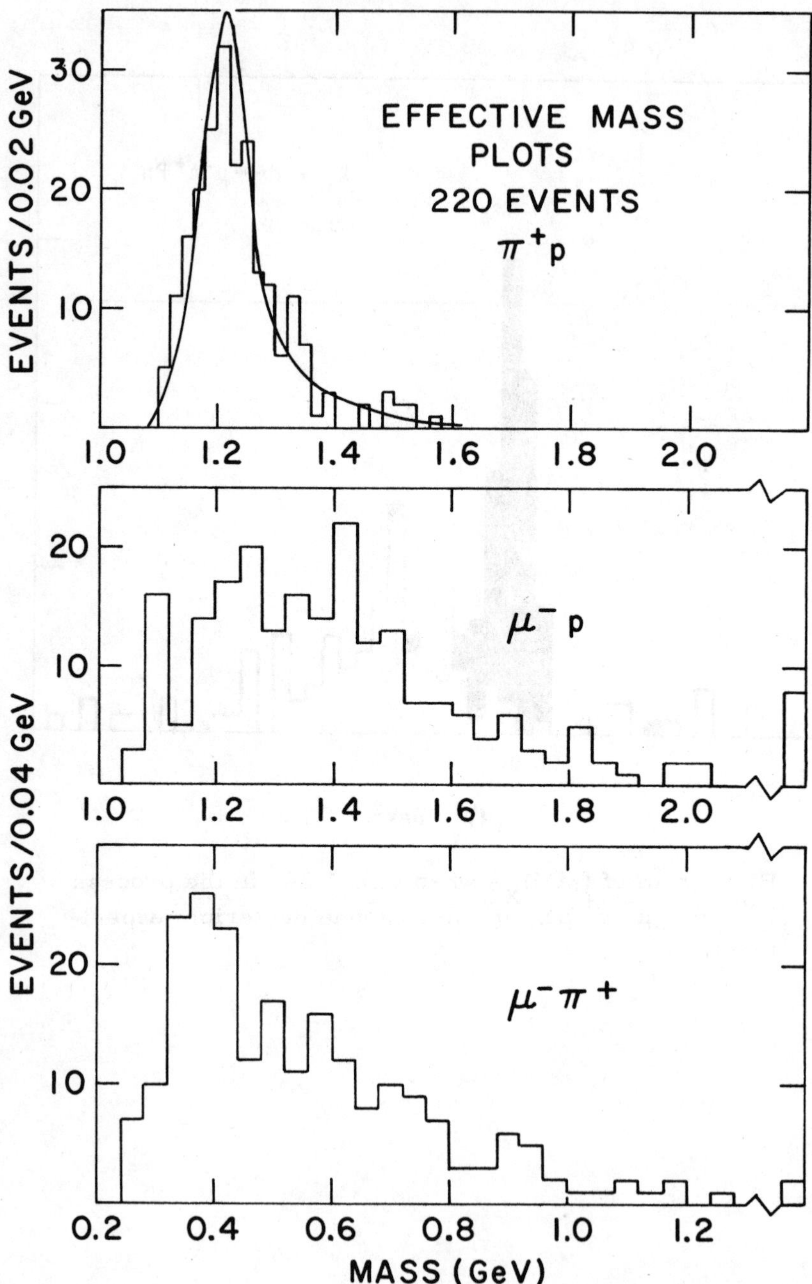

Fig. 18 Two-body invariant mass combinations for 220 $\nu p \rightarrow \mu^- \pi^+ p$ events in the Argonne hydrogen and deuterium experiments.

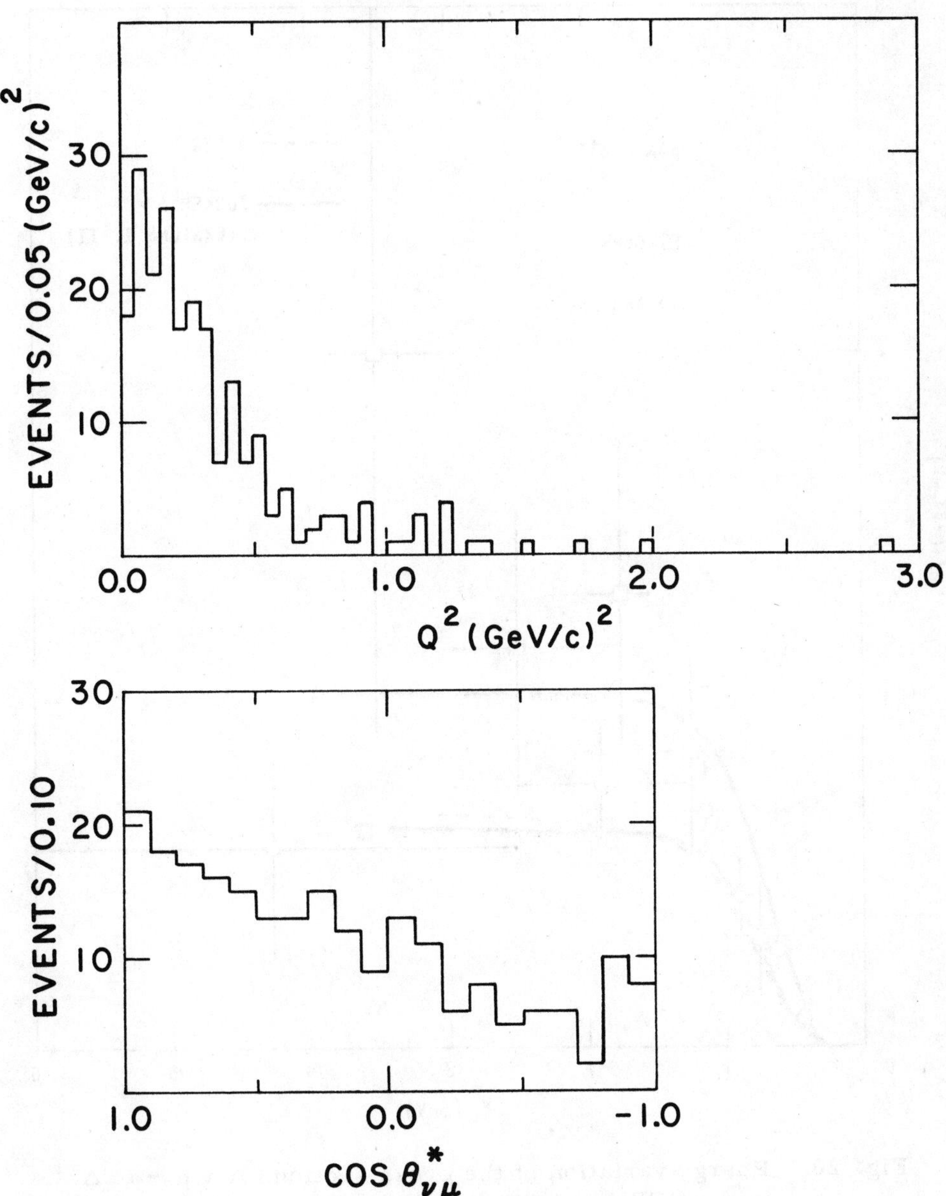

Fig. 19 Differential cross section for $\nu\, p \rightarrow \mu^- \Delta^{++}$ in the Argonne experiment plotted in terms of (a) q^2 and (b) $\cos\,\theta^*_{\nu\mu}$.

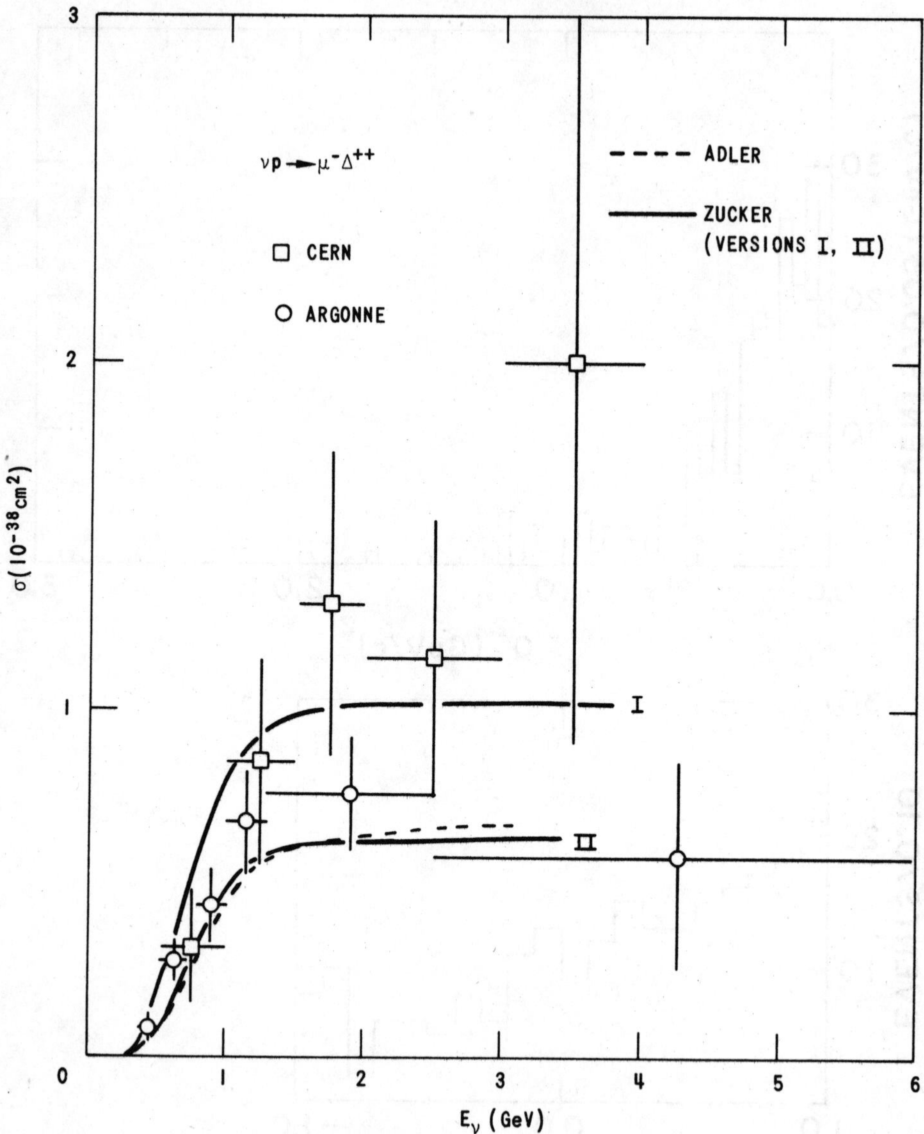

Fig. 20 Energy variation of the cross section for $\nu\, p \to \mu^- \Delta^{++}$ in the CERN and Argonne experiments. Curves correspond to models of S. Adler and P. Zucker.

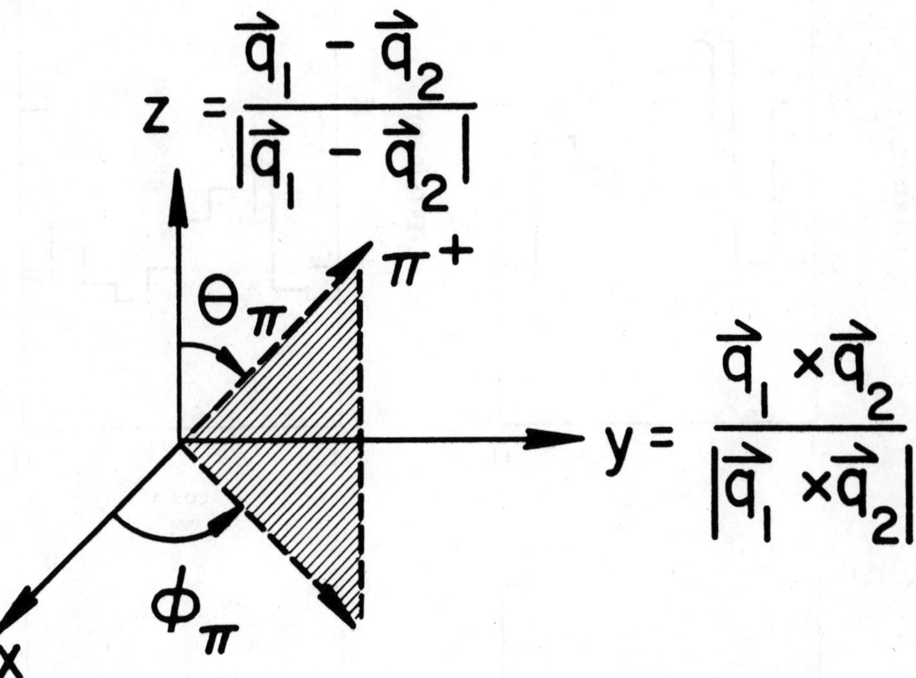

Fig. 21 Coordinate system for Δ^{++} decay in the reaction $\nu\, p \rightarrow \mu^{-}\, \Delta^{++}$.

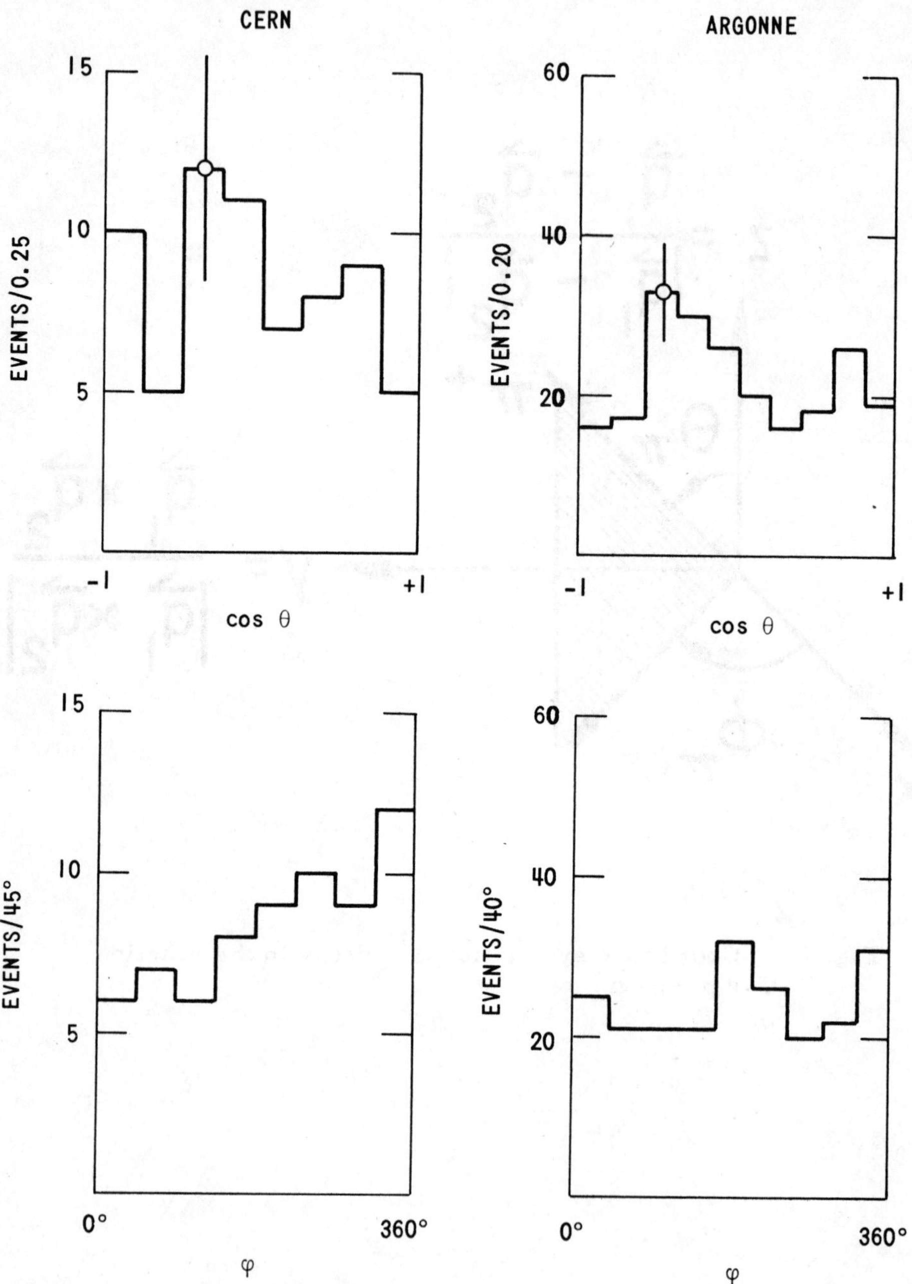

Fig. 22 Δ^{++} decay angular distributions in the CERN and Argonne experiments.

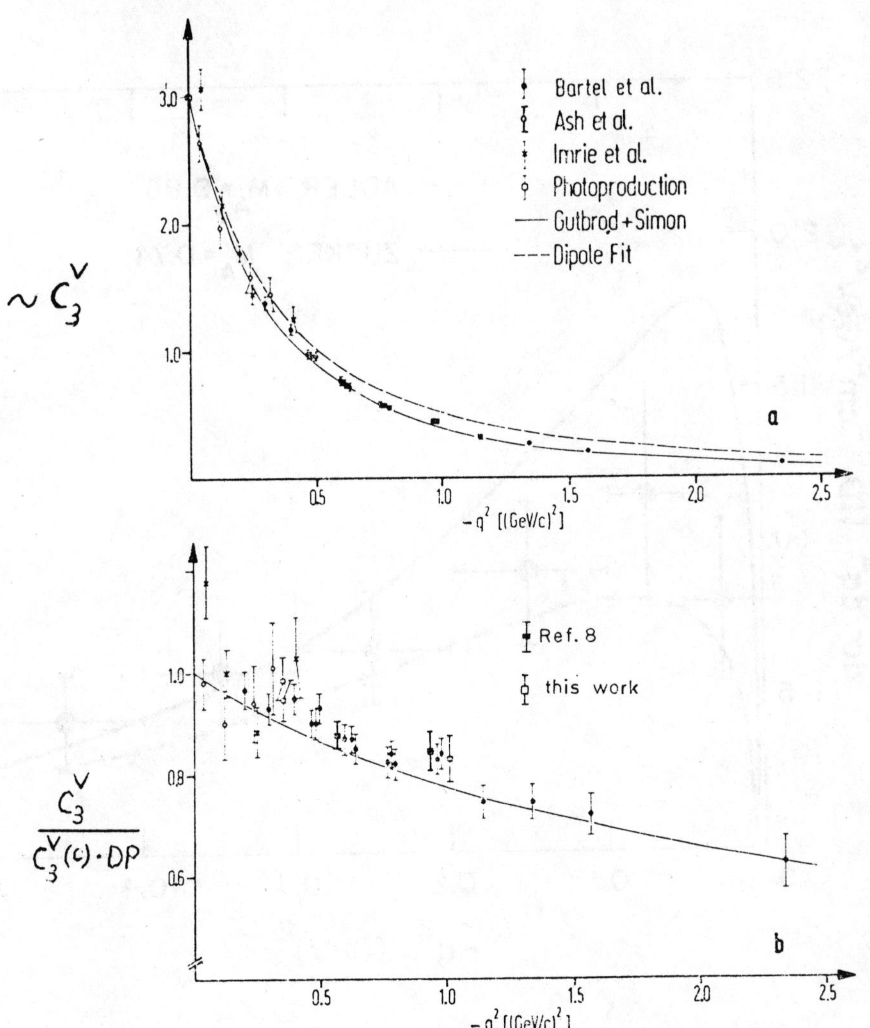

Fig. 23 Comparison of C_3^V with the dipole form factor measured in elastic electron scattering experiments.

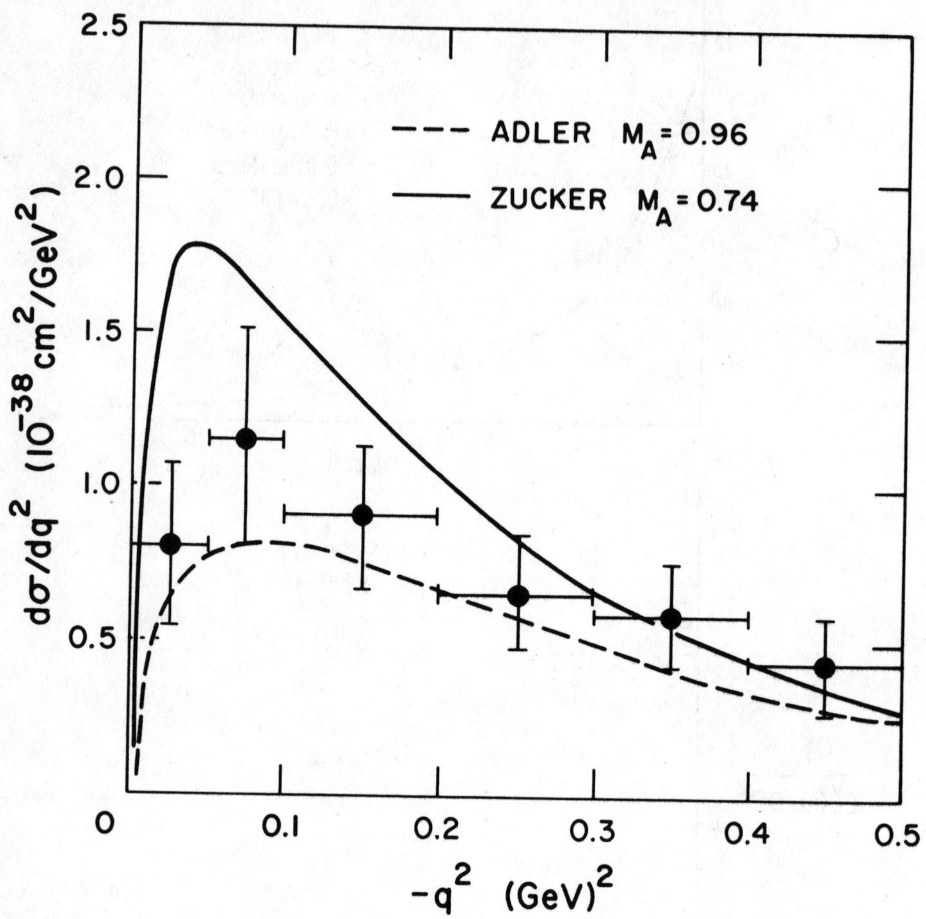

Fig. 24 Differential cross section for $\nu p \rightarrow \mu^- \Delta^{++}$ as measured in Argonne experiment.

NEUTRAL CURRENTS

A. Rousset
CERN, Geneva

ABSTRACT

This paper is an analysis of four neutrino experiments which are giving today positive evidence in favour of weak neutral current interactions : The search for muon neutrino elastic scattering on electrons in Gargamelle at CERN, the search for inclusive semi-leptonic neutrino and antineutrino reactions in Gargamelle at CERN, the search for the same reactions but at higher energies at NAL in the 1A experiment and the search for one pion channel in the Argonne 12 feet bubble chamber.

INTRODUCTION

"Do neutral currents exist or not ?" This paper is an attempt to give the best present answer to this question. The experiments which are giving a significant contribution to this question are critically discussed, from a strictly experimental point of view without any theoretical influence. The theoretical aspects are discussed after.

The first indication of neutral currents in neutrino interactions was reported by P. Musset, from the Gargamelle collaboration ation (1) at the Vittel Colloquium in May 1973. Muonless events were identified but the background analysis was not completed and only upper limits were given. It has to be noted that the principles of this experiment and preliminary but inconclusive results were presented (2) by A. Pullia already at the Balaton Conference in June 1972.

In January-March 1973, one candidate $\nu_\mu + e^- \rightarrow \nu_\mu + e^-$ was found in Gargamelle, presented at CERN and published (3) in Physics Letters 46 B, 1 p. 121, 3.9.73.

In July 1973, the observation in Gargamelle of neutrino-like interactions without muon or electron was definitely established, presented at CERN and published (4) in Physics Letters 46 B, 1, p. 138, 3.9.73. These results were presented at conferences held at SLAC (July 73), Berkeley (August 73), Bonn (August 73) and Aix-en-Provence (September 73).

Observation of muonless neutrino induced inelastic reactions was also presented at the same Bonn and Aix-en-Provence conferences by the NAL 1A experiment group and published (5) at a later date in Phys. Rev. Lett., A. Benvenuti et al., 32, 14, p. 800, 8.4.74.

A preliminary result from the Argonne bubble chamber was also presented at the Bonn Conference.

The review papers by G. Myatt (6) at the Bonn Conference, P. Musset (7) at the Aix-en-Provence Conference and H. Wachsmuth (8) at the Berkeley Conference give summaries of the experimental situation at that time.

The major part of this paper is devoted to new information which have appeared between September 73 and April 74. These are :

- New results from the Gargamelle experiment.
- New results from the 1A NAL experiment.
- New evidence from the Argonne 12 feet bubble chamber, presented at the APS Washington meeting (April 74).

The reader is supposed to have the knowledge of one of the papers 6, 7 or 8.

I) THE MUON ANTINEUTRINO SCATTERING ON ELECTRON IN GARGAMELLE

This interaction ($\bar{\nu}_\mu + e^- \rightarrow \bar{\nu}_\mu + e^-$) is characterized in Gargamelle by a single electron (e^-) originating in the liquid, emitted at small angle Θ_e with respect to the neutrino beam ($< 5^o$). The electron is expected to carry typically one third of the energy of the incident neutrino which is peaked between 1 and 2 GeV. The characteristics of the first event found in the antineutrino film in February 73 are presented in Table I. No event was found in the neutrino film.

The main source of background is the process $\nu_e + n \rightarrow e^- + p$ where the proton is not visible and the electron is emitted at small angle ($< 5^o$). This background has been determined in the neutrino film using the observed events of the type $\nu_\mu + n \rightarrow \mu^- (\Theta < 5^o) + p_+$ (not visible), and the estimated ν_e/ν_μ flux ratio (0.7 %), giving 0.3 ± 0.2 events. In the antineutrino film, the flux of ν_e is 10 times smaller and the number of expected background is then 0.03 ± 0.02 events.

The obvious criticism to this first result, is that one unique event cannot be taken as a definitive proof of something new. To get a more significant result the experiment with antineutrinos in freon was pursued using the increase of PS intensity given by the new injector. Two runs in November and December 73 were done with $5 - 6\ 10^{12}$ protons per pulse on the target. A second event was found in these pictures in January 74 (see Table I for the characteristics of this event). The running of the chamber and the scanning of the pictures are in continuous progress. With the present status of the scanning, the expected number of events at the minimum of the Weinberg curve (8, 9, 10) is 1.9 and the expected background 0.14 ± 0.09.

Table I - Characteristics of the electron events in Gargamelle

No and Lab.	Sign	Energy (MeV)	Θ_e (Degrees)
1) Aachen	-	385 ± 100	$1.4 {+1.6 \atop -1.4}$
2) Orsay	-	500 ± 120	2 ± 2
3) Milano	$\pm \gamma$	1500 ± 500	< 2
4) Paris	+	~ 2000	5.5
5) London	+	3000 ± 1000	$5 \pm .5$

The estimation of the background is checked by two direct observations :

1. The expected background in the neutrino film is $.3 \pm .2$ and no event was found. An underestimation of the background by a factor 7 is therefore excluded at 90 % confidence level.

2. Similar methods can be used to determine the number of expected e^+ coming from the reaction $\bar{\nu}_e + p \rightarrow e^+ (< 5^o) + n$ (non visible) and compared to the observed number. The ν_e/ν_μ flux ratio is still 0.7 % and 22 events of the type $\nu_\mu + p \rightarrow \mu^+ (< 5^o) + n$ (non visible) were found in a sample of the antineutrino film. Using these data the expected number of single positive electrons is 2 - 3 at the present stage of the scanning. 3 candidates were found : their characteristics are given in Table I. It has to be noticed that the energies of these 3 events are around 2 GeV or greater as expected in the $\nu_e + n \rightarrow e^+ + n$ with the cut at 5°.

The Milano event has an ambiguous sign because δ-rays and bremsstrahlung occur in the first centimetre of the track. It can also be a γ-ray, but the e^+ assumption is the most probable one. The Paris event has an angle larger than the 5° cut off. The London event has a vertex outside the fiducial volume, at the limit of the visible volume.

Therefore, among these 3 candidates only the first one would be retained if the cuts are strictly applied. The observed event number is not greater than the estimated number, and in conclusion the background in the negative electron is certainly not underestimated. The two relevant experimental numbers of the negative electron search in the antineutrino film are :

$$2 \text{ events} / 0.14 \pm 0.09 \text{ background}$$

The Poisson law gives a probability of about 2 % to explain the signal with the background. This experiment will be pursued in 1974 to double the number of pictures.

II) THE INCLUSIVE SEMI-LEPTONIC NEUTRAL CURRENT SEARCH IN GARGAMELLE

To detect a possible semi-leptonic neutral current interaction, according to the reaction ν_μ ($\bar{\nu}_\mu$) + N → ν_μ ($\bar{\nu}_\mu$) + Hadrons, all the detectable secondaries have to be identified as hadrons, to exclude a possible muon or electron : hadrons interacting in the liquid, clearly stopping protons, γ's coming from π^0 and giving electron pairs, decays of K^0 and Λ^0 ...etc... In the first published (4, 12) analysis 102 and 63 candidates (NC) were found in the neutrino and antineutrino pictures respectively. The absence of charged leptons in these events is particularly clear (except in the neutrino events where a small proportion of slow and non-decaying muons, less than 3 %, could be confused with short protons).

Neutrons simulate the same kind of events. Hence the various sources of neutrons were examined and only one source is found to be dangerous : the neutrons in equilibrium with neutrinos along the beam produced by neutrino interactions in the shielding and in the chamber. Such neutrons produce interactions in Gargamelle. If the source of the neutron is outside the visible volume (fig. 1 a), the interaction can be confused with a NC candidate; it will be a background event (B). If the source is visible inside the fiducial volume (fig. 1 b), the event is not confused as a NC candidate, it is taken as an "associated event" (AS) (AS). 15 and 12 AS events were found in the first analysis of the neutrino and and antineutrino pictures respectively. The number of AS events is proportional to the number of B events. The essential problem is to estimate the ratio B/AS.

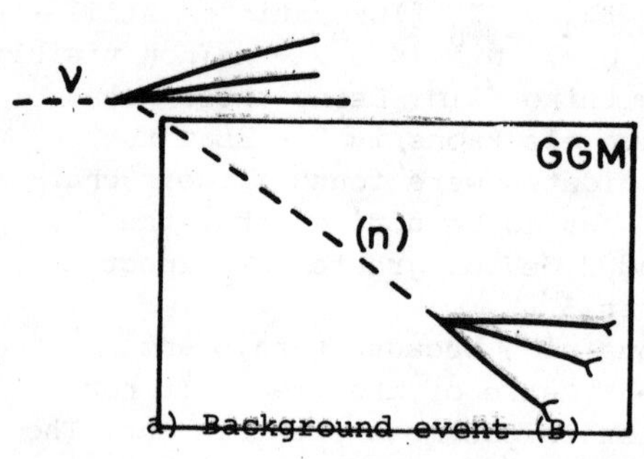

a) Background event (B)

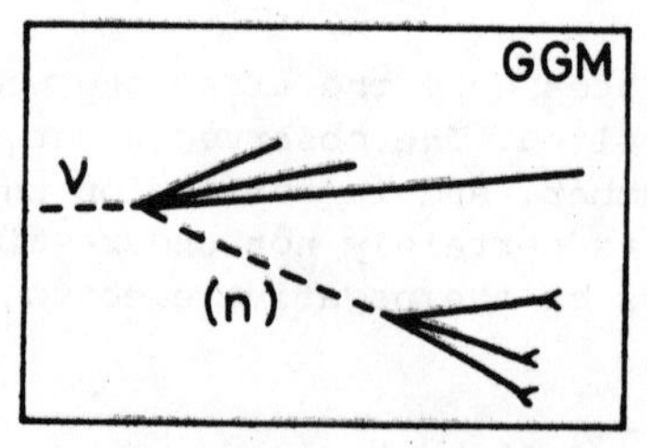

b) Associated event (AS)

Fig. 1 The neutron background

A very simple estimation of the ratio B/AS can be obtained using the pessimistic assumption that the neutrino flux density is uniform. Three equations can be written relating the various event rates in the fiducial volume of the chamber.

$$N = B + AS$$

where N is the total rate of neutron interactions satisfying the NC events criteria.

$$N = \alpha N_\nu \text{ (equilibrium neutrons-neutrinos)}$$

where N_ν is the rate of neutrino events and α is the proportion of neutrons able to produce an interaction satisfying the NC criteria, assuming an infinite potential length for their detection.

$$AS = \alpha N_\nu < p >$$

where AS is the rate of associated events, $< p >$ is the probability to detect a neutron interaction when the origin of the neutron is a neutrino interaction inside the fiducial volume.

$$\frac{B}{AS} = \frac{1}{< p >} - 1$$

The probability p is a function of the spatial distribution of the neutron interactions. Assuming that this distribution can be described by an exponential with a characteristic length λ,

$$p = 1 - e^{-\frac{L}{\lambda}}$$

where L is the potential length for the neutron detection.

This formula was completed by taking into account the hadron detection efficiency. Using as λ, the "effective mean free path", characteristic for the attenuation of the neutron flux in freon and quoted in the first publication (4) as $\lambda_e \sim 2.5\ \lambda$, the ratio B/AS was found to be 1.0, giving an estimation of 15 % neutron background in the NC events.

A more elaborate calculation was done introducing the following information :

- the radial distribution of the neutrino flux.
- the density distribution of matter around Gargamelle.
- the characteristics of a neutron cascade.
- the angular distribution of neutrons at production (in neutrino interactions).
- the energy spectrum of the neutrons.

The two first points are well determined. The neutron distributions were assumed to be the same as the distributions of protons emitted in neutrino interactions (angular distributions and energy spectrum). The characteristics of the neutron cascade were calculated by a Monte Carlo program using as input the elasticity curve of protons on nuclei (13) between 1 and 10 GeV.

The final result was B/AS = 0.6 ± 0.3 in agreement with the crude estimation.

A supplementary argument has been given to exclude the neutron as unique source of all the NC events. Assuming that all NC events are due to neutrons, the energy and angular distributions of these events are taken as energy and angular distributions for the neutrons. The resulting B/AS ratio was found to be 1.0 ± 0.25 which is still in strong disagreement with the result NC/AS $\sim$ 6.

Some obvious criticisms can be made to the preceding analysis :

- small statistics of the AS events.
- uncertainty of computed cascade processes of the neutrons.
- charge exchange of protons into neutrons is not taken into account.
- the neutron distributions are not directly measured.

The following presents some new development and gives answers to these criticisms.

1. New Statistics

The statistics of NC events and AS events in the neutrino film were nearly doubled. The results presented at the APS 74 meeting (24) are given in Table II. The numbers of events per film remain constant, confirming the published ratios.

Table II - New Statistics

		September 1973	April 1974
# NC / film	ν	$\frac{102}{111} = 0.92 \pm 0.13$	$\frac{191}{197} = 0.97 \pm 0.10$
	$\bar{\nu}$	$\frac{63}{276} = 0.23 \pm 0.012$	$\frac{70}{298} = 0.235 \pm 0$
#AS / film	ν	$\frac{15}{111} = 0.135 \pm 0.037$	$\frac{40}{277} = 0.144 \pm 0.025$
	$\bar{\nu}$	$\frac{12}{276} = 0.043 \pm 0.012$	$\frac{14}{328} = 0.042 \pm 0.012$

2. The Proton Experiment (preliminary results)

Pictures with incident protons at 4, 7, 12, 19 GeV/c were taken in March 74 to study the cascade process of protons. Assuming charge symmetry between neutron and proton on a nuclear target (Freon) the characteristics of the proton cascade can be used for the neutron cascade. The four following quantities were measured and compared to the Monte Carlo results :

a) The apparent mean free path λ_a, (fig. 2) defined as the average distance $< \ell >$ of the first interaction in which the proton deposits at least 150 MeV. Fig. 3 gives the results. This quantity is an important parameter in the determination of the neutron detection efficiency.

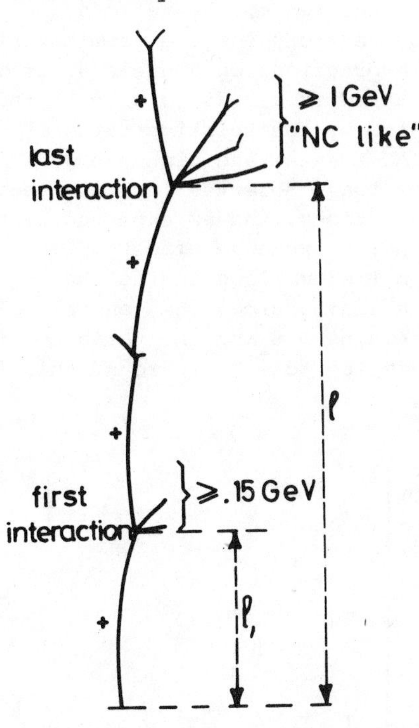

Fig. 2 The proton experiment

$\lambda_a = < \ell_1 > \quad \lambda_c = < \ell >$

b) The cascade length λ_c (fig. 2) defined as the average distance of the last interaction of the cascading proton which would simulate a NC event (visible deposited energy larger than 1 GeV). The cascade length (fig. 3) is slightly larger than λ_a at low energy (4 GeV) and increases with energy. In the range of the AS event energies, the cascade length is about 1.2 m. If such a length is used to characterize the spatial distribution of the AS events then the ratio B/AS is equal to 0.7.

c) The proportion of protons which interact after the first interaction at least once simulating a NC event. This proportion which increases with proton energy (fig. 4) is directly compared with the Monte Carlo prediction; the agreement is good.

d) The proportion of neutron interactions which come from proton neutron charge exchange and simulate NC events. Again the comparison with the Monte Carlo is good (fig. 4 b).

More detailed information on the inelasticity and neutron energy measurement will be extracted in the future from this proton experiment. It already provides an excellent check of the details in the Monte Carlo estimation.

3. The Proton-Neutron Charge Exchange

The other difficulty in the neutron background estimation is the knowledge of the neutron production characteristics. In the first publication (4) the energy spectrum and the angular distribution was varied to check this influence on the B/AS ratio. Here, the proton-neutron charge exchange is included in the calculation. The protons are indeed a source of neutrons by charge exchange in the iron, and this source is not entirely detected in the AS events. Charge exchange is included in the Monte Carlo program, as well as more refined elasticity curves for various neutron energies. The resulting ratio B/AS is given in fig. 5, as a function of the ratio n/n + p of neutrons emitted from neutrino interactions on a nucleus. It was varied from 0.25 to 0.75 as indicated by the relative ratio of (AS-events) and high energy protons, observed in neutrino interactions, and as expected by the charge exchange effect on the initial nucleon inside the nuclear source. The central value remains 0.6 and the variation is limited to $\pm$ 20 % around this value.

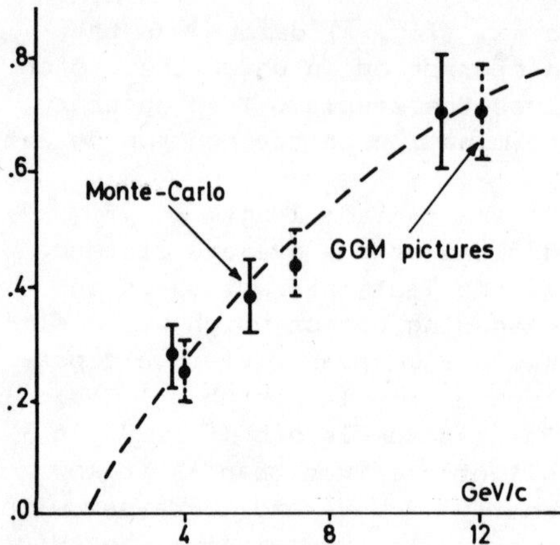

Fig. 4 (a) Fraction of protons interacting more than once and simulating an NC event

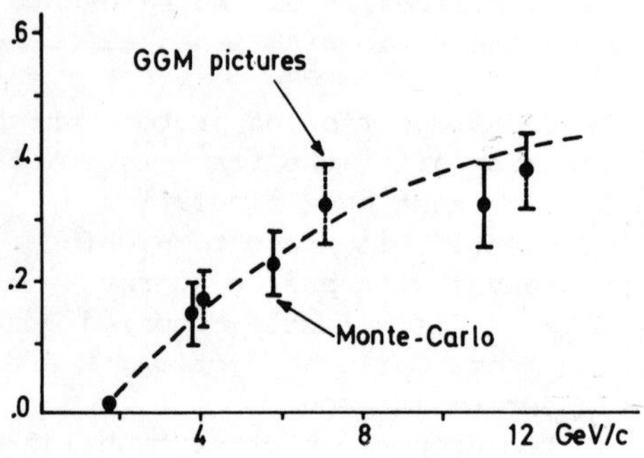

Fig. 4 (b) Fraction of secondary neutrons simulating NC events

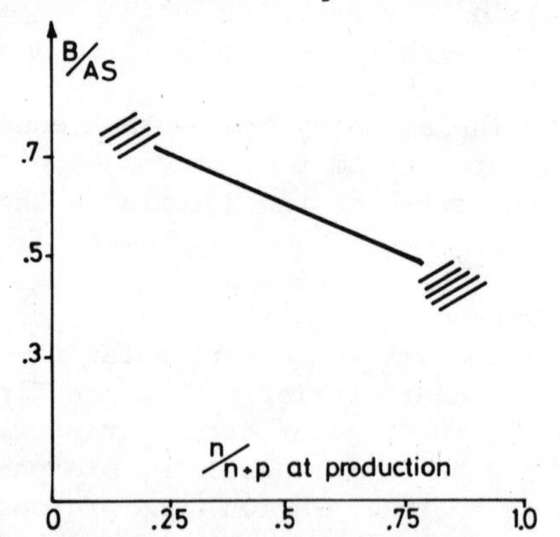

Fig. 5 Neutron and proton dependence of the B/S ratio

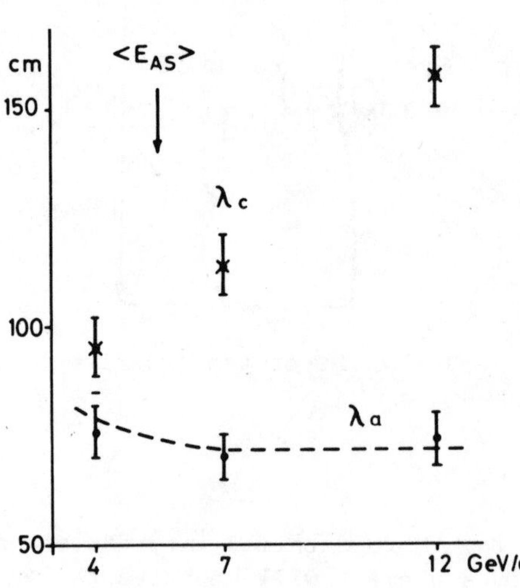

Fig. 3 Apparent mean free path and cascade length

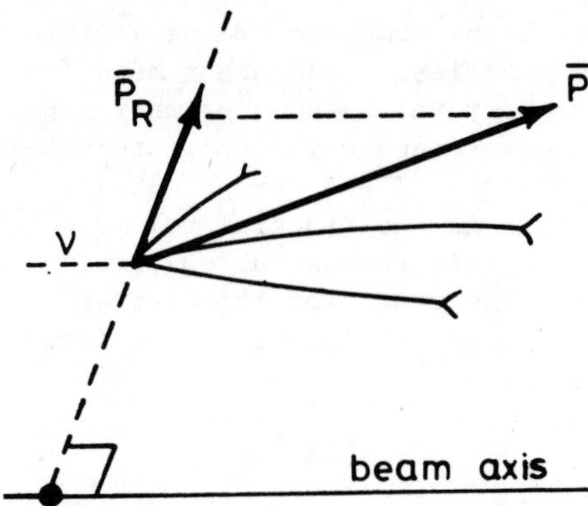

Fig. 6 Definition of the radial momentum $\bar{p}_R$

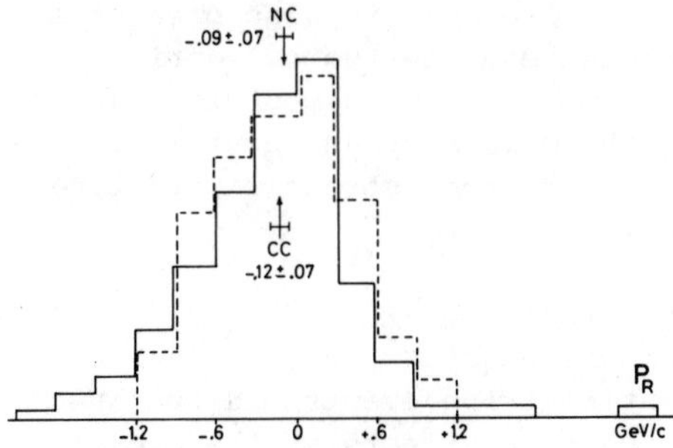

Fig. 7 The radial momentum distribution in the neutrino events

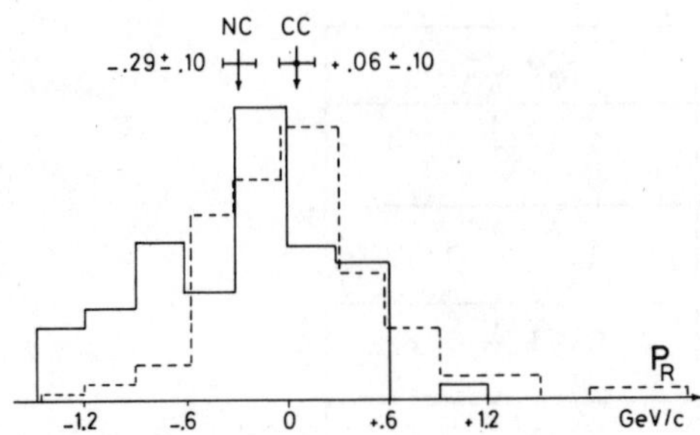

Fig. 8 The radial momentum distribution in the antineutrino events

4. The radial Momentum Distribution

The radial momentum P_R is defined as the projection of the reconstructed momentum onto the transversal axis passing through the vertex of the event (fig. 6).

For the neutrino or antineutrino events the average P_R is expected to be negative and small, of the order of 50 MeV, due to the effect of the hadron detection efficiency. For a neutron entering from the side the radial momentum is expected to be large, of the order of .7 GeV/c. Therefore the P_R distribution is a test to discriminate neutrino events from neutron events. Fig. 7 shows the P_R distribution of the CC and NC events with neutrino. The two distributions are compatible. Fig. 8 shows the P_R distribution of antineutrino events. There is an apparent disagreement, which can be interpreted partially as a statistical fluctuation: $< P_R >_{CC}$ which is expected slightly negative is **slightly positive. The** combined neutrino and antineutrino event distribution gives

$$< P_R >_{CC} = - 40 \pm 60 \text{ MeV}$$

$$< P_R >_{NC} = - 180 \pm 60 \text{ MeV}$$

$$< P_R >_{NC} - < P_R >_{CC} = - 140 \pm 80 \text{ MeV}$$

$< P_R >_{CC}$ is compatible with the expected effect. The difference between NC and CC is compatible with a contamination of about 15 % neutrons, and clearly not compatible with only neutrons.

5. The apparent mean free path of the primary neutral particle

Assuming that the direction of the incident neutral particle can be reconstructed from the

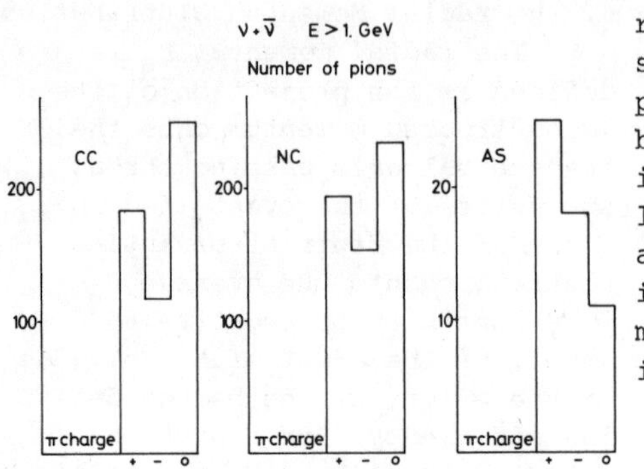

Fig. 9 Pion charge distribution

resulting momentum of the visible secondaries, an apparent mean free path for the unknown primary can be determined using the interaction length ℓ and the potential length in the fiducial volume along this reconstructed direction. The result of the Bartlett method on the increased statistics is :

$$\left(\frac{1}{\lambda}\right)_{NC} = 0.24 \pm 0.11$$

as compared to

$$\left(\frac{1}{\lambda}\right)_{CC} = 0.20 \pm 0.10$$

applying this same method to the hadronic part of the CC events.

The small systematic effect on the CC events is also present in the NC events, but the good agreement between the two categories shows that the primary particles are compatible with neutrinos and cannot be particles with mean free path smaller than about 4 m, which correspond in freon to at least 4 apparent strong interacting mean free paths (90 % confidence level).

6. Analysis of the Pion Charge Distribution

Fig. 9 shows the observed distributions of π^+, π^-, π^0 in the NC, CC and AS events. Neutrino and antineutrino events are combined with equal weight to compensate for possible effects of the muon charge. Table III gives the ratios $\pi^0/(\pi^+ + \pi^-)$, π^0/π^- and the relevant differences.

Table III - Pion Charge Distributions ($\nu + \bar{\nu}$)

	$\pi^0/\pi^+ + \pi^-$	π^0/π^-
CC	0.83 ± 0.08	2.14 ± 0.24
NC	0.68 ± 0.06	1.72 ± 0.29
AS	0.26 ± 0.09	0.61 ± 0.27
NC - CC	$- 0.15 \pm 0.10$	$- 0.42 \pm 0.37$
NC - AS	$+ 0.42 \pm 0.11$	$+ 1.11 \pm 0.40$

The π^0 proportion in the NC events and in the AS events is significantly different (~ 4 standard deviations). The difference with CC events is smaller, and compatible with an admixture of neutron background. This observation favours a different origin to the NC events and to the AS events and similar origin to the CC events.

These charge distributions cannot be directly compared to other measurements, e.g. in $e^+ e^-$ annihilations at SPEAR or in neutron-proton interactions, because the secondary interactions inside the nucleus modify the charge distribution. The future propane experiment will give better determination of the pion charge distribution.

In conclusion, after 6 months of supplementary analysis, the Gargamelle results are confirmed mainly on the following points :

- The NC and AS rates are doubled and the ratios NC/CC and AS/NC are not significantly modified.
- The neutron cascade process is experimentally studied using the proton pictures in Freon and the measurements agree with the Monte Carlo estimations.
- The π charge distributions show an agreement between NC and CC events, and a significant difference between NC and AS events.

New ratios will be given at the London Conference (July 74) when the complete analysis will be done on the new statistics, but the previous numbers can be confirmed.

$$\left(\frac{NC}{CC}\right)_{\nu} = 0.23 \pm 0.04 \qquad \left(\frac{NC}{CC}\right)_{\bar{\nu}} = 0.43 \pm 0.12$$

A total of 260 muonless events (ν and $\bar{\nu}$) are presently identified in Gargamelle with an estimated neutron background of 30 ± 15 events. Such a result is highly significant (15 standard deviations).

III) THE SEARCH FOR INCLUSIVE SEMI-LEPTONIC NEUTRAL CURRENTS IN THE EXPERIMENT 1A AT NAL

1. The First Experiment

The detector is a calorimeter filled with 70 tons of scintillating mineral oil, split into four sections of 4 segments each, (Fig. 10 a). The energy of the hadronic shower of a neutrino interaction is measured by the total pulse height coming from all segments. The accuracy is of the order of 15 %. The muon is identified and measured in a spectrometer made of magnetized iron, split into four sections, 1.5 m long each. The particles, hadrons and muons, are localized by spark chambers between two successive sections. A counter A in front of the calorimeter, in anticoincidence eliminates the incoming charged particles. Two counters B and C are used to sign a muon.

A neutral current candidate (NC) is defined by the following requirements : $\bar{A}$, E, $\bar{C}$ (and SC5). E is a selection of an energy deposition in the calorimeter larger than a value E.

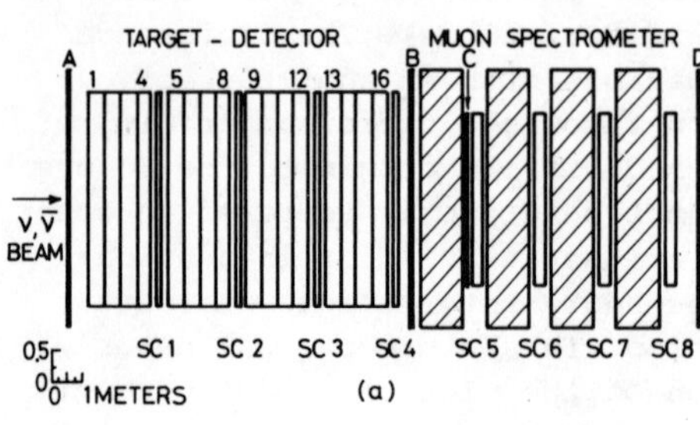

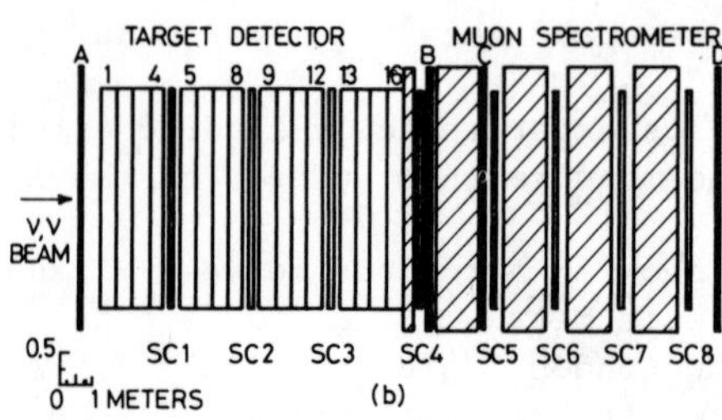

Fig. 10 NAL 1A experiment set-up

A charged current candidate (CC) is defined by : $\bar{A}$,E,B,C or SC5.

The event vertices are selected in the following fiducial volume : from element 7_2 to 12 and in a square 2.4 x 2.4 m^2 in transverse section. The elements 1 to 6 are eliminated to avoid a possible neutron contamination and to eliminate events with a small muon detection efficiency. The events in elements 12 to 16 have not a good measurement of the hadronic shower. 93 CC and 76 NC events were found. A subtraction on the NC events has to be done to take into account the event with a muon missing the C counter. The angular distribution is taken as measured in the detector using azimuthal symmetry of an event and adding the information of the CC events produced in the first iron block to increase the solid angle of the muon detection. The correction for undetected muons is about 50 % of the initial signal. The final result given at the Bonn Conference was R_1 = (NC/CC) corrected = 0.28 $\pm$ 0.09. The beam was a non-focused and non separated beam produced by protons of 300 and 400 GeV incident on the target. It is a natural mixture of neutrinos (80 %) and antineutrinos (20 %). The most important criticism to this result is the problem of the small muon detection efficiency which requires a large correction. An improved analysis of this correction was done after the Bonn Conference and is now published (5). It gives the result : R_2 = 0.23$\pm$0.08.

2. The Second Experiment

Two important modifications were done to increase the muon detection efficiency (fig. 10 b).

1. The addition of 35 cm of iron immediately downstream of the calorimeter to filter the muons. The B counter and the SC4 spark chamber are now used to identify a muon.
2. The increase of the area of the counter C and the exchange of narrow gap spark chambers with large gap spark chambers.

The muon detection efficiency is therefore increased to about 85 %. But a new correction is now necessary: the hadrons can penetrate the 35 cm of iron and simulate muons by a pulse in the C counter or a track in SC4. The probability of this effect, the punch-through probability ε_p is determined from the CC events with muons well identified in the C counter and with an additional track in B counter and SC4 spark chamber. ε_p is therefore measured in function of the vertex position in the calorimeter and in function of the hadronic energy E (Fig. 11 a and b). The average value of ε_μ is determined from the observed distribution and assuming the azimuthal symmetry of an event. Nevertheless 4 % of the events are never detected, with a muon angle larger than 0.5 radian. A correction based on a theoretical model (scaling) has been done. The check of the ε_μ efficiency on the observed events is shown in Fig. 11 c.

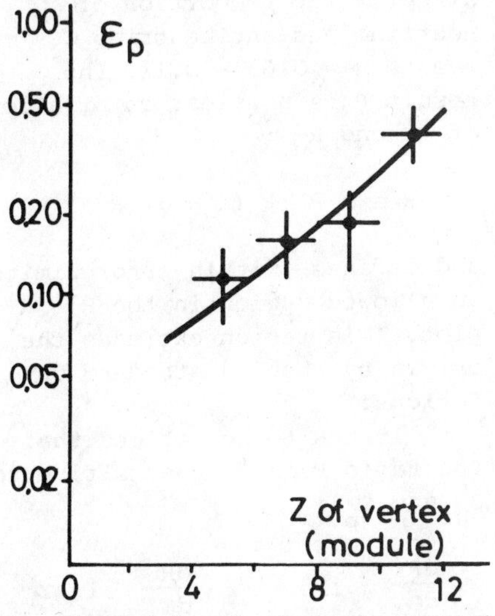

Fig. 11 (a) Punch-through correction versus the vertex position

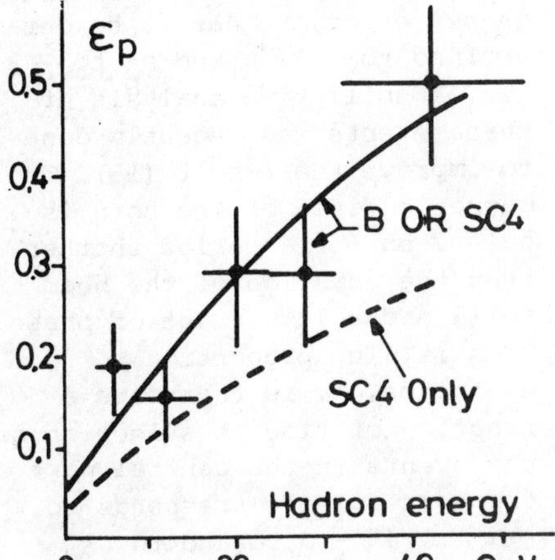

Fig. 11 (b) Punch-through correction versus the hadronic energy

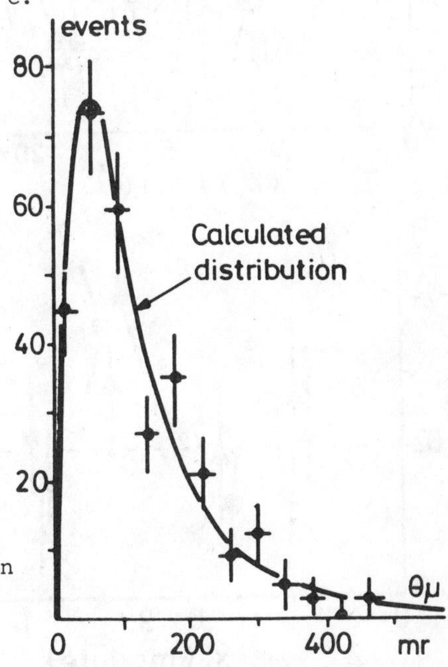

Fig. 11 (c) Muon angular distribution

If R_m is the measured ratio of events without a muon and with a muon, the corrected ratio R is

$$R = \frac{NC}{CC} = \frac{(\varepsilon_\mu + \varepsilon_p - \varepsilon_\mu \varepsilon_p)(1+R_m) - 1}{1-\varepsilon_p (1+R_m)}$$

The ratio R has been studied as a function of the vertex position (Fig. 12 a and b) and as a function of the E cut off (Fig. 12 c). It remains compatible with a constant value. The global result is (14) :

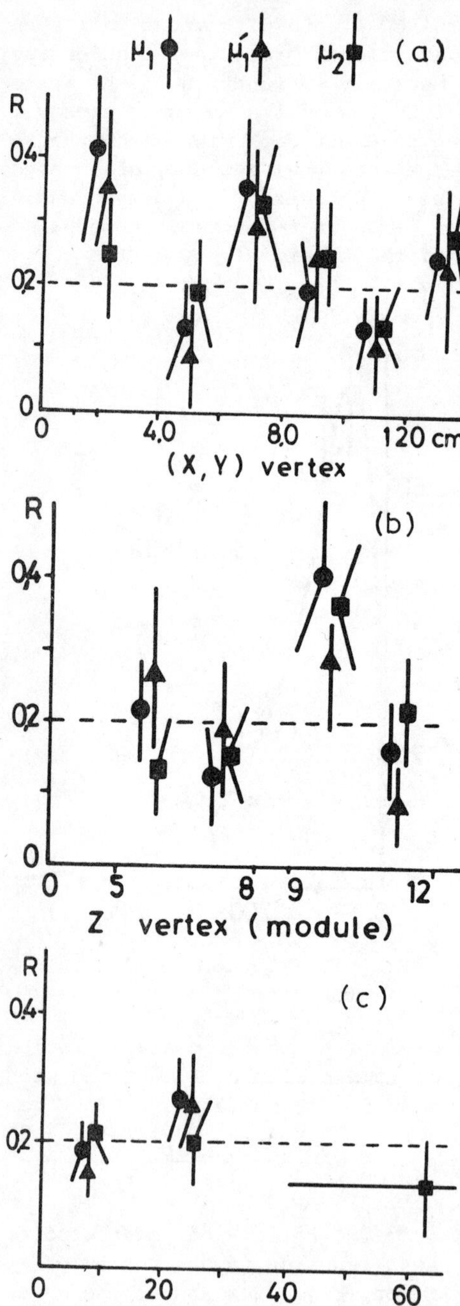

Fig. 12 Position and hadronic energy dependence of the result

$$R = 0.20 \pm 0.05$$

The incoming flux is a mixture of neutrinos and antineutrinos enriched in antineutrinos by the operation of a magnetic horn. The measured proportion of CC events with a negative muon and a positive muon gives directly the proportion of neutrino and antineutrino CC events $\alpha = 0.63 \pm 0.11$. The result R is a linear combination of R_ν and $R_{\bar{\nu}}$:

$$R = \alpha\, R_\nu + (1 - \alpha)\, R_{\bar{\nu}}$$

and defines within error limits an allowed region in the R_ν - $R_{\bar{\nu}}$ plot. This region excludes the origin by about 4 standard deviations.

It has to be noticed that the ratio R is very sensitive to ε_μ and ε_p :

$$\frac{dR}{d\varepsilon_\mu} \sim 1.5 \qquad\qquad \frac{dR}{d\varepsilon_p} \sim 1$$

To be compatible with the quoted error, ε_μ has to be determined to $\pm$ 2 % and ε_p to $\pm$ 3 %.

A spill time analysis of these events was recently done to improve the result (15). The magnetic field of the horn is pulsed on for a period shorter than the duration of the beam spill and a little out of phase with it. The proportion of neutrinos, α, is therefore a function of time. A selection of the events in the central part of the horn pulse corresponds to $\alpha = 0.45 \pm 0.06$ and to $\alpha = 0.74 \pm 0.06$ for the events outside this central part (fig.13).

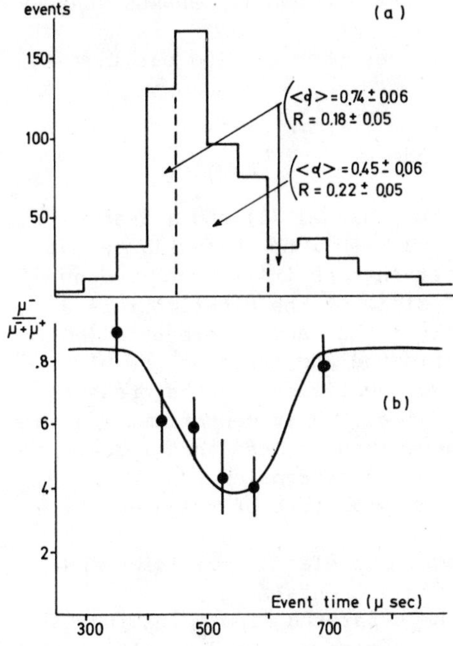

Fig. 13 Spill time analysis

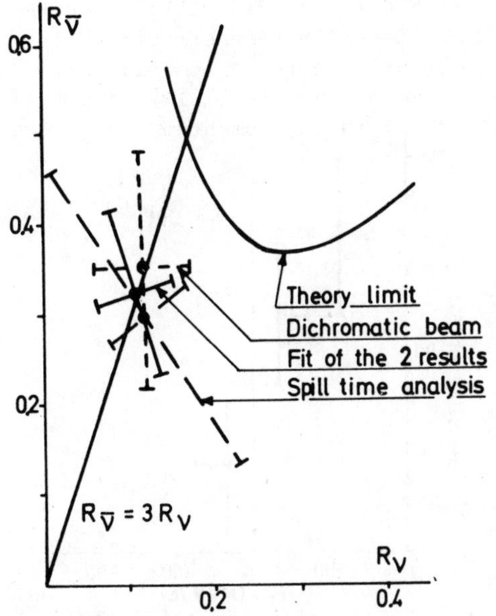

Fig. 14 NAL 1A experiment results

The corresponding ratios are found to be :

$$R = 0.18 \pm 0.05 \quad (\alpha = .74 \pm .06)$$

$$R = 0.22 \pm 0.05 \quad (\alpha = .45 \pm .06)$$

This new information reduces the allowed region in the R_ν - $R_{\bar{\nu}}$ plot to an ellipse (fig. 14) favouring a value $R_{\bar{\nu}}$ larger than R_ν.

3. The Third Experiment

The same experimental set-up was recently used in the dichromatic beam. The charge separation of the π and K parents at 140 GeV makes quasi pure neutrino or antineutrino beams (98 and 88 % respectively). The energy of the neutrino is about 50 GeV. On the basis of a few muonless events, the two ratios $R_{(\nu)}$ and $R_{(\bar{\nu})}$ were independently determined and found to be :

$$R_{(\nu)} = 0.13 \pm 0.06 \quad R_{(\bar{\nu})} = 0.34 \pm 0.12$$

and a global fit of the second and the third experiment gives the final values of :

$$R_{(\nu)} = 0.11 \pm 0.05 \quad R_{(\bar{\nu})} = 0.32 \pm 0.09$$

All the results of the experiment 1A are plotted in (figure 14). The origin (absence of neutral currents) is excluded by about 5 standard deviations.

IV) THE SEARCH FOR ONE PION CHANNEL IN THE ARGONNE 12' BUBBLE CHAMBER

A search for two specific channels has been carried out in the Argonne 12' bubble chamber (16).

$$\nu p \to \nu n \, \pi^+ \qquad (1)$$

$$\nu p \to \nu p \, \pi^0 \qquad (2)$$

Only a single π^+ is detected in the channel (1) and a proton with one γ-ray in the channel (2). These reactions obviously cannot be identified with the kinematic constraints as it is usually done in a hydrogen bubble chamber. The principle of the experiment is a "counting" experiment, as in Gargamelle : the candidate events (NC) are accumulated, the background is estimated and the final result comes from the subtraction of these two quantities. In the present experiment, 800'000 pictures were analysed, 2/3 in deuterium, 1/3 in hydrogen. The typical energy of the neutrinos is less than 1 GeV.

The identified π^+'s are divided into 3 categories :

1. The stopping π^+, with the clean signature of the successive decays, $\pi^+ \to \mu^+ \to e^+$ (10 events).

2. The interacting π^+ on a proton. The elastic collision and the π mass can be checked by kinematics (4 events).

3. The leaving π^+ which has to be separated from a leaving proton or from an incoming π^-. This separation can be done efficiently up to 400 MeV/c by range-curvature relation. The separation from incoming π^- is done by comparison between the measured momenta at the beginning and at the end of the visible track (fig. 15). 4 events survive finally as good candidates. The identification efficiency is determined applying exactly the same criteria on the π^+ coming from the very well fitted reaction $\nu + p \to \mu^- + p + \pi^+$. (Fig. 16) is a clear justification of the cut at 400 MeV/c in the momentum.

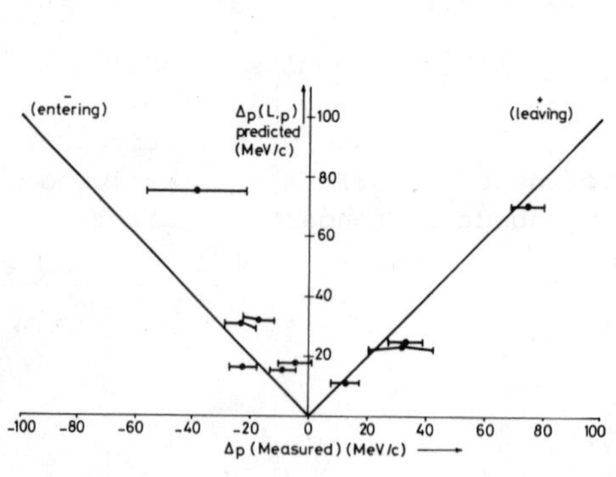

Fig. 15 Separation of the outgoing tracks from incoming tracks

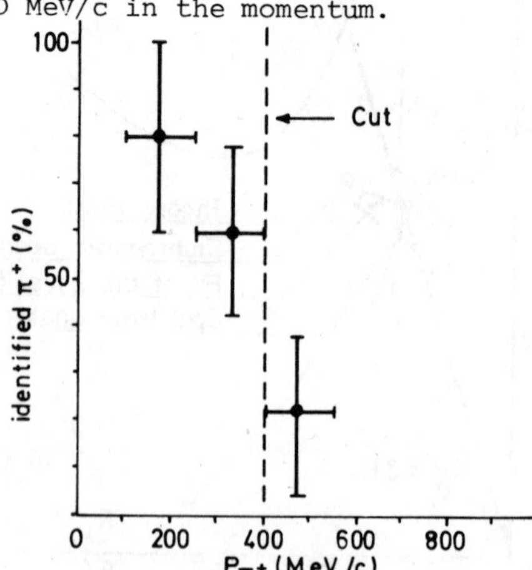

Fig. 16 π identification efficiency using the leaving π's in $\nu p \to \mu^- \pi^+ p$

Some cuts are applied to select the $p\gamma$ events :
$\pm 60^{\circ}$ in the dip and azimuthal angles of the proton; all the $\mu^- p \pi^+$ are inside the square defined by these cuts.

< 1 GeV/c in the momentum of the proton, which is practically the end of the proton spectrum in the $\mu^- p \pi^+$ events. 7 events survive these cuts. An 8th event has an ambiguous vertex, the γ-ray pointing equally well to the two extremities of the proton track. In one case the event is inside the $\pm 60^{\circ}$ square box, in one case it is outside!

The efficiency to detect the $p\gamma$ event is measured in an exposure of the chamber to a neutron beam. A search for $np \rightarrow p\, p\, \pi^-$ (3 mb) and $np \rightarrow n\, p\, \pi^0$ (8 mb) has given the ratio

$$\frac{p\gamma \text{ events}}{p\, p\, \pi^- \text{ events}} = 0.19 \pm 0.04$$

Many sources of backgrounds were considered:

The neutrons can produce π^+ according to the reaction : $n + p \rightarrow n\, n\, \pi^+$ which simulates a NC candidate. This background can be easily measured using the charge symmetry reaction: $n + p \rightarrow p\, p\, \pi^-$ (12 events). In fact the exact symmetry of charge produces a symmetric π^+ in the centre of mass; this effect has been studied and found to be negligible in the detection efficiency. The complication of the proton spectator in the deuterium has also been studied.

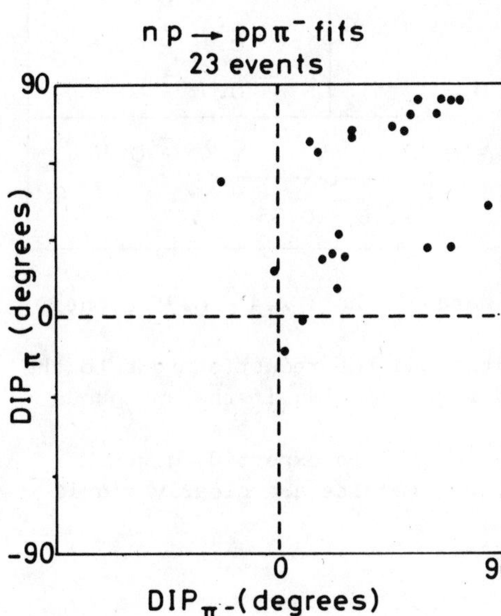

Fig. 17 Dip angles of the neutron and of the π^- in the $pp\pi^-$ events

A very interesting feature of the background is shown in (fig. 17), which is a scatter plot of the neutron events in function of the dip angles of the π^- and of the reconstructed neutron. Most of the neutrons are coming from the top of the chamber and the π^- has a certain memory of the neutron direction. A cut at 0° in dip angle of the π^- eliminates 90 % of the neutron background. Therefore this cut is applied to the events to optimise the signal/background ratio, even if 50 % of the signal disappears.

Another source of background is the single pion-photo-production by γ-rays produced above the chamber in the magnet from cosmic muons. A cut is therefore applied on events closer than 20 cm to a cosmic ray, and the measured spectrum of cosmic γ-rays was used to estimate the remaining background.

The $\bar{\nu}$ contamination in the beam is also a source of background in the events $\bar{\nu} + p \rightarrow \mu^+ + n$, because the leaving μ^+ cannot be separated from a leaving π^+. The background was estimated using the $\nu n \rightarrow \mu^- p$ and appropriate form factors.

Incoming π^- giving charge exchange and γ materialisation can be confused with an outgoing proton and the event taken as a $p\gamma$ event. The background is evaluated using the elastic scattering $\pi^- p \rightarrow \pi^- p$ events, which are easily identified.

The coincidences of a γ-ray and a one-prong was determined with two samples of stray rays and isolated one-prong.

The final estimation of background with different sources is given in Table IV.

Table IV - Background Summary

	π^+ events	$p\gamma$ events
Neutron	0.55 ± 0.55	0.91 ± 0.45
γ production	0.38 ± 0.16	0.07 ± 0.03
K^o	0	0
$\bar{\nu}p \rightarrow \mu^+ n$	0.05 ± 0.02	0
$\pi^- p \rightarrow \pi^o n$	0	0.36 ± 0.16
Random pointing	0	0.06 ± 0.06
	0.98 ± 0.57	1.40 ± 0.48
	2.38 ± 0.75	

The total background is estimated to be : 2.4 ± 0.75 events

Table V gives the NC candidates and the reductions due to the successive cuts. The final number is 13 (or 14, if the ambiguous event is added).

The Poisson law applied to 13 when 3 is expected, gives a probability of the order of 10^{-4}. The results are clearly significant.

Table V - Neutral Current Candidates

	H^2	D^2	Total	After cuts Cosmic	Dip
Stopping π^+	2	8	10	7	6
π^+ scatter	3	1	4	3	1
leaving π^+	3	1	4	3	0
π^+ total			18	13	7
$p\gamma$	3	4(5)	7(8)	6(7)	
π^+ and $p\gamma$			25(26)	13 (14)	

V) CONCLUSION FROM AN EXPERIMENTALIST'S POINT OF VIEW

- Four different experiments give evidence for new phenomena which cannot be explained by known processes :

The two electron events in Gargamelle are incompatible with the expected background at the level of 1 to 2 % in probability.

The 260 muonless events in Gargamelle (neutrino and antineutrino) are incompatible with the 30 expected events from neutron background. The significance is strongly depending upon the error on the neutron background. The quoted number B/AS = 0.6 ± 0.3 corresponds to about 15 standard deviations.

The muonless events in the NAL 1A experiment cannot be explained by a lack of muon detection. The experimental significance is about 5 standard deviations.

The 13 events in the Argonne 12' bubble chamber are incompatible with background at the level of 10^{-4} in probability.

- The neutrino beam energies are very different in all these experiments :

0.5 - 1.5 GeV in the Argonne experiment.
2 - 10 GeV in the CERN experiment.
5 - 50 GeV in the NAL experiment.

- The reactions studied are also very different :

purely electronic process in Gargamelle.
inclusive reactions in Gargamelle and in the NAL 1A experiment.
two specific channels in the Argonne experiment.

- The sources of background are very different :

  neutrons coming from the hadron beam at Argonne.
  neutrons in equilibrium with neutrinos in Gargamelle.
  undetected muons in the 1A experiment.
  ν_e contamination of the $\bar{\nu}_\mu$ beam in the electron search in Gargamelle.

It is very unlikely that all the four experiments are wrong!

It is very unlikely that four new processes are necessary to explain the four experimental results. Therefore it is reasonable to look for a unique explanation to the four experiments.

VI) SOME POSSIBLE INTERPRETATIONS

Before a discussion of the interpretation with neutral currents, some possible explanations, more or less esoteric, are examined here.

Neutral semi-strong Interacting Particles

The experiments detecting only hadrons in the final state cannot exclude the possible existence of neutral semi-strong interacting particles in equilibrium with the neutrinos. The Gargamelle experiment requires a mean free path longer than 4 m. An argument against this hypothesis is the relative constant value of NC/CC from .5 to 50 GeV.

Decaying Neutral Particles

The detection of evaporation prongs in most of the interactions in Gargamelle eliminates the possible existence of neutral long lived particles in equilibrium with neutrinos.

Production of heavy Leptons by Neutrinos

The hadronic events could be interpreted by a production of heavy leptons, decaying into hadrons and neutrinos. A variation of the cross-section would be expected as a function of the neutrino energy or as a function of the energy transfer to the hadrons, because threshold effects would appear when the available energy is just above the mass of the heavy lepton. Such effects are not significantly observed.

VII) THE NEUTRAL CURRENT INTERPRETATION

The theory of charged currents in weak interactions remained unsatisfactory during many years because it was not renormalizable. Recently, Salam, Weinberg and t'Hooft have demonstrated that a renormalizable theory can be elaborated if neutral currents exist, or heavy leptons, or both. These ideas are very attractive and today many theories or models are proposed. In this paper, only the comparison with the Weinberg theory is done because the experimental results are so preliminary, that we are not able to discriminate between most of the proposed models. The Weinberg model introduces only one parameter, the angle Θ_w, defined by the relation $tg\Theta = g'/g$ where g and g' are the coupling constants of the isovector and isoscalar field respectively. All the cross-sections can be computed as a function of the unique parameter $\sin^2\Theta_w$. All the available experimental data are analysed to test if they are compatible or not with a unique value of the parameter $\sin^2\Theta_w$. This analysis has been already presented in the review papers (6, 7, 8); therefore the following discussion is restricted to new results, but all the previous constraints on $\sin^2\Theta_w$ are quoted :

- $\nu_\mu e^-$ and $\bar{\nu}_\mu e^-$ scattering. The Gargamelle results give the following constraint on the Weinberg angle :

  $0.1 < \sin^2\Theta_w < 0.5$ at 90 % confidence level.

  A lower limit of the $\nu_\mu e^-$ cross section can be estimated at $0.05.10^{-41}$ $cm^2 . E_{\bar{\nu}}$ (GeV).

- $\bar{\nu}_e e^-$ scattering. The Savannah nuclear reactor experiment was interpreted in the Weinberg theory as giving the following limit (17).

$$\sin^2\Theta_w < 0.3$$

The uncertainties on the neutrino flux make this constraint probably weaker.

- $\nu_\mu N \to \nu_\mu$ + hadrons. (Fig. 18) shows the Gargamelle and the NAL 1A results on the same R_ν, $R_{\bar{\nu}}$ plot. The two results are compatible within 1.5 standard deviations. From this relative agreement between two measurements at very different energies, it may be concluded that the new phenomenon "scales" with energy as well as the charged current cross-sections.

The results are compared with the Weinberg theory, using the Paschos-Wolfenstein (23) analysis in which the contribution of the iso-scalar term is neglected. The energy cut-offs and the hadron detection efficiencies introduce different but small corrections to the theoretical curve. The corrected curve applied to the Gargamelle experiment; the corrections in the NAL 1A experiment are, in principle, smaller. A value of $\sin^2\Theta_w$ between 0.3 and 0.5 would not be incompatible with the present data.

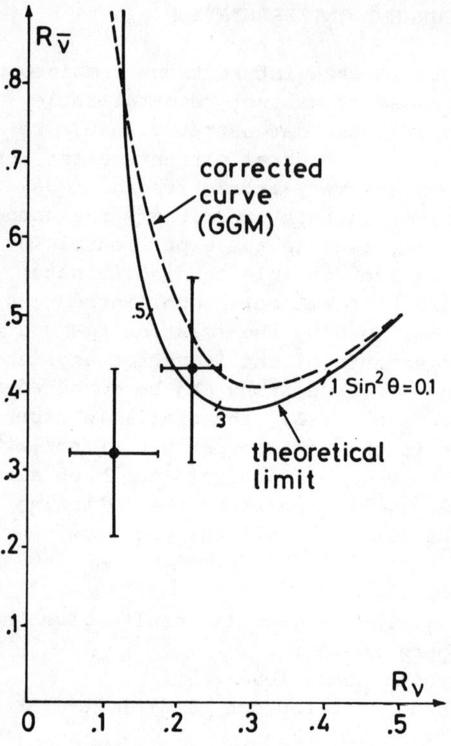

Fig. 18 Gargamelle and NAL 1A results

- single π^0 production. Old results have given lower limits on the ratio $\frac{\sigma(\nu n \to \nu n \pi^0) + \sigma(\nu p \to \nu p\ \pi^0)}{2\ \sigma(\nu n \to \mu^- p\ \pi^+)}$ < 0.21 in Gargamelle (19) and < 0.14 in spark chambers (18). The interpretation in the Weinberg model is not easy (6). The large values of $\sin^2\Theta_W$ are favoured up to a lower limit around 0.3.

The new results from Argonne gives the ratio $\frac{\sigma(\nu p \to \nu p\ \pi^0)}{\sigma(\nu p \to \mu^- p\ \pi^+)}$ $= 0.48 \pm 0.24$ which is not incompatible with the Weinberg model (19) but does not give a constraint on $\sin^2\Theta_W$.

- single π^+ production. The new result from Argonne gives the ratio $\frac{\sigma(\nu p \to \nu p\ \pi^+)}{\sigma(\nu p \to \mu^- p\ \pi^+)} = 0.17 \pm 0.07$. This ratio favours the low values of $\sin^2\Theta_W$, but do not give an upper limit.

ν_μ ($\bar{\nu}_\mu$) Inél. GGM - 1A

νp (CERN)

ν_μ ($\bar{\nu}_\mu$) e^- (GGM)

$\bar{\nu}_e$ e^- (Savannah)

0 0.2 0.4 0.6 0.8 1.0

Fig. 19 Limits on $\sin^2\Theta_W$

- νp scattering. An old result from CERN gives the small constraint $\sin^2\theta_W < .9$ at 90 % confidence level (22).

- $\bar{\nu}_e d \rightarrow \nu_e pn$. As a preliminary result reported by Gürr an upper limit on the cross section of this process is given to be 6 times the V-A expected cross section.

Conclusions : All the present experimental results are compatible with the existence of neutral weak currents, as described in the Weinberg theory with a value of $\sin^2\theta_W$ around 0.3 - 0.5. Fig. 19 is a resumé of the interpretations of the experimental results. The results are also compatible with many other models, but significant comparisons will be possible only when more accurate experimental results will be available.

Other arguments favour the neutral current interpretations :

- The apparent "scaling" of the results from very low energy (.5 GeV) to high energy (50 GeV).

- The very similar behaviour of the hadronic showers in the CC events and in the NC events in the Gargamelle experiment: the energy distribution, the angular distribution, the pion charge distribution.

A definite proof of the neutral current will be very difficult because the outgoing neutrino cannot be detected and identified. Nevertheless, the neutral currents appear today to theoreticians as the most suitable interpretation of the results obtained by the experimentalists.

REFERENCES

1. P. Musset, Etats des résultats expérimentaux sur la recherche des courants neutres. Colloque de Vittel de Mai 1973. Journal de Physique - Supplément 11 - 12, Tome 34, p. C3-1, 1973.
2. A. Pullia, Gargamelle collaboration, Search for Neutral Currents in Gargamelle, Proceeding of the Balaton Conference (June 1972). Neutrino 72.
3. F.J. Hasert et al., Gargamelle collaboration, Search for Elastic Muon-Neutrino Electron Scattering, P.L. 46 B, 1, p. 12, 3.9.73.
4. F.J. Hasert et al., Gargamelle collaboration, Observation of Neutrino-like Interactions without Muon or Electron in the Gargamelle Neutrino Experiment, P.L. 46 B, 1, p. 138, 3.9.73.
5. A. Benvenuti et al., Observation of muonless Neutrino-Induced inelastic Interactions, P.R.L., 32 - 14, p. 800, 8.4.74.
6. G. Myatt, Neutral Currents, International Symposium on Electron and Photon Interactions at high Energies, Bonn, August 27 - 31st 1973.
7. P. Musset, Neutrino Interactions, Aix-en-Provence International Conference on Elementary Particles, 6 September 1973.
8. H. Wachsmuth, Aug. 1973 Review of latest Results from high Energy Neutrino Experiments, Conference of the American Physical Society Div. Particles and Fields, Berkeley, Aug. 13 - 17, 1973, Conference Proceedings No 14, p. 2 - 20.
   see also :
   H. Wachsmuth, Latest Results from Gargamelle Neutrino Experiments, Proceedings of Summer Institute on Particle Physics at SLAC, SLAC 167, Vol. II, p. 235.
9. A. Salam and J.G. Ward, P.L. 13, p. 168, (1964).
10. S. Weinberg, P.R.L. 19, p. 1264, (1967).
11. G. t'Hooft, P.L. 37 B, p. 195, (1971).
12. F.J. Hasert et al., Observation of Neutrino-like Interactions without Muon or Electron in the Gargamelle Neutrino Experiment, TC-L/Int. 74-1, 7.1.74, to be published in Nuclear Physics B.
13. M.E. Law et al., L.R.L. 80 (1972).
14. B. Aubert et al., Further Observations of muonless Neutrino induced inelastic Interactions (March 74), to be published in P.R.L.
15. B. Aubert et al., Measurement of Rates for muonless deep inelastic inelastic Neutrino and Antineutrino Interactions, (April 74), to be published in P.R.L.
16. S. Barish et al., Search for neutral Current in the 12 Feet Argonne Bubble Chamber (one Pion Production). A.P.S. Meeting in Washington, (April 74).
17. F. Reines, Topical Meeting on Weak Interactions, Trieste, (June 1973).
18. W. Lee, Phys. Letters, 40 B, 423, (1972).
19. P. Musset, Bull. Am. Phys. Soc. 18, 73, (1973).

20. C.H. Albright, B.W. Lee, E.A. Paschos, L. Wolfenstein, P.R. D7, 2220, (1973).
22. D.C. Cundy et al., P.L. 31 B, 479, (1970).
23. E.A. Paschos and L. Wolfenstein, P.R. D7 91 (1973).
24. D. Haidt, Contribution to the A.P.S. Washington 74 Meeting.

TOTAL AND DIFFERENTIAL CROSS-SECTIONS IN DEEP INELASTIC NEUTRINO SCATTERING*

F. J. Sciulli
California Institute of Technology, Pasadena, Calif. 91109

Invited talk given at the 1974 International Conference on Neutrino Physics and Astrophysics, April 1974, at Downingtown, Pennsylvania.

CONTENTS

INTRODUCTION

It has been recognized for some time that the study of neutrino interactions at high energies will provide an opportunity to study the weak interactions in a completely unexplored regime. Extrapolation of the V-A theory for leptons up to very high energy creates theoretical paradoxes that can only be resolved by modifying present theory. The prediction and observation of scaling phenomena in electron-nucleon and neutrino-nucleon scattering provides further stimulus to such studies, because with scaling the lepton-lepton paradox also applies to the lepton-nucleon case. In addition, ideas derivative to the scaling phenomenon (parton models, light-cone algebra, etc.) present a context within which the scaling functions may be interpreted. In many instances, these interpretations are extremely physical and intuitive, and so very appealing.

Since the theoretical framework for the treatment of deep-inelastic scattering is being discussed by Prof. Lee at this conference, I will confine myself to a short description of the relevant formula, with some attention to the assumptions involved. The major part of this talk will concern itself with the existing experimental data.

* Work supported in part by the U. S. Atomic Energy Commission.

THEORETICAL EXPECTATIONS (BEST GUESSES)

Assumption 1: V-A and Scaling

The Theoretical expectations for the scattering of neutrinos from nucleons revolve around two major assumptions: V-A coupling for the weak interaction, and the scaling hypothesis for hadron coupling. The former assumption is very well tested by the weak decays of metastable particles, with energy releases of a few hundred MeV. The scaling hypothesis was originally found to work in the measured electromagnetic cross-sections

$$e^- + N \to e^- + \text{hadrons}$$

made at SLAC.

In this picture, the incident neutrino of energy E_ν emits a virtual boson propagator and μ^- at the upper lepton vertex. This propagator carries laboratory energy $\nu = E_\nu - E_\mu$, and invariant 4-momentum transfer $q^2 = -Q^2 = (E_\nu - E_\mu)^2 - (\vec{p}_\nu - \vec{p}_\mu)^2$. At momentum transfers where the rest mass (Λ) of this propagator becomes important ($Q^2 \approx \Lambda^2$), an apparent violation of scaling will be observed.

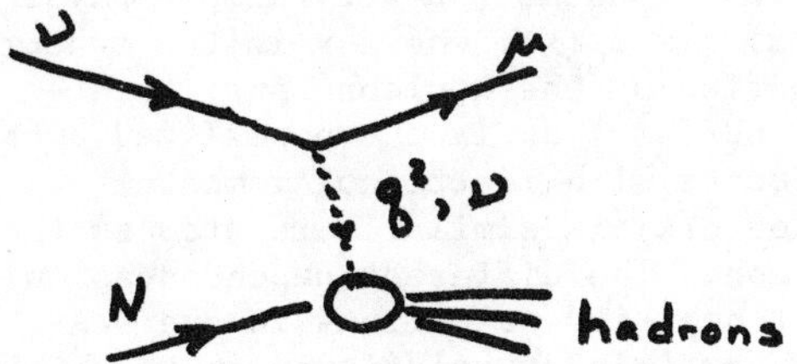

The scaling (dimensionless) variables are defined as $x = \frac{Q^2}{2M\nu}$ and $y = \nu/E_\nu$ with M = nucleon (target) mass. At high energies, the scaling variables have a kinematic range $0 < x < 1$, $0 < y < 1$. In kinematic regions where $Q^2 \to \infty$, $\nu \to \infty$, but the ratio remains finite, one expects that the differential cross-section for the process $\nu_\mu + N \to \mu^- + \text{hadrons}$ will be of the form

$$\frac{d\sigma^{\nu,\bar{\nu}}}{dxdy} = \frac{G^2 M E_\nu}{\pi} \left\{ F_2(x)\,(1-y) + 2x F_1(x)\,\frac{y^2}{2} \mp x F_3(x)\, y(1-\frac{y}{2}) \right\}$$

where F_1, F_2, F_3 are 3 structure functions describing the hadronic vertex. The important point is that the (V-A) + scaling hypotheses require the cross-section to depend linearly on neutrino energy times some function of the dimensionless parameters, x and y. Some additional dependence on Q^2, ν, or E_ν would indicate a breakdown of either V-A, or scaling, or both. For example, the finite mass propagator might result in the modification:

$$\frac{d\sigma^{\nu,\bar{\nu}}}{dxdy} \to \frac{d\sigma^{\nu,\bar{\nu}}}{dxdy} \; \frac{1}{(1 + Q^2/\Lambda^2)^2}$$

Assumption 2: Callen-Gross Relation

If we postulate that neutrino scattering from nucleons behaves as if the nucleons consisted of predominantly spin 1/2 constituents, then the Callen-Gross relation would hold: $F_2(x) = 2x\, F_1(x)$. The differential cross-sections take a somwhat simpler form:

$$\frac{d\sigma^{\nu}}{dxdy} = \frac{G^2 ME_{\nu}}{\pi} \left\{ q(x) + \bar{q}(x)\,(1-y)^2 \right\} \tag{1}$$

$$\frac{d\sigma^{\bar{\nu}}}{dxdy} = \frac{G^2 ME_{\nu}}{\pi} \left\{ q(x)\,(1-y)^2 + \bar{q}(x) \right\} \tag{2}$$

where $q(x) = \dfrac{F_2(x) - x\,F_3(x)}{2}$ $\quad \bar{q}(x) = \dfrac{F_2(x) + x\,F_3(x)}{2}$

In the parton model, these functions have a very simple physical interpretation. For example, $q(x) = x\, f_q(x)$, where x is the fraction of the total nucleon momentum carried by the nucleon constituent which engages in the scattering and $f_q(x)\,dx$ is the normalized probability for finding quark constituents with fractional momentum between x and $x+dx$. The function $\bar{q}(x)$ is similarly related to the anti-quark component of the nucleon. The different dependences on y are a reflection of the different helicity selections in the V-A theory between particle and anti-particle at velocities approaching c.

Assumption 3: Small Anti-quark Component

If we assume $\bar{q}(x) \ll q(x)$ for $x > 0$, the differential cross-section becomes as follows:

$$\frac{d\sigma^{\nu}}{dxdy} \simeq \frac{G^2 ME_{\nu}}{\pi}\, q(x) \tag{3}$$

$$\frac{d\sigma^{\nu}}{dxdy} \simeq \frac{G^2 ME_{\nu}}{\pi}\, q(x)\,(1-y)^2 \tag{4}$$

with $q(x) = F_2(x) \propto F_2^{ed}(x)$, neglecting the effect of strange quarks. Since the x-distribution relates to nucleon constituents, we expect the same shape to occur in neutrino scattering off equal admixtures of neutrons and protons as occurs in inelastic electron scattering from deuterium. The normalization of the two functions is different, however, because the electron scattering depends on the charge of the constituents while neutrino scattering is independent of electric charge.

Total cross-sections:

Integrating equations (1) and (2) over y, we obtain

$$\sigma^{\nu} = \frac{G^2 ME_{\nu}}{\pi} \int \left(q(x) + \frac{1}{3}\, \bar{q}(x) \right) dx$$

$$\sigma^{\bar{\nu}} = \frac{G^2 ME_{\nu}}{\pi} \int \left(\frac{1}{3}\, q(x) + \bar{q}(x) \right) dx$$

Thus, for $\int \bar{q}\, dx \ll \int q\, dx$

$$\frac{\sigma^{\bar{\nu}}}{\sigma^{\nu}} \simeq \frac{1}{3}\left(1+2.7 \frac{\int \bar{q}\, dx}{\int q\, dx} \right)$$

and the ratio of neutrino to anti-neutrino cross-sections is identically 1/3 in the limit of no anti-quark component in the nucleon. This cannot be exactly true, since we expect that $\bar{q}$ (0) = q (0) at exactly x = 0, to satisfy the sum rule: $\int \frac{\bar{q}(x) - q(x)}{x} dx = 3.$ Away from x = 0, however, we might expect to be dominated by the three valence quarks. Deviations from a ratio of 1/3 provide a measure of this anti-quark component. The CERN-Gargamelle experiment, for example, finds

$$\frac{\sigma^{\bar{\nu}}}{\sigma^{\nu}} = 0.38 \pm .02,$$ indicating perhaps a small, but finite, anti-quark component.

Independently of how large the anti-quark component is, the sum of the cross-section is

$$\sigma^{\nu} + \sigma^{\bar{\nu}} = \frac{4}{3} \frac{G^2 M}{\pi} E_{\nu} \int_0^1 (q(x) + \bar{q}(x)\ )\, dx$$

$$= \left\{ \frac{4}{3} \frac{G^2 M}{\pi} \int F_2(x)\, dx \right\} E_{\nu}$$

$$= (\alpha^{\nu} + \alpha^{\bar{\nu}})\, E_{\nu}$$

Here $\int F_2(x)\, dx$ is the fraction of the nucleon momentum carried by non-strange quarks. In addition, this integral is related to the similar quantity $\int F_2^{ed}(x)\, dx$ measured in electron-deuteron scattering by the mean-square charge of the constituents (with small corrections for strange quarks and neutron-proton excess).

The best quark-parton model prediction, using the SLAC data, is $\alpha^{\nu} + \alpha^{\bar{\nu}} = 1.10 \times 10^{-38}$ cm^2/GeV, with 54% of the momentum carried by the non-strange quarks. The measured value from CERN is

$$\alpha^{\nu} + \alpha^{\bar{\nu}} = (1.02 \pm .10) \times 10^{-38}\ \text{cm}^2/\text{GeV}$$

EXPERIMENTAL DATA

1. CERN-Gargamelle

Historically, the first indications that the whole idea of scaling was valid came from the SLAC-MIT electron scattering data. At that time, it was a mild surprise that the idea worked at energies as low as 16-18 GeV. It was even more surprising when the neutrino data from CERN-PS showed the same qualitative behavior. This data lies predominantly at energies $E_\nu < 5$ GeV, although a smaller number of events exist up to 11 GeV. In the SLAC-MIT data, scaling (or energy independence) holds when the data is plotted as a function of the scaling variable x, for $Q^2 > 1\ (\text{GeV}/c)^2$ and for final-state hadron states with invariant mass $W > 2\ \text{GeV}/c^2$. It was found, largely empirically, that scaling held in the region of much smaller Q^2, when plotted as a function of the variable

$$x' = \frac{Q^2}{2M\nu + M^2}$$

Of course, they are equivalent as $\nu \to \infty$ (i.e. $\nu \gg M$).

CERN, at even lower energy than SLAC, finds a universal shape for $F_2(x)$ only at their highest energies. On the other hand, their data plotted as a function of x′ shows a qualitative universality at all energies. This is shown in 4 energy intervals in figures 1 and 2. The qualitative agreement with the smooth curve, the $F_2^{ed}(x')$ structure function, is apparent. This is true even at the very lowest energies, where a substantial fraction of the data consists of psuedo-elastic events: $\nu_\mu + n \to \mu^- + p$. For this analysis, the Callen-Gross relation, $F_2(x) = 2x F(x)$, was assumed.

Figures 3a and 3b show the distributions $q(x')$ and $\bar{q}(x')$, obtained by using the sum and difference of neutrino and anti-neutrino differential cross-sections, integrated over y. Note first that, as expected, $\bar{q}(x) \ll q(x)$ for $x > 0$. At the same time, the distributions meet at $x = 0$, as anticipated by the Gross-Llewelyn Smith sum rule.*

Figures 4 and 5 show the y-dependence of the neutrino and anti-neutrino data, respectively, in several energy ranges. These distributions are distorted at low energy by two factors: (1) at small y, there is a substantial contribution from psuedo-elastic events; (2) at large y, there is an observable kinematic limit due to the low energy incident neutrino. For the neutrino case (fig. 4), these effects combine to give y-distributions which appear to fall as y increases rather than be flat. For the anti-neutrino case, they appear more like expected for a substantial quark component of the nucleon.

Figure 6 shows the ratio of the total cross-sections of anti-neutrino and neutrino data, as a function of incident neutrino energy. The data are consistent with a ratio independent of energy, with average value

$$\frac{\sigma^{\bar{\nu}}}{\sigma^{\nu}} = 0.38 \pm .02.$$

---

* Whether these distributions should really be interpreted as anti-quark component at the smallest values of x′ is open to question. As $x' \to 0$, a scattering constituent of finite rest mass will have very low velocity and the helicity selection rule which provides the y-dependence discussed previously does not exactly hold.

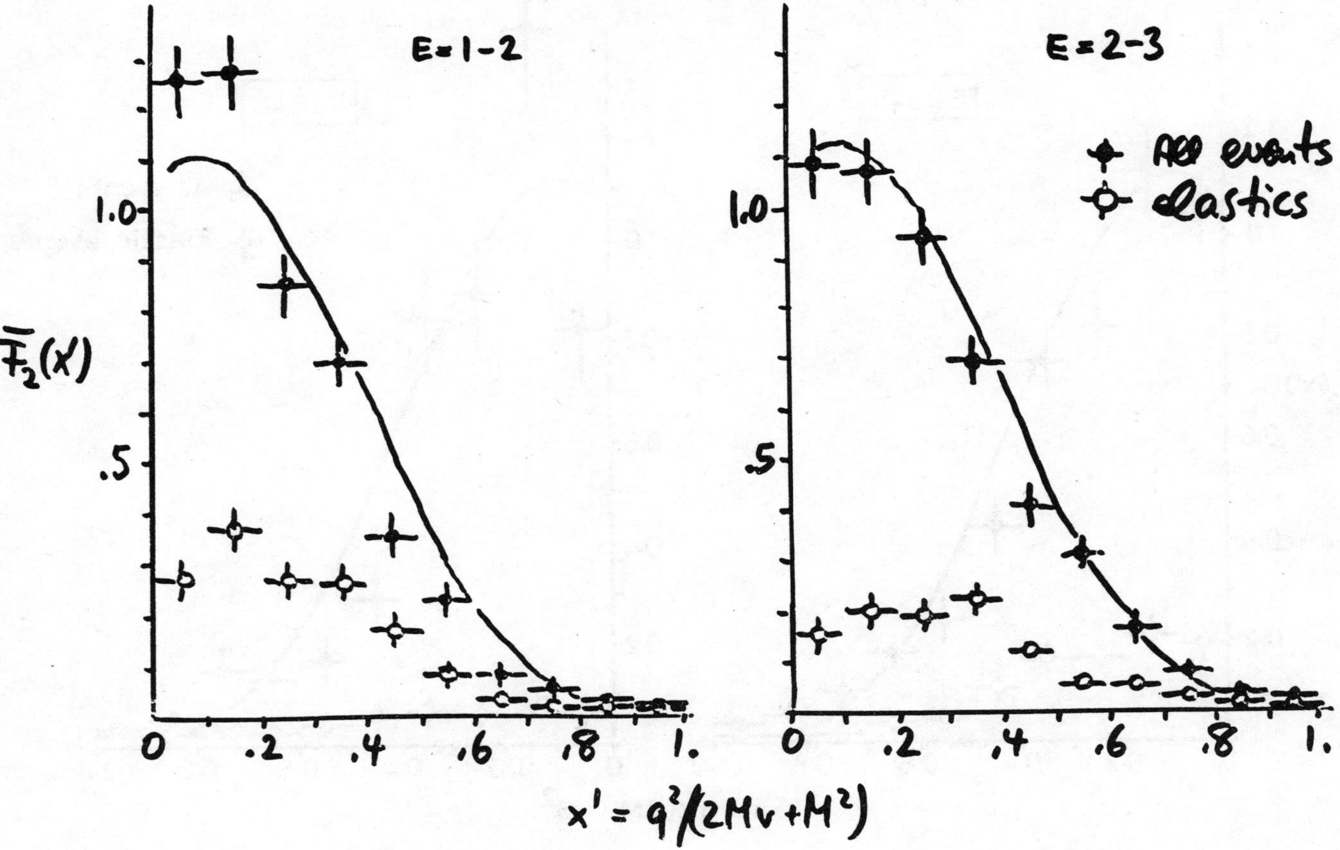

Figure 1

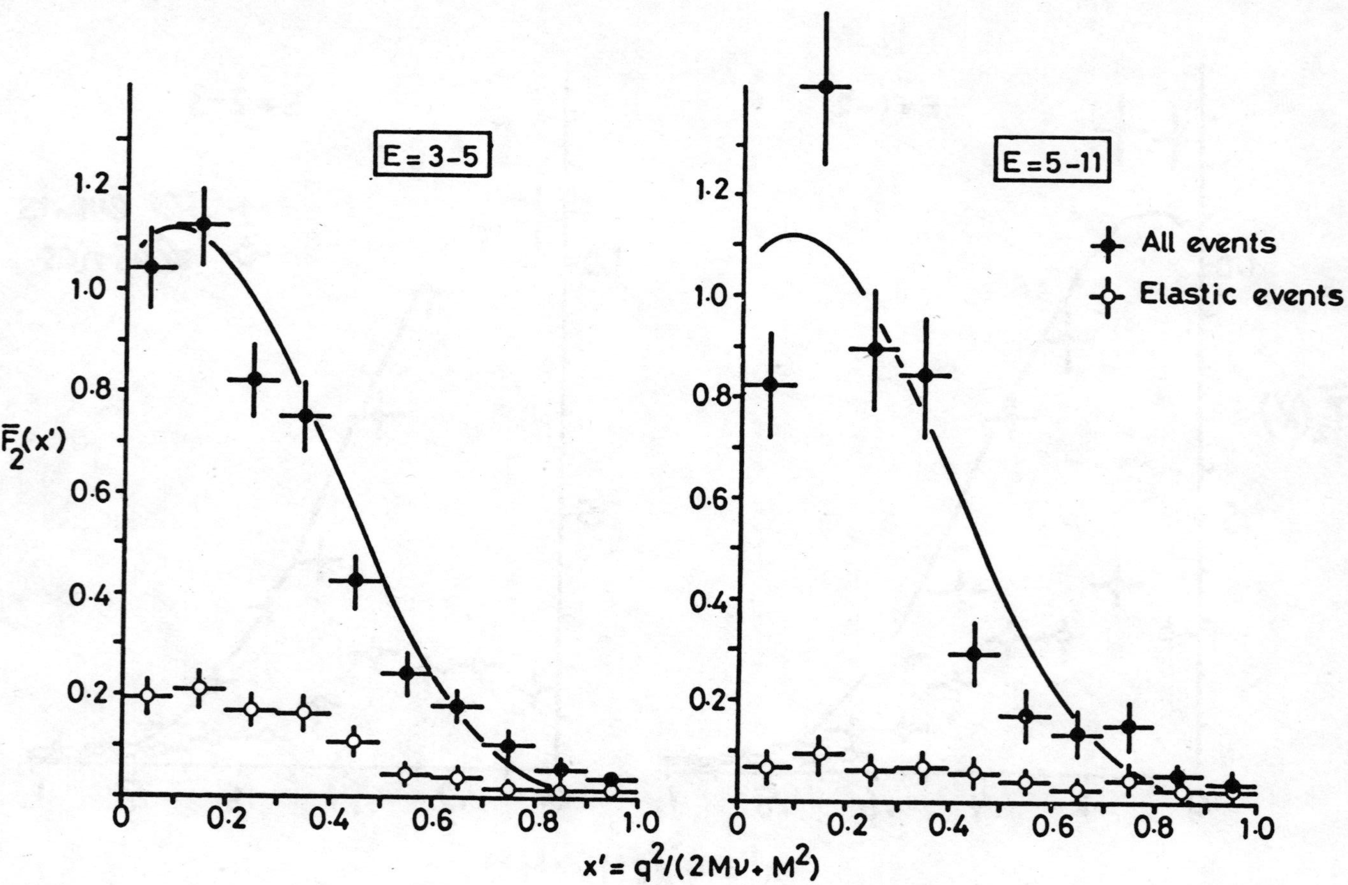
E=3-5
E=5-11
All events
Elastic events
$\bar{F}_2(x')$
$x' = q^2/(2M\nu + M^2)$
1.2
1.0
0.8
0.6
0.4
0.2
0
0.2
0.4
0.6
0.8
1.0

Figure 2

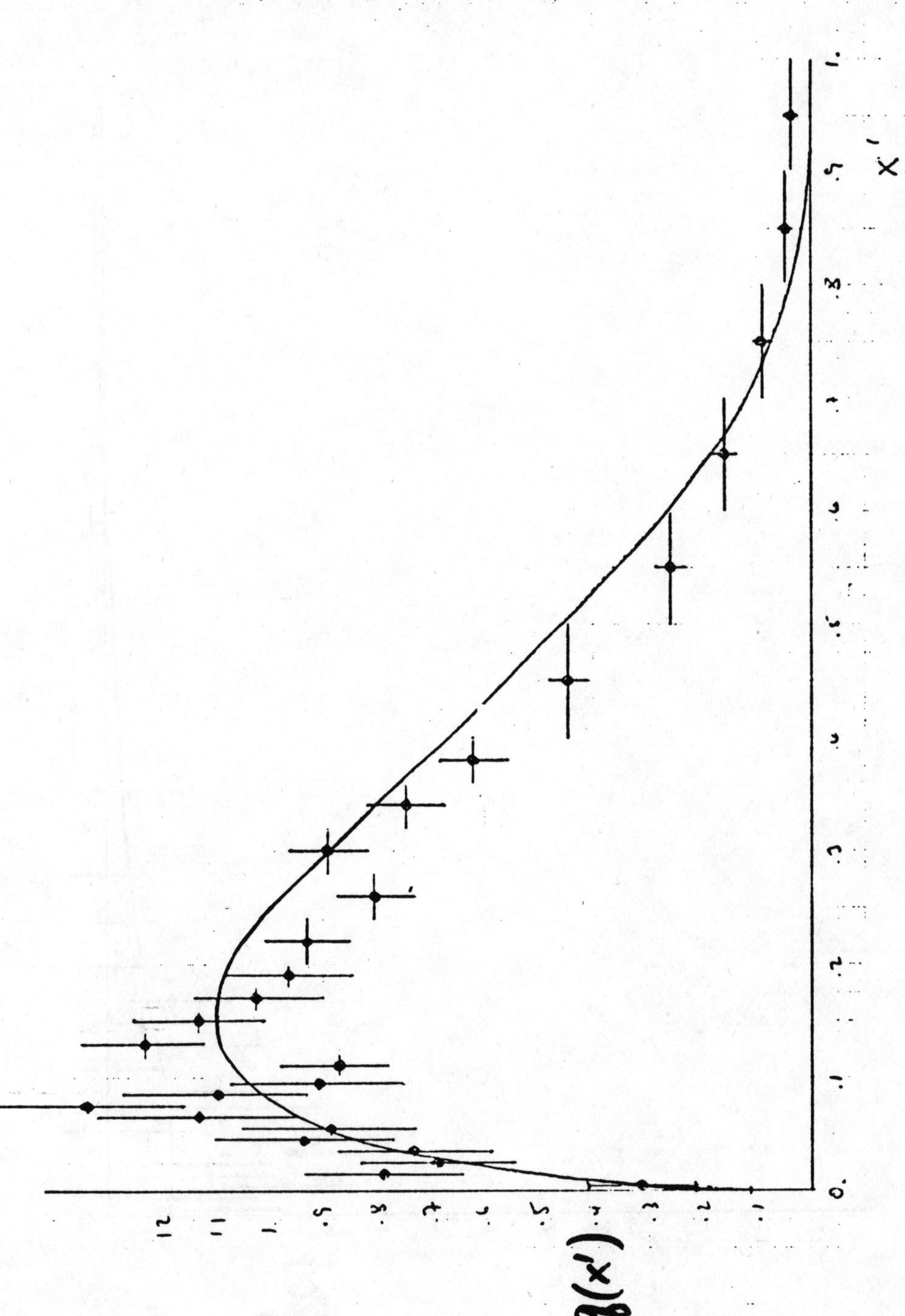

Figure 3a

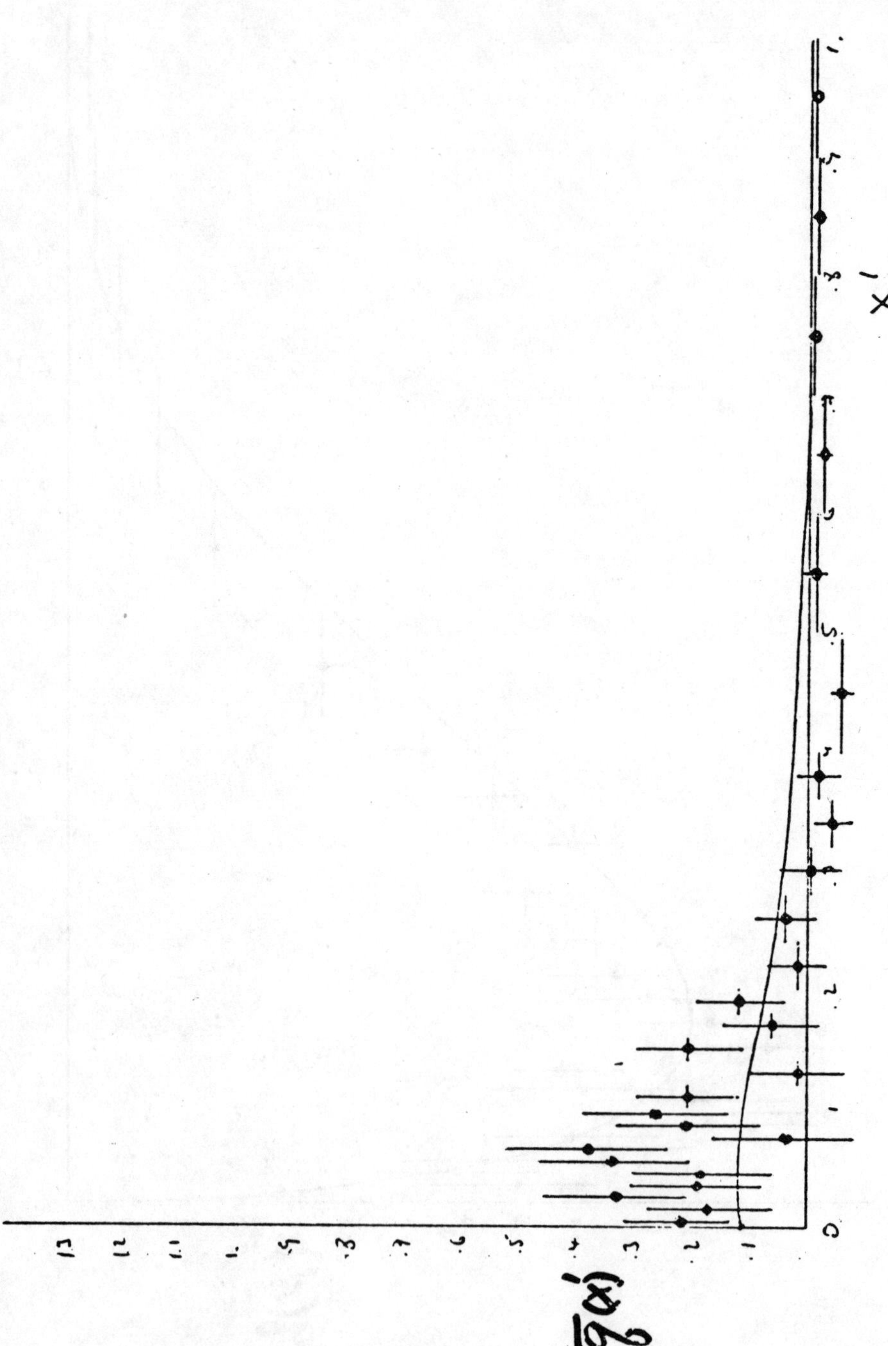

Figure 3b

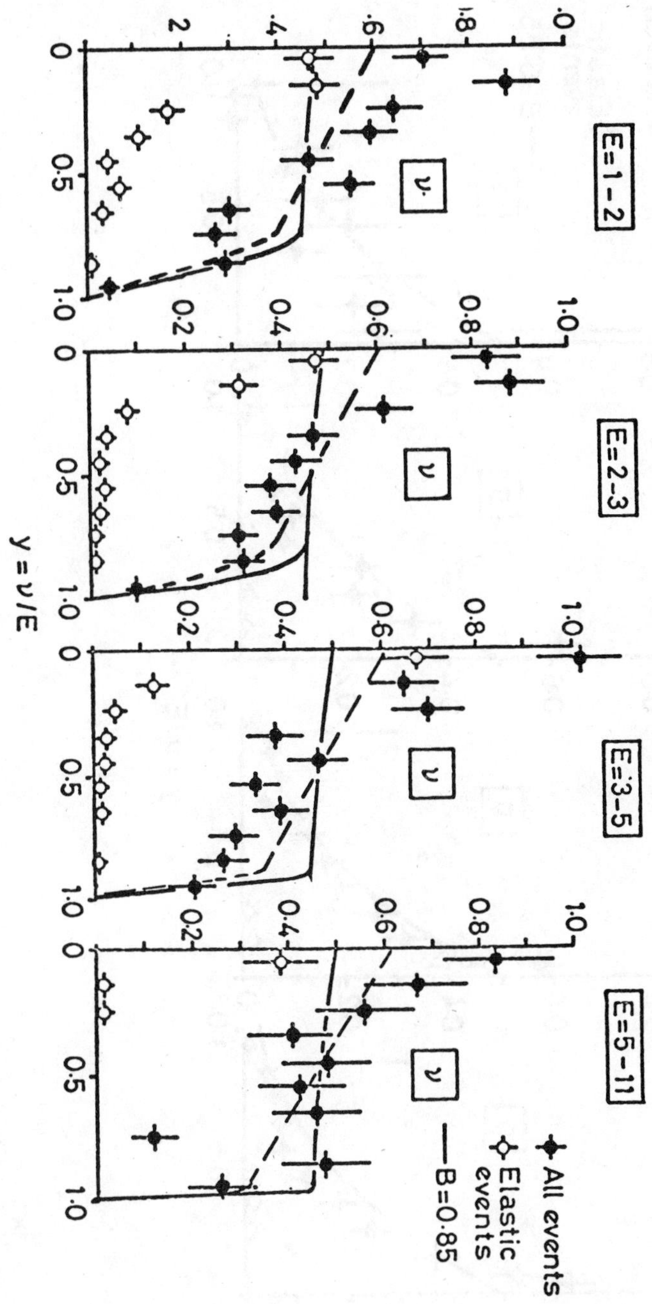
E=1-2
E=2-3
E=3-5
E=5-11
ν
y = ν/E
All events
Elastic events
B=0.85

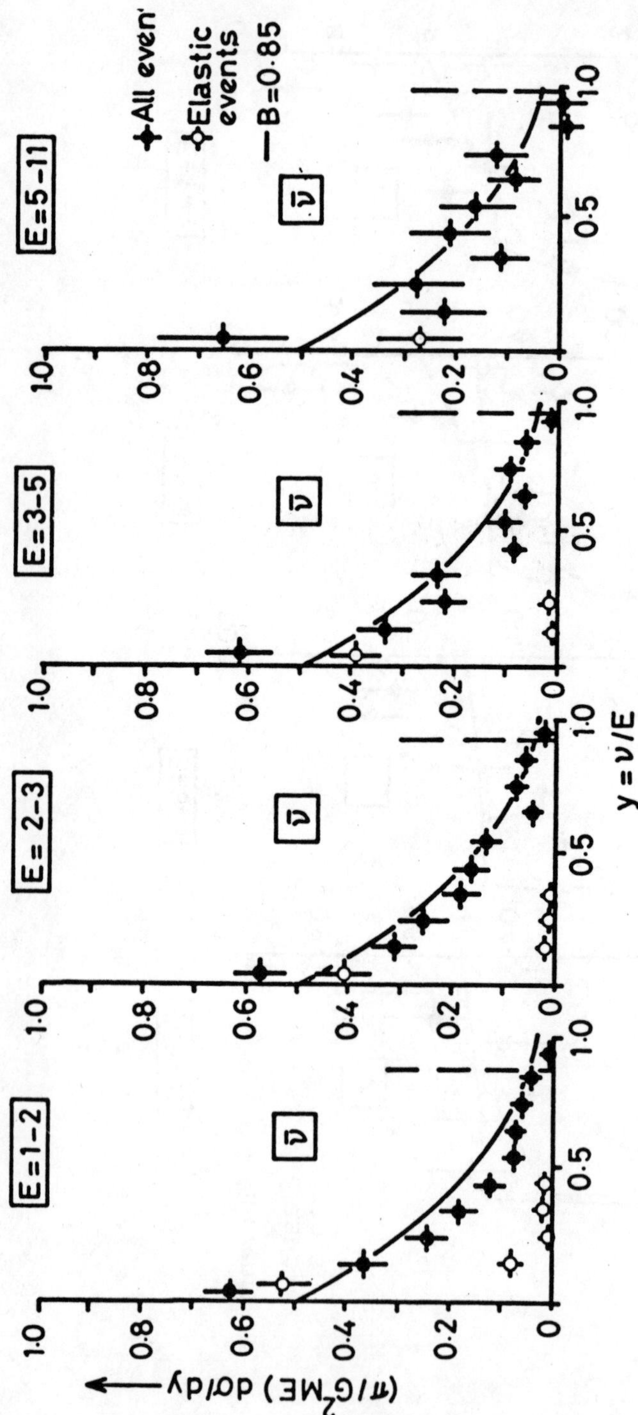
E=1-2
E= 2-3
E=3-5
E=5-11
$\bar{\nu}$
Elastic events
B=0.85
$y = \nu/E$
$(\pi/G^2ME)\ d\sigma/dy$

Figure 5

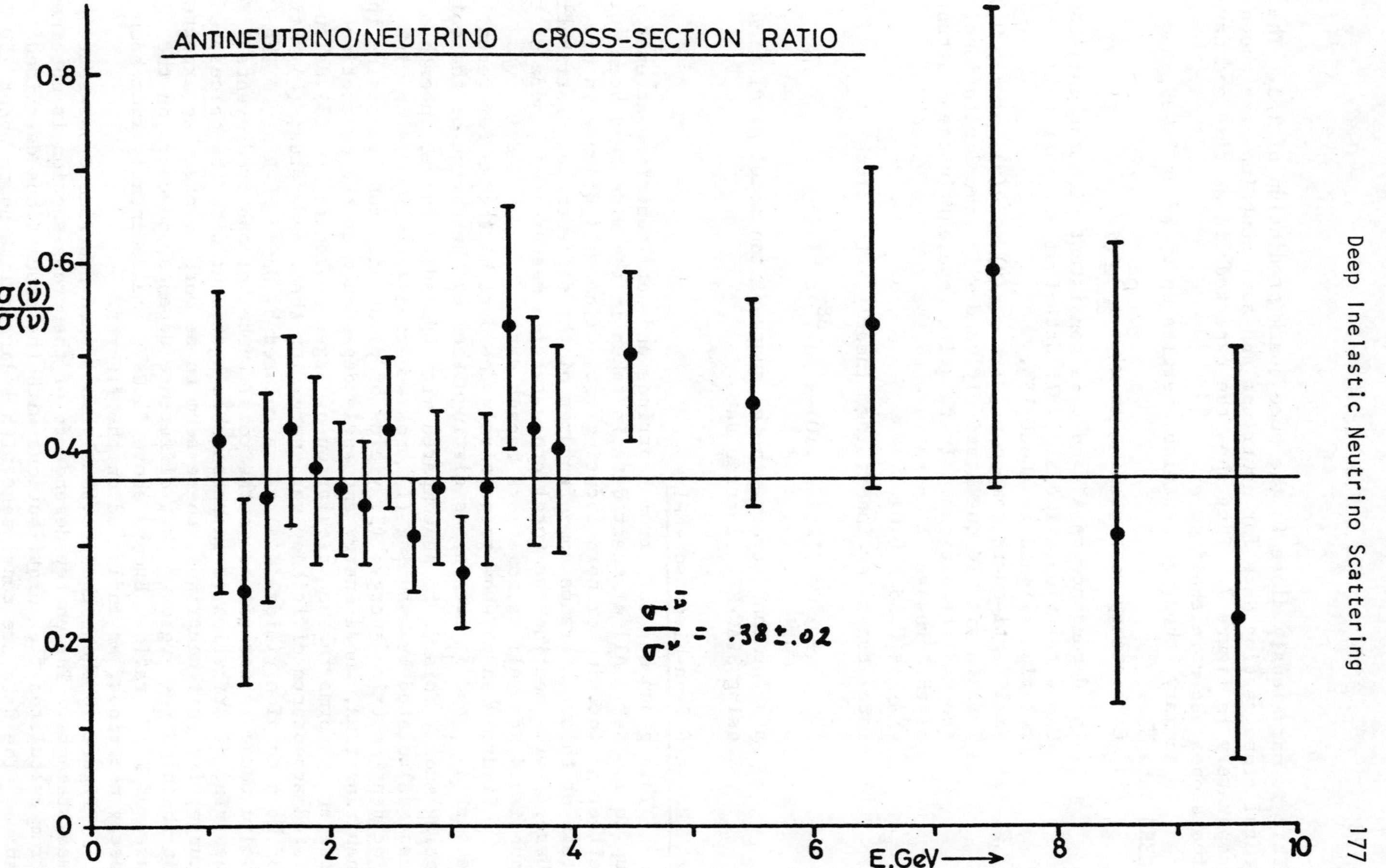
ANTINEUTRINO/NEUTRINO CROSS-SECTION RATIO
$\frac{\sigma(\bar{\nu})}{\sigma(\nu)}$
0.8
0.6
0.4
0.2
0
0
2
4
6
8
10
E, GeV →
$\frac{\sigma^{\bar{\nu}}}{\sigma^{\nu}} = .38 \pm .02$

Figure 6

This is marvelously close to the pure quark prediction of 1/3. The total cross-section data for neutrinos and anti-neutrinos are shown separately in figure 7. They show the expected linear rise and have the slopes shown in the figure.

In summary, then, at neutrino energies up to about 5 GeV, the CERN data show:

(1) Consistency with scaling in $x' = \frac{Q^2}{2M\nu + M^2}$.

(2) The dependence on x′ and y is consistent with the Callen-Gross relation. The SLAC-MIT data indicate that this relation is good to about 15%.

(3) Small anti-quark component, i.e. $\bar{q}(x) \ll q(x)$ for $x \gtrsim 0.05$. This is visible qualitatively, and with some limitations, in the y-distributions. It is quite apparent in the constancy of the cross-section ratio with the average value $\sigma^{\bar{\nu}}/\sigma^{\nu} = 0.38 \pm 0.02$.

(4) Total cross-section rising linearly, with slope

$$\alpha^{\nu} + \alpha^{\bar{\nu}} = (1.02 \pm .10) \times 10^{-38}\ \mathrm{cm}^2/\mathrm{GeV},$$

a value consistent with the quark-parton model prediction using SLAC-MIT electron data.

2. Harvard-Penn-Wisconson-NAL

This group has data from neutrinos and anti-neutrinos at energies up to 150 GeV. All of their data is taken in the wide band beam, with either no focusing or horn focusing. D. Cline will discuss in his talk at this conference some features of the differential distributions. Here, I will mention the total cross-section measurements, some of them published, as well as some more recent data.

Figures 8 and 9 show respectively the data histogram for recently observed ν_μ and $\bar{\nu}_\mu$ events, as distinguished by the charge of the outgoing muon. This is to be compared with the solid curve, in each case calculated by assuming (1) cross-section linearly rising with incident neutrino energy; (2) yields of pions and kaons supplying the neutrino flux, whose energy and angle dependence at the production target conforms to the calculation of Hagedorn-Ranft; (3) calculated acceptance efficiency vs. energy for their apparatus; (4) ratio of kaon to pion yields (K/π) as calculated by Hagedorn-Ranft. The solid curve is normalized to the total number of observed events. The neutrino distribution is in reasonable agreement with the calculated curve; for anti-neutrinos, there seem to be more events than expected at the higher energies. This difference depends, however, on the assumed K^-/π^- ratio. Events above 75 GeV come dominantly from kaon decay neutrinos, as indicated on the figures.

Figure 10 shows the previously published neutrino cross-section measurements. The energy dependence of the cross-section is obtained from calculated flux distributions with the assumptions described above. The data are consistent with a linear rise up to about 150 GeV (figure 10a). They obtained a numerical measure of the slope of the cross-section by utilizing "quasielastic" events for normalization. Figure 10c shows the distribution in W^2 for observed events with

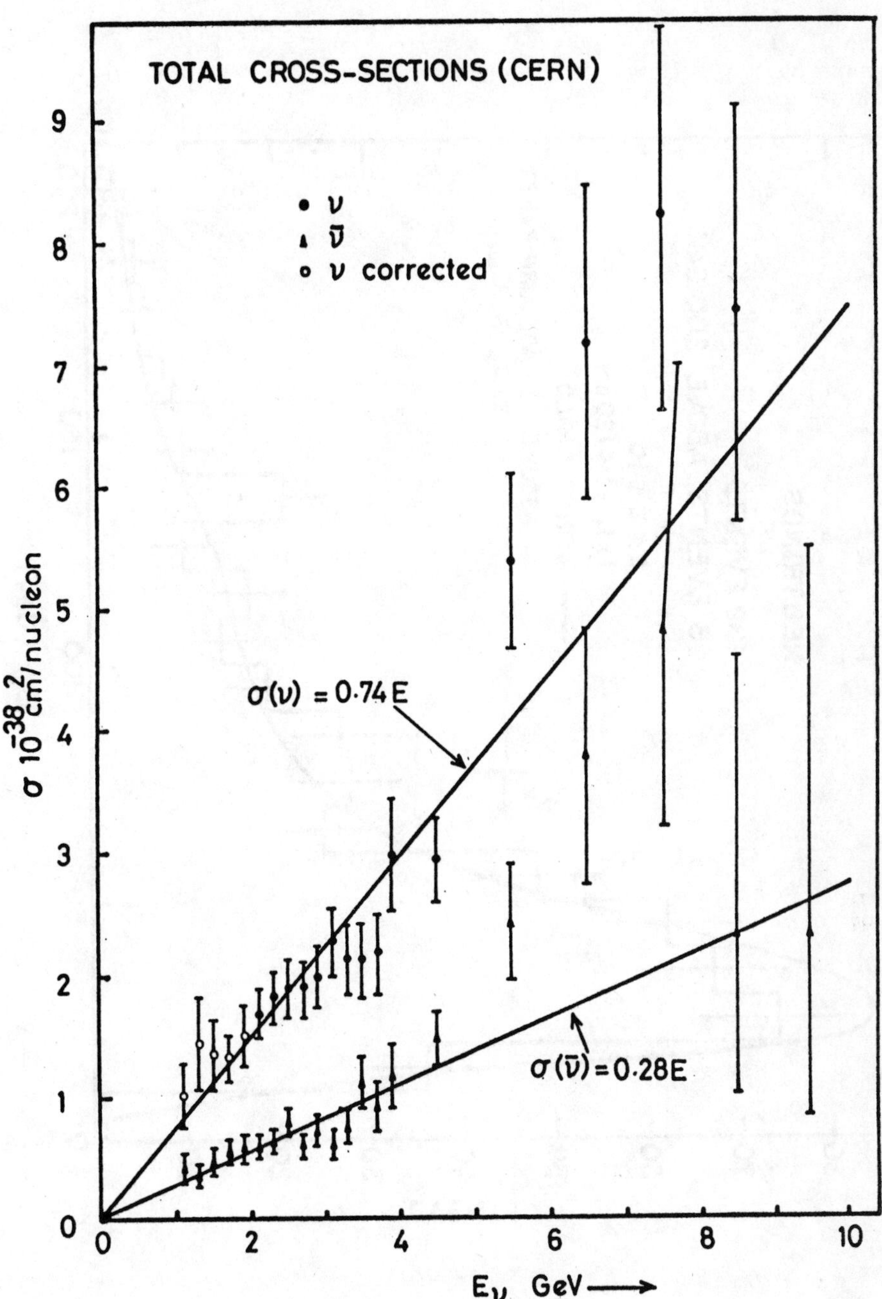
TOTAL CROSS-SECTIONS (CERN)
• ν
▲ ν̄
o ν corrected
σ(ν) = 0.74E
σ(ν̄) = 0.28E
σ 10^-38 cm^2/nucleon
Eν GeV
0 1 2 3 4 5 6 7 8 9
0 2 4 6 8 10

Figure 7

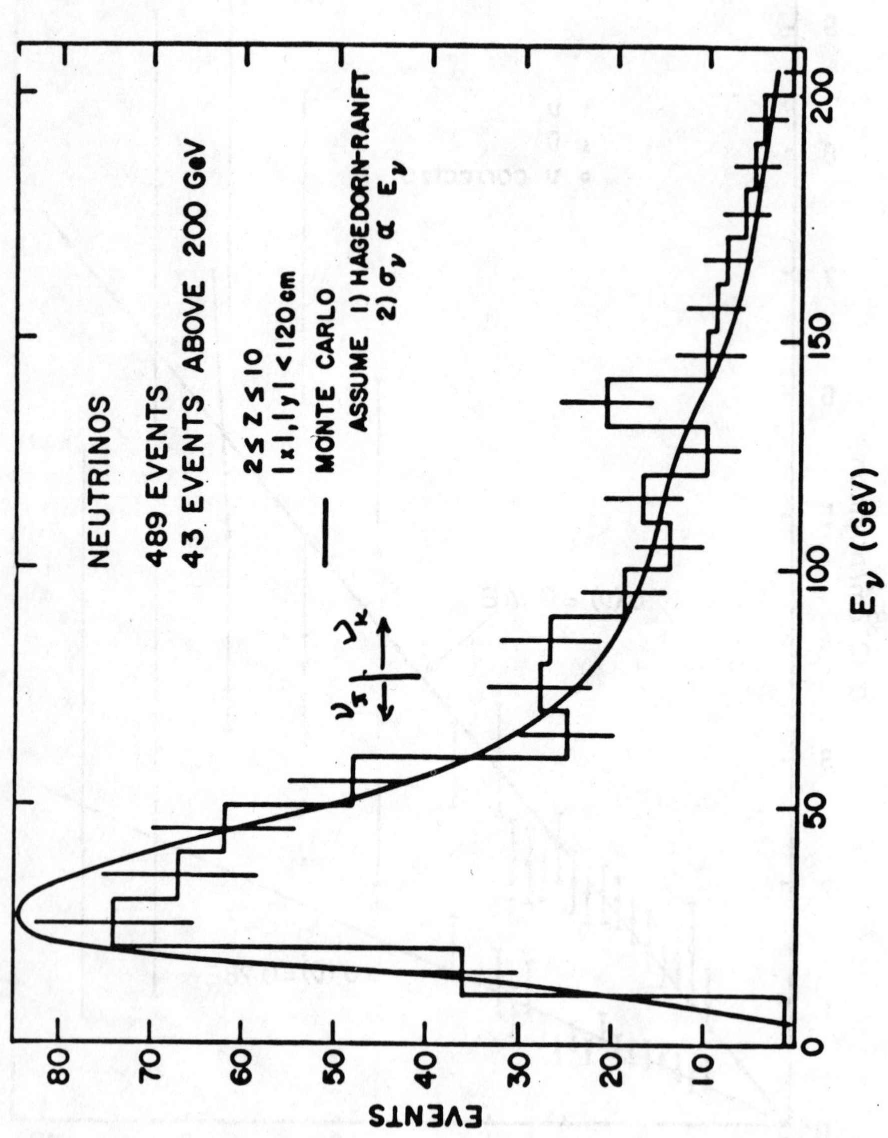
NEUTRINOS
489 EVENTS
43 EVENTS ABOVE 200 GeV
$2 \leq Z \leq 10$
$|x|,|y| < 120$ cm
— MONTE CARLO
ASSUME 1) HAGEDORN-RANFT
2) $\sigma_\nu \propto E_\nu$
ν_π
ν_K
EVENTS
80
70
60
50
40
30
20
10
0
50
100
150
200
E_ν (GeV)

Figure 8

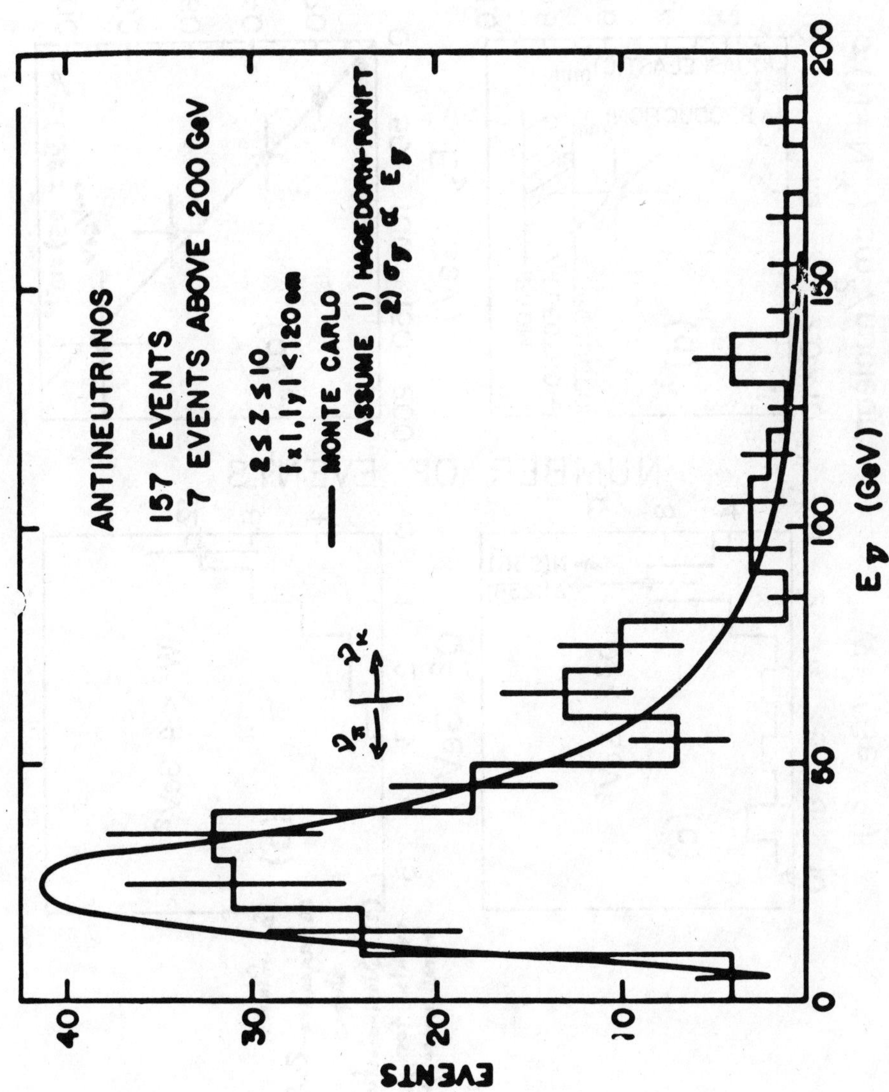
ANTINEUTRINOS
157 EVENTS
7 EVENTS ABOVE 200 GeV
$2 \leq Z \leq 10$
$|x|, |y| < 120$ cm
MONTE CARLO
ASSUME 1) HAGEDORN-RANFT
2) $\sigma_{\bar{\nu}} \propto E_{\bar{\nu}}$
ν_π
ν_K
EVENTS
40
30
20
10
0
50
100
150
200
$E_{\bar{\nu}}$ (GeV)

Figure 9

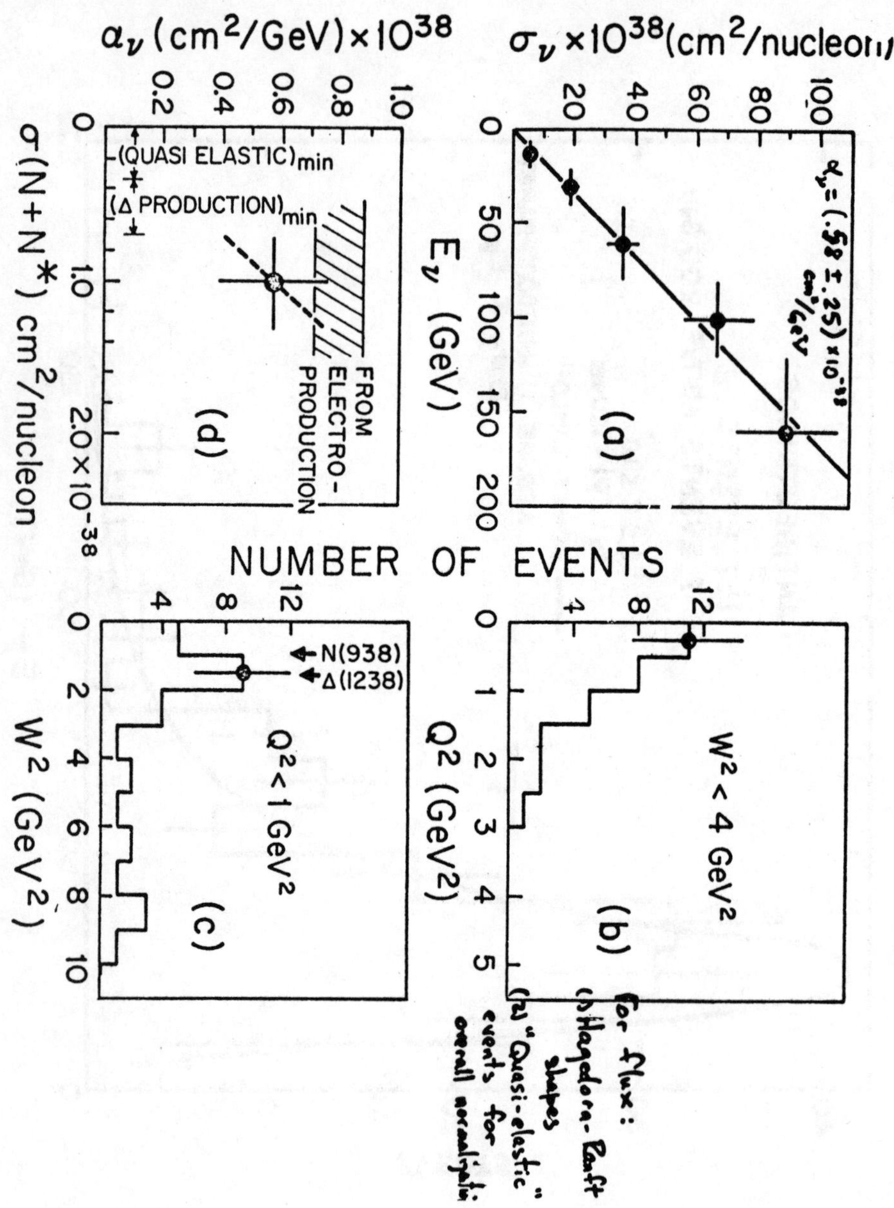

σ_ν ×10^38 (cm²/nucleon)
E_ν (GeV)
(a)
(b)
W² < 4 GeV²
Q² (GeV²)
NUMBER OF EVENTS
α_ν (cm²/GeV) ×10^38
(QUASI ELASTIC)_min
(Δ PRODUCTION)_min
FROM ELECTRO-PRODUCTION
(d)
σ(N+N*) cm²/nucleon
N(938)
Δ(1238)
Q² < 1 GeV²
(c)
W² (GeV²)

Figure 10

$Q^2 < 1\ GeV^2$. The enhancements visible at small W is assumed to come from the production of nucleon resonances. It is expected that these partial cross-sections will be very constant with energy. (High energy provides more available Q^2, but these exclusive channels should be damped at the higher Q^2 by form factors.) Using the best values for the quasi-elastic cross-section available from lower energy neutrino experiments, the authors quote for the neutrino slope:

$$\alpha^{\nu} = (0.58 \pm 0.25) \times 10^{-38}\ cm^2/GeV.$$

Figure 10 d shows the dependence of this result on the assumed quasi-elastic cross-section.

Figure 11 shows the ratio of the anti-neutrino to neutrino cross-sections vs. energy. The data are shown up to 75 GeV; higher energies require reliable estimates of the kaon yields. Because this measurement depends critically on the production ratio of π^- and π^+, from which the observed neutrinos are obtained, figure 11 b shows the dependence of the result on the assumed flux ratio. The data are consistent with a cross-section ratio independent of energy up to 80 GeV, with an average value

$$\frac{\sigma^{\bar{\nu}}}{\sigma^{\nu}} = 0.34 \pm 0.08.$$

Summarizing the cross-section results of this group in the energy range $0 < E_{\nu} < 150$ GeV:

(1) The observed events plotted against energy are consistent with the Hagedorn-Ranft flux estimate and a linearly rising cross-section.

(2) Using the Hagedorn-Ranft flux estimate, they find that $\sigma^{\bar{\nu}}/\sigma^{\nu} = 0.34 \pm 0.08$ and independent of energy up to $E_{\nu} = 80$ GeV.

(3) With quasi-elastic events setting the overall normalization, $\alpha^{\nu} = (0.58 \pm 0.25) \times 10^{-38}$, or $\alpha^{\nu} + \alpha^{\bar{\nu}} = (0.78 \pm 0.34) \times 10^{-38}$ cm^2/GeV.

This is about one standard-deviation lower than the prediction of the quark-parton model.

3. Caltech-NAL

The Caltech-NAL experiment utilizes a narrow band beam to obtain its neutrinos. This involves a momentum-analyzed hadron beam traveling in the forward direction toward the apparatus. The small angle decays then give neutrinos into the apparatus as shown schematically in figure 12 for an idealized case of a monochromatic pion and kaon beam. The two energy bands of neutrinos correspond to the decays of the pion and kaons, respectively.

In the real beam at NAL, the momentum spread of the hadrons is about ± 18%. Events wherein the muon and hadron energies are both detected allows us to reconstruct the incident neutrino energy to about 25%. The experimental spectrum for 1522 observed ν events are shown in figure 13. Even with the overall resolution in detection and beam

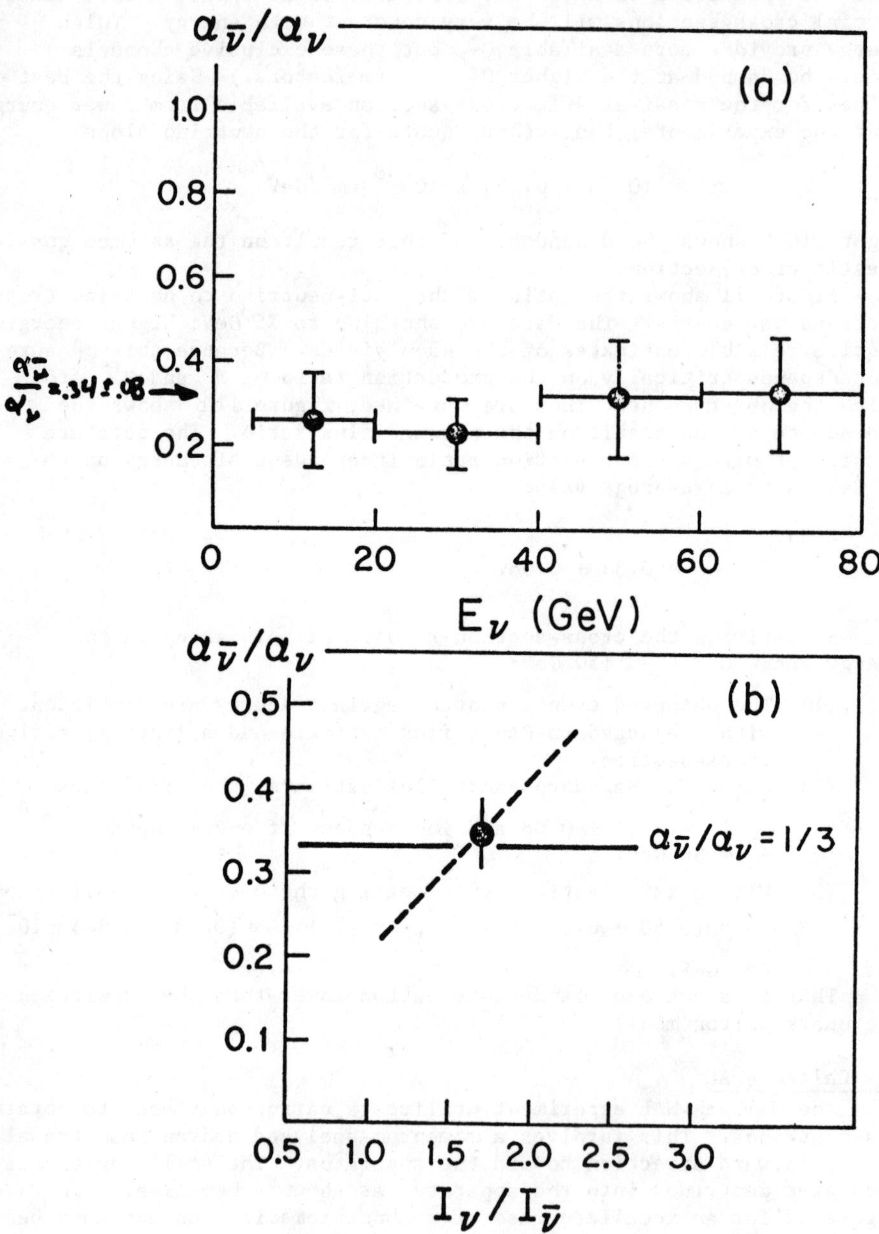

$a_{\bar{\nu}}/a_{\nu}$
(a)
1.0
0.8
0.6
0.4
0.2
0
20
40
60
80
E_{ν} (GeV)
$a_{\bar{\nu}}/a_{\nu}$
(b)
0.5
0.4
0.3
0.2
0.1
$a_{\bar{\nu}}/a_{\nu} = 1/3$
0.5
1.0
1.5
2.0
2.5
3.0
$I_{\nu}/I_{\bar{\nu}}$

Figure 11

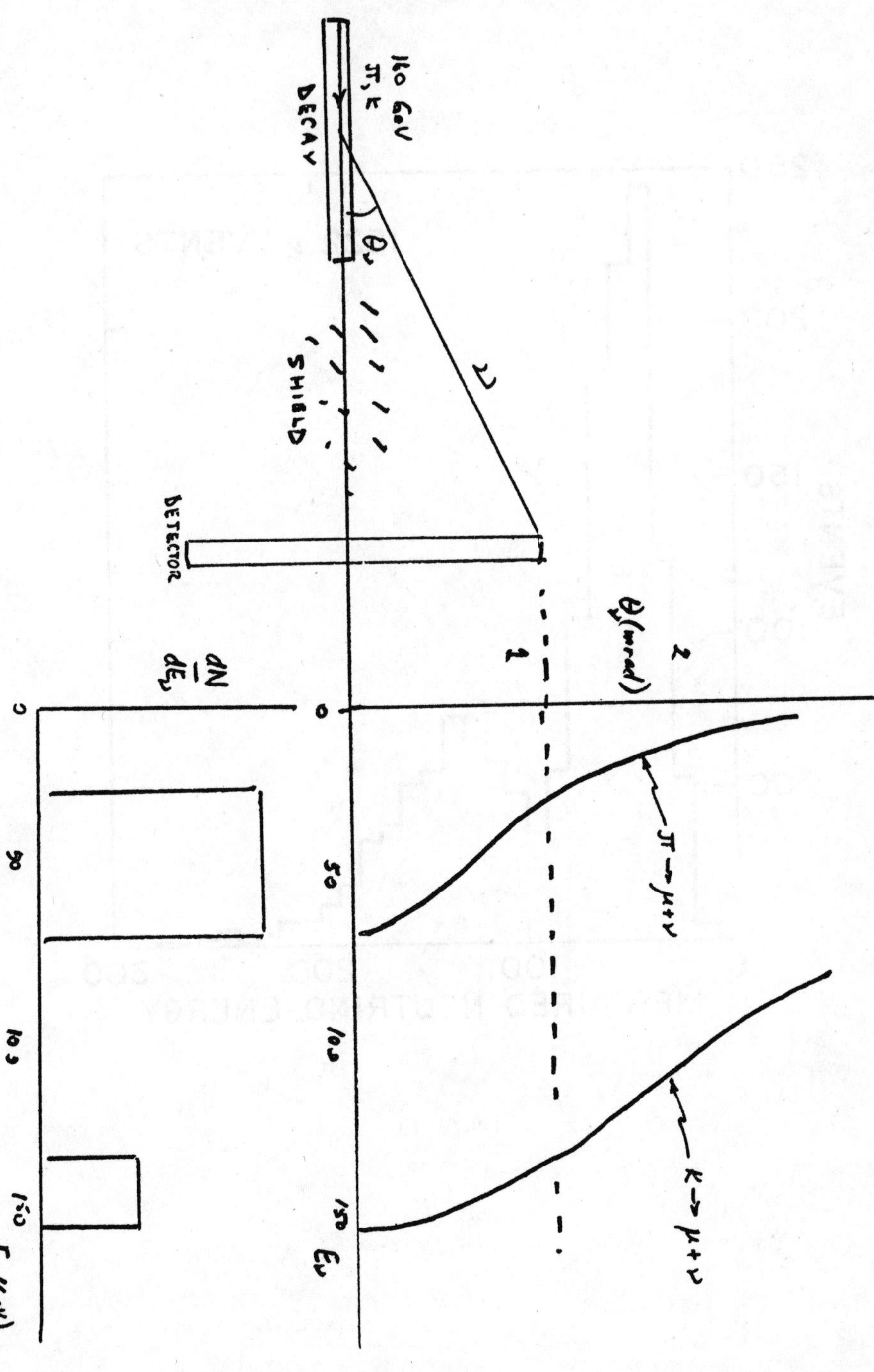
160 GeV
π, K
DECAY
θ_ν
ν
SHIELD
DETECTOR
θ_ν(mrad)
1
2
0
50
100
150
E_ν
π → μ + ν
K → μ + ν
$\frac{dN}{dE_\nu}$
0
50
100
150
E_ν (GeV)

Figure 12

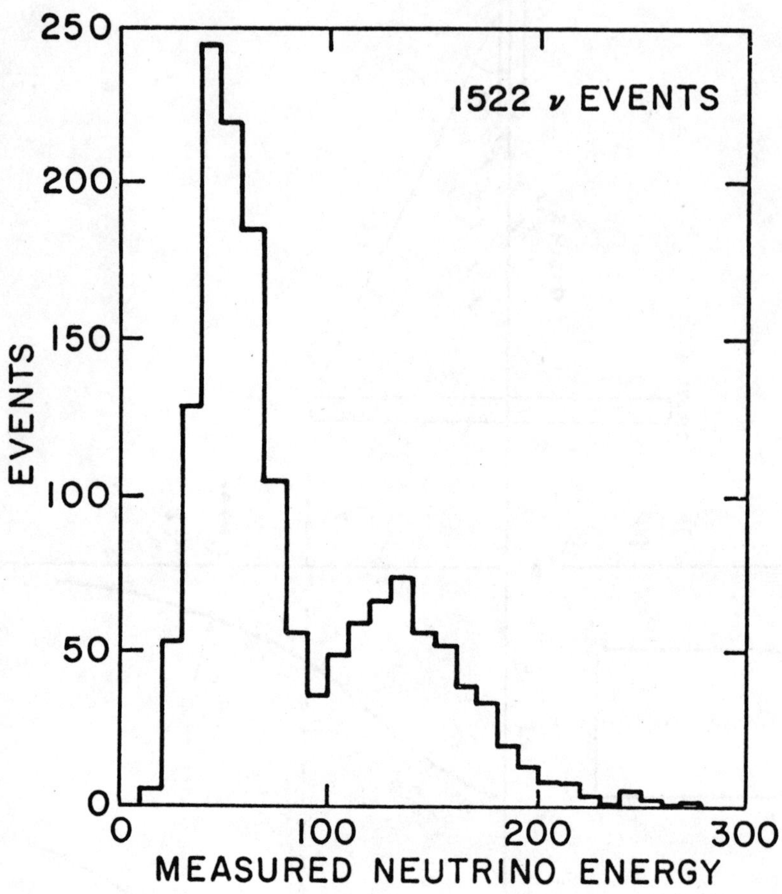
250
200
150
100
50
0
EVENTS
1522 ν EVENTS
0
100
200
300
MEASURED NEUTRINO ENERGY

Figure 13

presently available, the two peak structure expected for this kind of beam is observed.

The detection apparatus is depicted in figure 14. Neutrinos interact in the target calorimeter, consisting of 160 tons of 1.5m x 1.5m steel plate, with scintillation counters and spark chambers distributed throughout. The produced muon is observed in spark chambers, and it its angle is small enough, it enters an iron-core spectrometer magnet. The angle of bend gives the muon momentum.

Measurement of the hadron energy is accomplished with the scintillation counters located throughout the target. The hadronic cascade developing downstream of the interaction point is sampled in the steel. The total pulse height is then proportional to the total energy in the hadron shower.

We estimate our present resolutions as follows:

E_μ : statistically ~ 21%, systematically ~ 5%

E_μ : statistically 15-30%, systematically ~ 10%.

a. Differential Distributions

(1) x-distributions: Precise measurement of x-distributions really require extremely good resolutions. Resolution effects will change the shape of observed x-distributions. Figure 15 shows just such an effect. The curve labelled $F_2(x)$ is the structure function measured by SLAC-MIT with electrons. When we fold in the experimental resolutions for this experiment, we find the expected experimental curve labelled $[F_2(x)]_{smeared}$. If we further fold in the acceptance of the apparatus, we see the lower curve.

Therefore, it should be emphasized that an $F_2(x)$ structure function of the same form as that obtained at SLAC will result in an experimental distribution that is sharply peaked near x = 0, and with a somewhat different falloff, coming from effects of resolution and acceptance.

Figure 16 shows the observed x-distribution for 1027 events inside our fiducial volume where both the muon and the hadron energies are observed. These data were taken with a mean hadron beam setting of 170 GeV, and an observed neutrino energy distribution as shown in figure 13. The solid curve is the expected distribution, assuming the SLAC $F_2^{ed}(x)$ structure function and with resolutions and efficiencies folded in. The agreement with the expected distribution is quite good. There are no statistically significant differences observable between the x-distributions for pion neutrinos and kaon neutrinos.

(2) Q^2-distributions: The distribution in Q^2 corrected for efficiency for all events is shown in the figure 17. The expected distribution in Q^2 assumes the flat y-distribution as well as $F_2^{ed}(x)$ from SLAC. This curve is labelled $\Lambda = \infty$ in the figure. The multiplicative propagator term $[1+Q^2/\Lambda^2]^{-2}$, gives a much steeper falloff. For Λ = 10 GeV, for example, the calculated curve, normalized to the data, falls below the data at all points below the first. Figure 18 shows the confidence level for a fit to this propagator term. Above $\Lambda = 15$ GeV/c^2, we have no sensitivity; the fit is equally likely at $\Lambda = 10.3$. Therefore, we place a 90% confidence on a propagator mass for the term we have included: $\Lambda > 10.3$ GeV/c^2.

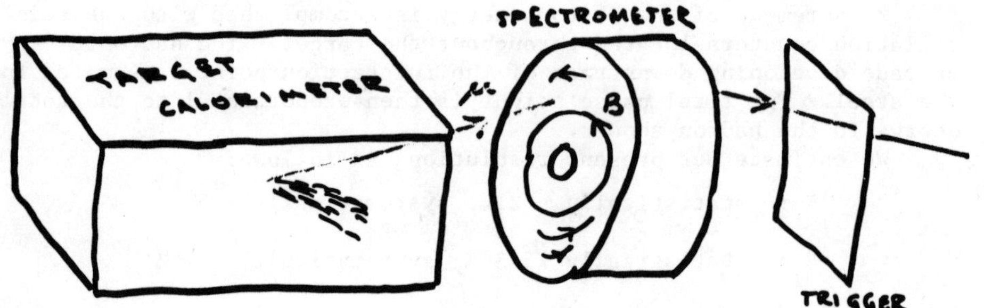

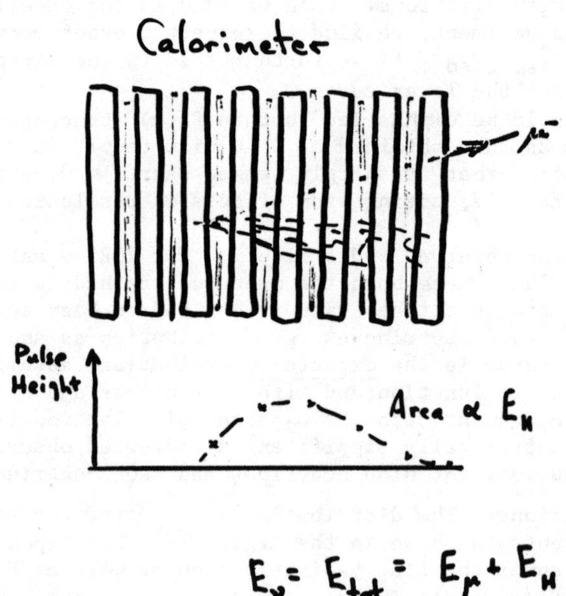

$$E_\nu = E_{tot} = E_\mu + E_H$$

Figure 14

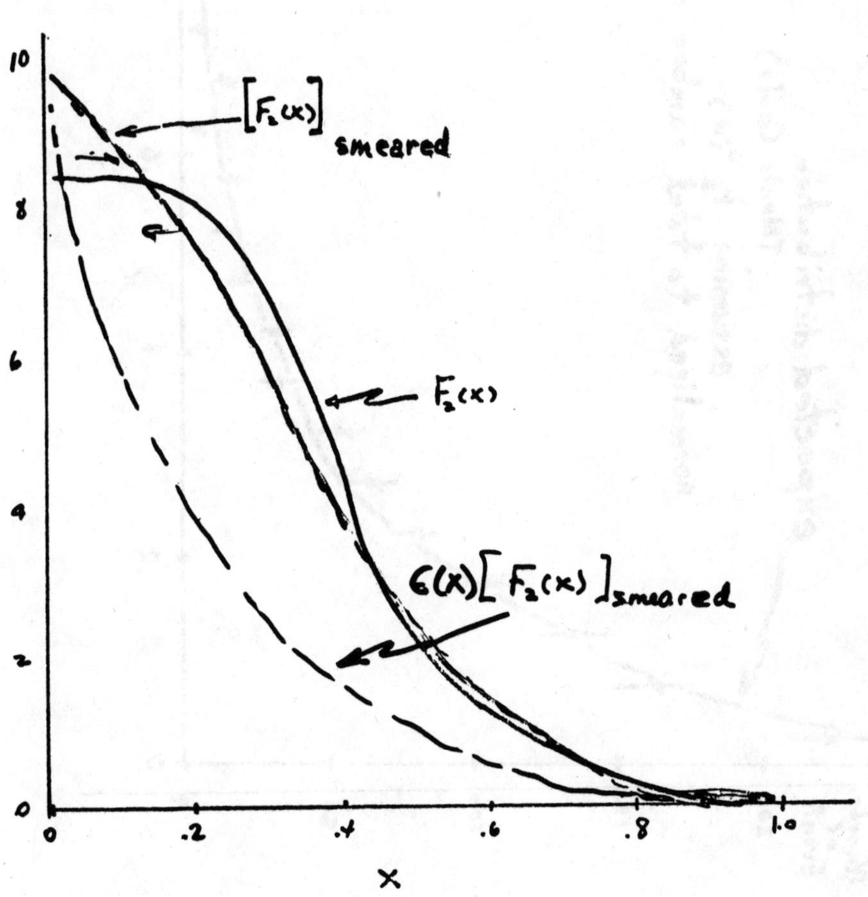
Effect of resolution
on $F_2(x)$
10
8
6
4
2
0
0
.2
.4
.6
.8
1.0
x
$[F_2(x)]$ smeared
$F_2(x)$
$G(x)[F_2(x)]_{smeared}$

Figure 15

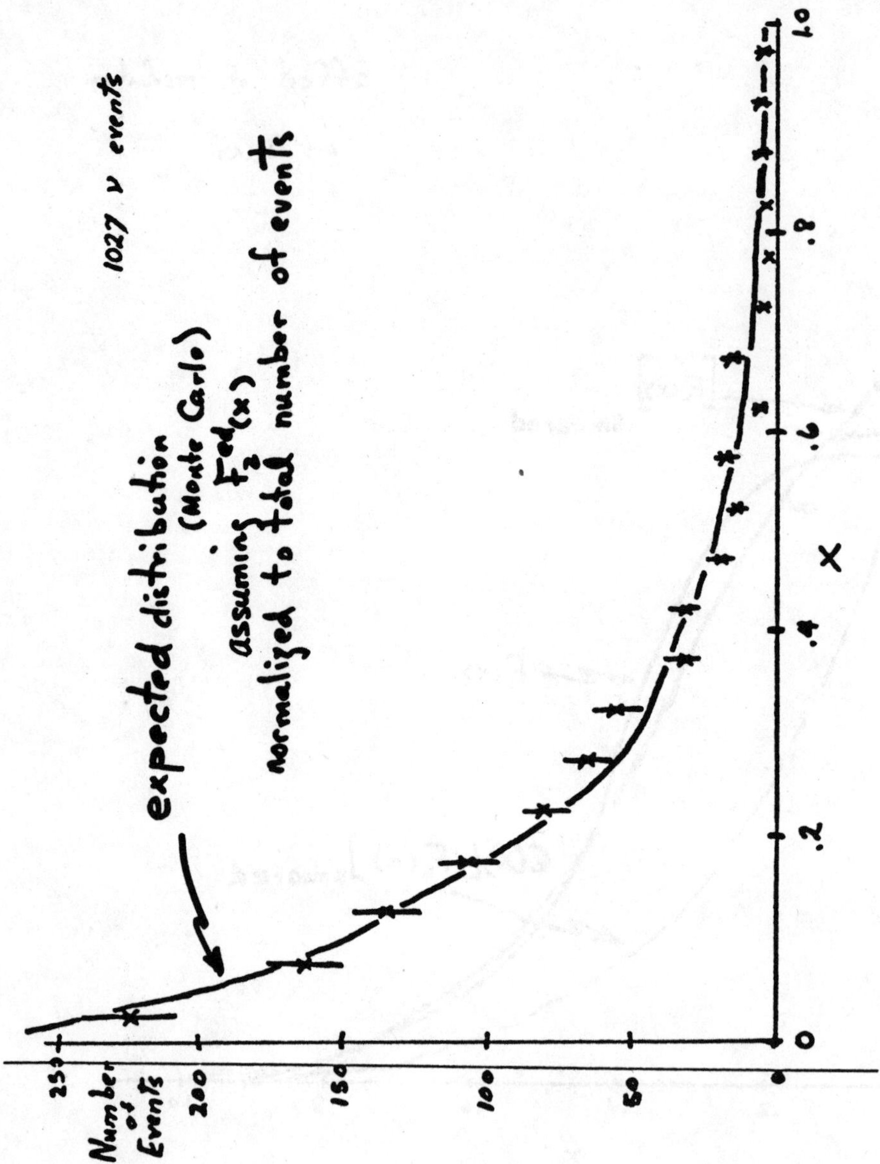
expected distribution
(Monte Carlo)
assuming $F_2^{ed}(x)$
normalized to total number of events
1027 ν events
Number of Events
250
200
150
100
50
0
0
.2
.4
.6
.8
1.0
X

Figure 16

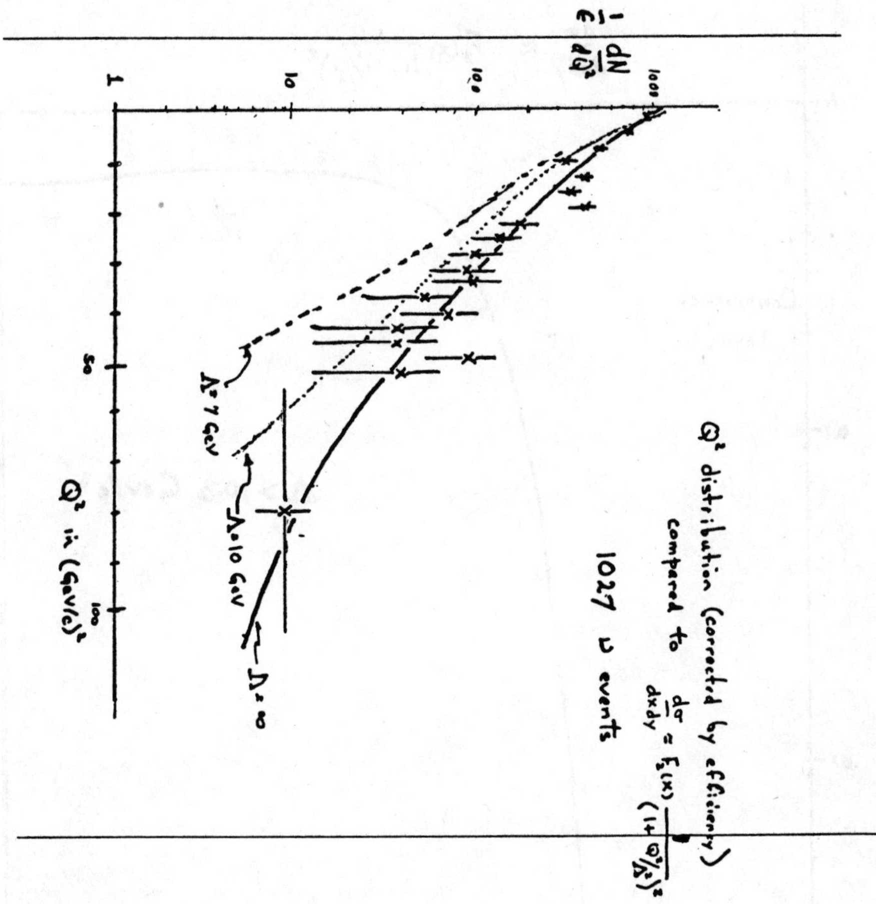

Q² distribution (corrected by efficiency)
compared to dσ/dxdy = F₂(x) 1/(1+Q²/Λ²)²
1027 ν events
dN/dQ²
1000
100
10
1
50
100
Q² in (GeV/c)²
Λ = 7 GeV
Λ = 10 GeV
Λ = ∞

Figure 17

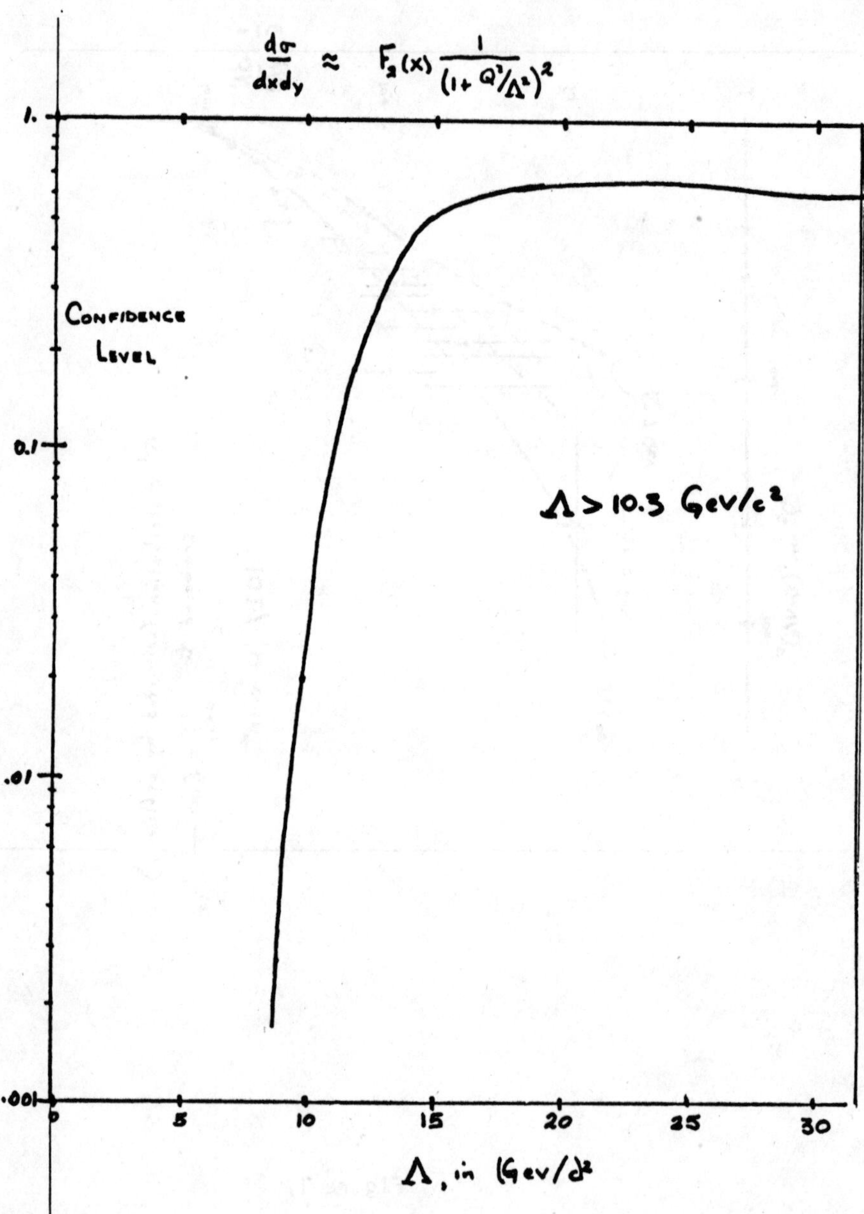

$\frac{d\sigma}{dxdy} \approx F_2(x)\frac{1}{(1+Q^2/\Lambda^2)^2}$
1.
Confidence Level
0.1
.01
.001
0
5
10
15
20
25
30
$\Lambda > 10.3$ GeV/c²
Λ, in (GeV/c)²

Figure 18

(3) y-distributions: The y-distributions are more effected by the acceptance function than by resolution. The smooth curve shown in figure 19 a is the expected y-distribution observed in our apparatus for ν interactions: effectively, it corresponds to the efficiency of the apparatus in y. The data are also shown with the appropriate statistical error. Figure 19 b shows the same distribution in a log plot. The systematically high points for $y > 0.6$ occur in a region where the efficiency has fallen below 10%. Below this value, there are no systematic departures from the expected flat distribution for neutrinos.

Figure 20 a shows all the neutrino events, with $y < 0.6$, corrected for efficiency and plotted vs. y. The data are consistent with the expected flat distribution. To obtain a numerical estimate of this consistency, we have fit to a function of the form

$$\frac{dN}{dy} = C\left[1 + a(1-y)^2\right]$$

where C is constrained by the overall normalization, and a is a free parameter. It could be thought of as representing an average anti-quark component in the nucleon. Figure 20 b shows the result:

$$a = +.05 \begin{array}{l} +.25 \\ -.17 \end{array}$$

consistent with zero, and consistent with the 5% average anti-quark component found in the low energy CERN data.

Summarizing these preliminary results of the Caltech-NAL group on the unnormalized distributions:

(1) The data with muon traversing the magnet are consistent with the $F_2^{ed}(x)$ shape observed by the SLAC-MIT group.

(2) The same data fit the expected distribution in Q^2 without an additional propagator term. The data are inconsistent with propagators of masses less than 10.3 GeV/c^2. That is, $\Lambda > 10.3$ GeV/c^2 (90% confidence).

(3) From 1027 ν events with measured muon momentum, a fit to

$$\frac{dN}{dy} = C\left[1 + a(1-y)^2\right], \text{ it is found}$$

$$a = +.05 \begin{array}{l} +.25 \\ -.17 \end{array}$$

consistent with the expected flat y-distribution for ν events.

(4) There appears no obvious dependence on E_ν in either x or y distributions for neutrinos.

b. Preliminary Total Cross-sections

I am going to report some very preliminary cross-section data taken in the NAL narrow band beam. This represents a measurement which is completely internal; that is, all relevant normalizations are measured directly in the same experiment. One nice feature of a normalized measurement is that it involves two separately measured numbers (the number of observed events and the incident flux), which

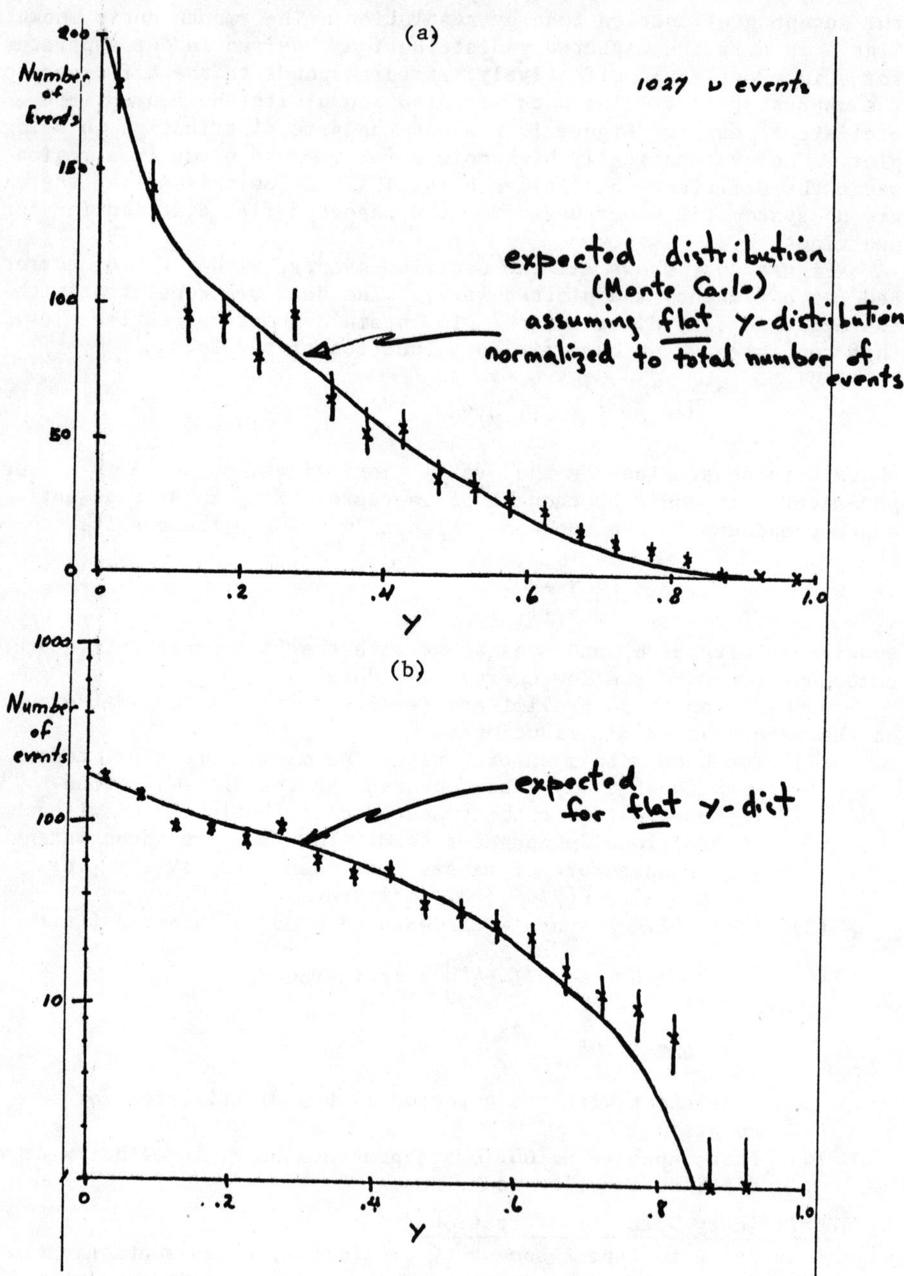
(a)
Number of Events
1027 ν events
expected distribution (Monte Carlo) assuming flat y-distribution normalized to total number of events
Y
(b)
Number of events
expected for flat y-dist
Y

Figure 19

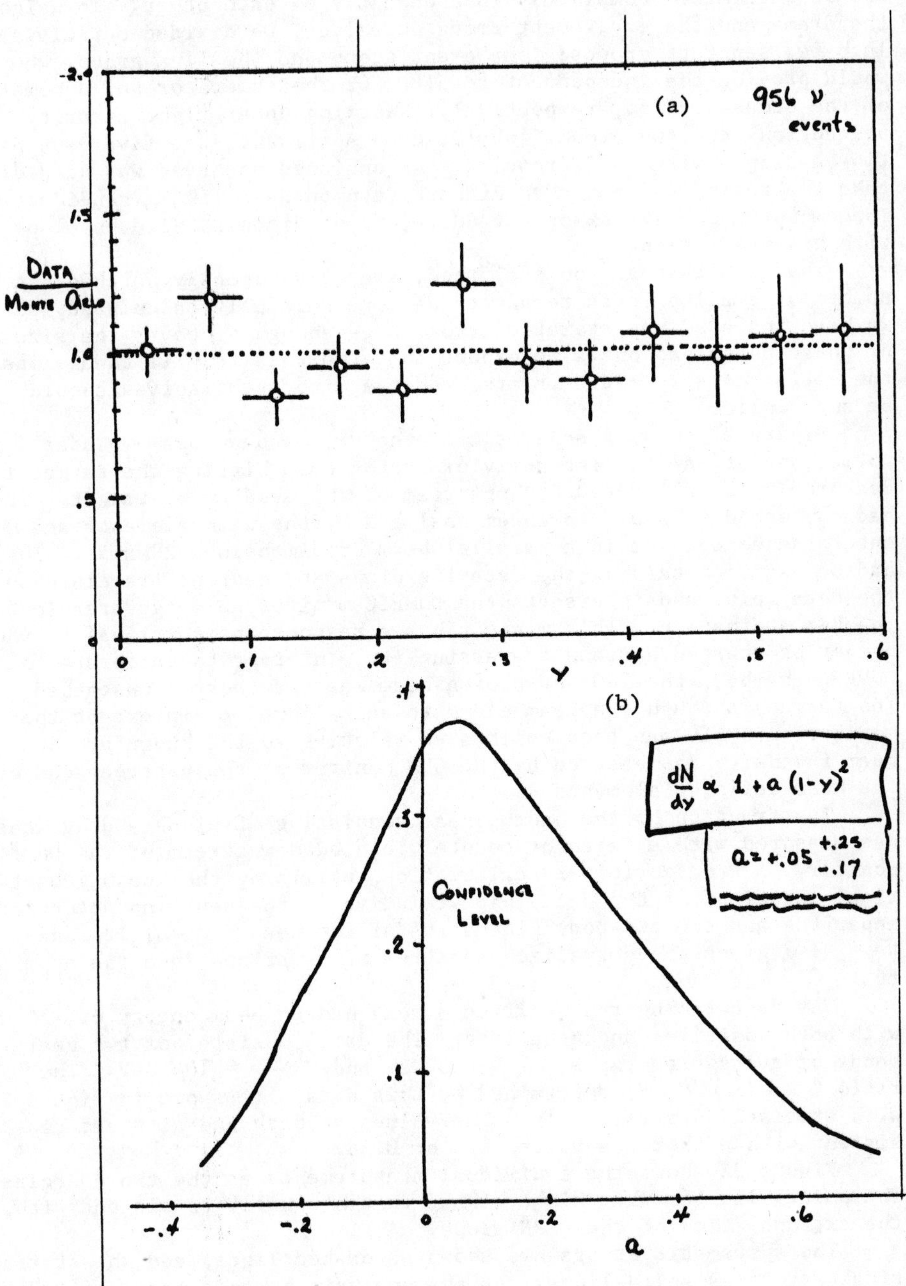
2.0
(a)
956 ν events
1.5
DATA / MONTE CARLO
1.0
.5
0
0
.1
.2
.3
.4
.5
.6
y
.4
(b)
$\frac{dN}{dy} \propto 1 + a(1-y)^2$
$a = +.05 {}^{+.25}_{-.17}$
.3
CONFIDENCE LEVEL
.2
.1
-.4
-.2
0
.2
.4
.6
a

Figure 20

can be determined completely independently of each other. To heighten the drama and the excitement among ourselves, we divided ourselves into two separate groups: the event group and the flux group, who would provide the independent results for the numerator and denomator of the cross-section, respectively. Working under tight security precautions the two groups submitted on April 22, just five days ago, two envelopes with their results. An unbiased observer was needed to take the ratio, so we picked Richard Feynman as a likely candidate. To demonstrate his lack of prejudice, he provided a third envelope with his prediction.

The results that you see, then, are only recently hatched. A number of small effects remain to be more carefully calculated, so we have included a systematic error large enough to cover the size of these perturbations. We expect that the results will change inside the estimated systematic errors, and the errors themselves should become smaller.

Figure 21 shows a schematic of the narrow band beam. Under normal operation, the intensity of proton beam hitting the target is measured by a calibrated SEM upstream of the production target. The hadron secondaries are momentum analyzed by the beam elements and enter the decay pipe in a parallel beam of dimensions 2" x 5". The hadron beam, containing the decaying pions and hadrons travels down the decay pipe and intersects the DUMPIC monitor, a large area ion chamber at the end. This monitor has a response proportional to the number of charged hadrons traversing it. In order to calibrate the DUMPIC chamber, the 200 GeV proton beam was transported unscathed (no target) through the beam elements and allowed to intercept the monitor. DUMPIC was then calibrated relative to the known proton beam intensity as measured by the SEM monitor at the upstream end of the beam defining elements.

The fraction of the hadron beam consisting of pions and hadrons was measured with a Cerenkov counter located downstream of the DUMPIC monitor. A straightforward calculation, utilizing the known geometrical locations of the decay pipe relative to the neutrino detection apparatus and the two-body kinematics of the decays $K \rightarrow \mu + \nu$ and $\pi \rightarrow \mu + \nu$ gives the normalized spectrum of neutrinos into the apparatus.

The data-taking run utilized a mean hadron beam energy of 120 GeV with both positives and negatives. The data consisted of two energy bands of average energies $\langle E_\nu \rangle = 38$ GeV and $\langle E_\nu \rangle = 104$ GeV. The ratio $R = \sigma(\bar{\nu}) / \sigma(\nu)$ determined by this data, is shown in figure 22, with statistical errors only. The values at both energies are consistent with either the values 1/3 or 0.38.

Figure 23 shows the individual measurements at the two energies. The points lie somewhat higher than, though usually consistent with, the extrapolation of the CERN slope.

The systematic errors are shown as dashed lines, and the statistical errors as solid lines. As the analysis becomes more refined, we expect the best value to move but to stay inside the dashed lines, and the systematic errors to become smaller. The statistical errors will not be helped by refinements of analysis. We conclude that the cross-section is indeed rising as the energy is increased, and that the rate

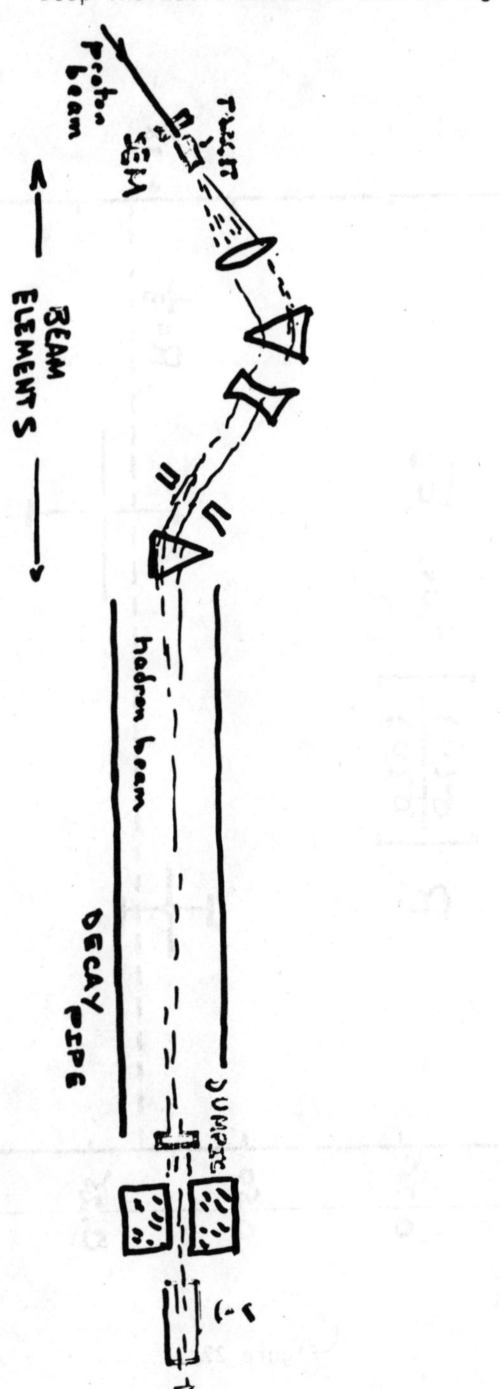
BEAM SCHEMATIC (not to scale)
proton beam
SEM
BEAM ELEMENTS
hadron beam
DECAY PIPE

Figure 21

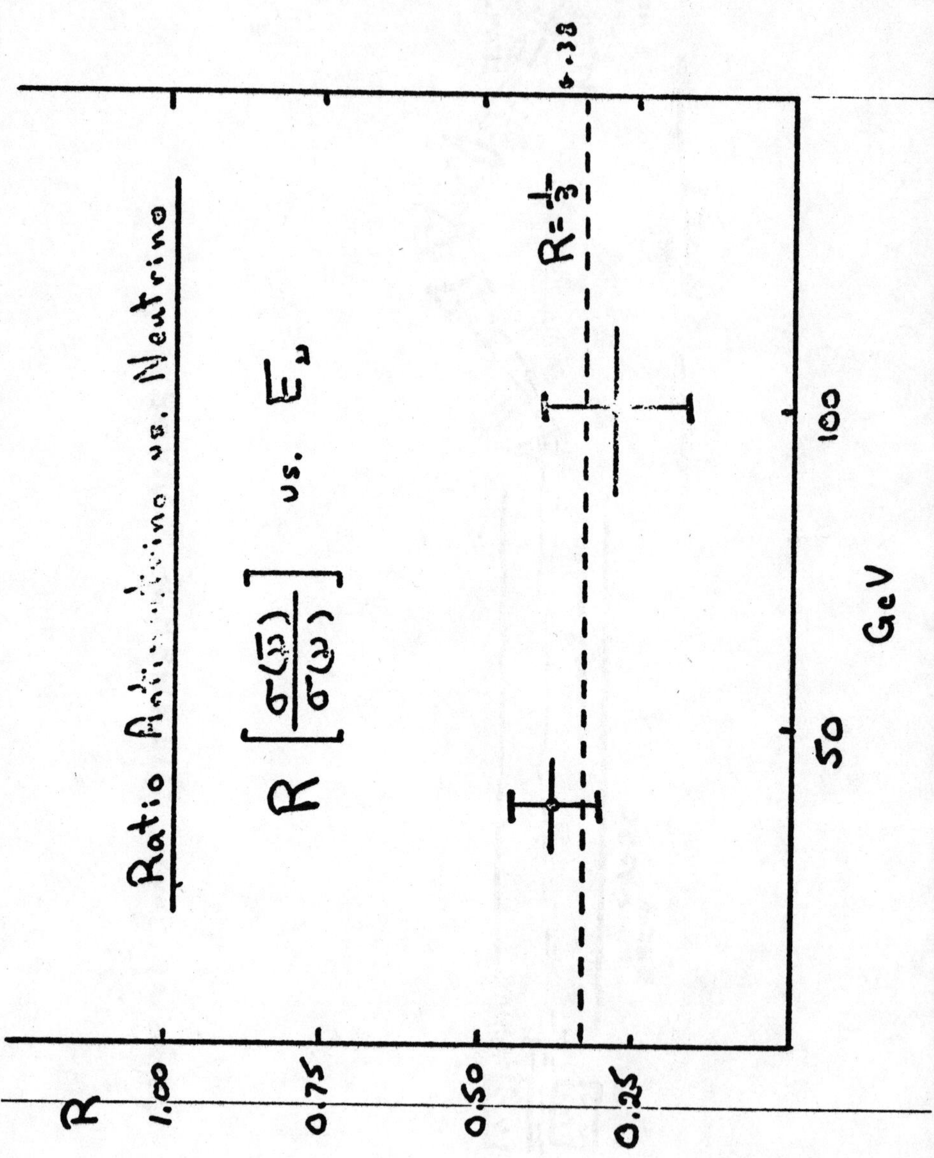

Figure 22

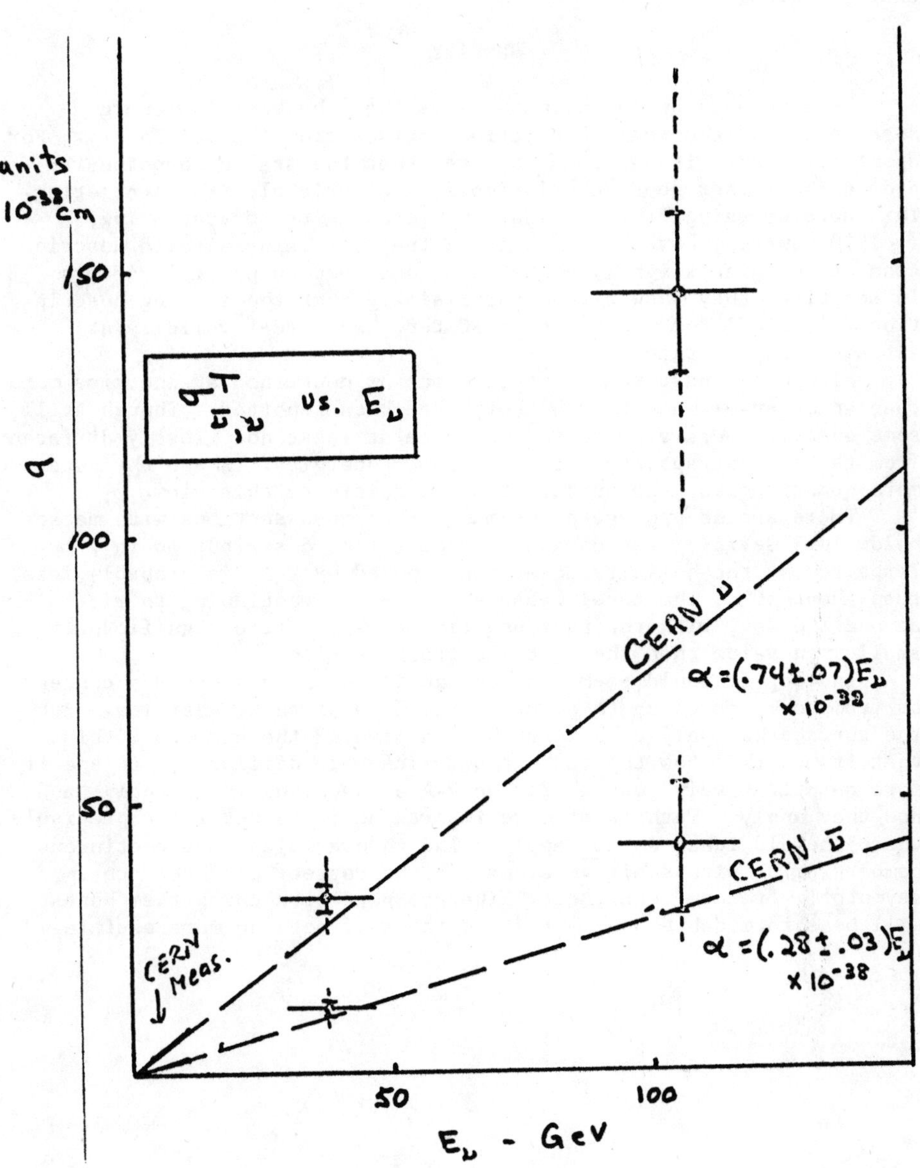
units
10^{-38} cm
150
100
50
σ
$\sigma^T_{\bar{\nu},\nu}$ vs. E_ν
CERN ν
α = (.74 ± .07) E_ν
× 10^{-38}
CERN $\bar{\nu}$
α = (.28 ± .03) E_ν
× 10^{-38}
CERN Meas.
50
100
E_ν - GeV

Figure 23

of rise is at least as fast as observed by Gargamelle-CERN in the low energy region.

SUMMARY

We have been aware for many years that the best low energy description of the weak interactions arises from the V-A Theory. For about six years, it has been observed that the scaling hypothesis provided a rather good description of inelastic electron scattering. This data emerging from CERN-Gargamelle on neutrino scattering, ($\sim$1970 onwards) have shown that V-A Theory remains a valid description at energies slightly higher than observed in particle decays. In addition, they show rather surprisingly that the scaling description (ala SLAC) for the hadronic states, has a real validity at extremely low energies.

Within the past year, interactions of neutrinos at energies more than an order-of-magnitude higher, have been reported. Though still some qualitative, we perceive a picture at least not grossly different from the one extrapolated from the lower energies. Where the data are more quantitative, the numbers tend to reinforce this view.

There are no propagator terms in the cross-sections with masses below 10.3 GeV/c^2. The deviation from a flat distribution in y (expected by the helicity selection imposed by V-A) is probably less than about 30%. The total cross-sections are continuing to rise beyond 100 GeV, with the anti-neutrino cross-section significantly smaller in value than the neutrino cross-section.

Indeed, it would seem that enough is known to provide a context, or framework, in which to describe the data at these energies. But the surface has only been scratched in finding the answers within that framework. Now the task is becoming more difficult: to specify more quantitatively just where the V-A and scaling ideas apply, and how they apply. Perhaps of more importance is to determine precisely where the old ideas do <u>not</u> apply. The theoreticians are continuously bombarding us with their versions for the correct picture: charm, asymptotic freedom, etc., etc. The prospect of keeping them honest will be a formidable one, but I can think of nothing more exciting.

Experimental Search for New Particle Production in High Energy Neutrino Interactions*

David B. Cline
University of Wisconsin
Madison, Wisconsin 53706

CONTENTS

1. A SELECTED LIST OF POSSIBLE NEW PARTICLES

A common and useful procedure in particle physics is the invention of new particles to explain force laws, conservation laws or in an attempt to unify seemingly diverse phenomena. A less common practice has been the discovery of these particles. Nevertheless the few spectacular cases where the particles did exist continues to make even skeptics somewhat optimistic. It seems that at present there exists an overly abundant number of suggested new particles and any attempt to summarize the speculated spectrum of particles has to be selective.

The focus of this review talk concerns the experimental signature for new particle production and the present status of the experimental evidence available from high energy neutrino collision data. We also discuss the different experiments required to make a definitive search for new particles independent of the theoretical estimates of the branching fraction into specific channels. It is also necessary to review the expectations for the scattering distributions in the absence of new particle production in order to access expected background. Early data from NAL experiments E21 and E1A are shown where relevant and the implications of a possible scaling breakdown indicated by some of these data is discussed.

The use of neutrino beam to produce new particles has many advantages over other projectiles in that a weakly interacting particle is bombarding a strongly interacting particle and therefore the production of

TABLE 1

EXPERIMENTAL SIGNATURE FOR NEW HADRON PRODUCTION

PARTICLE	REASON FOR SUSPECTED EXISTENCE	PRODUCTION AND DECAY	SIGNATURE AND BACKGROUND
Charm	Suppress $K_L^o \to \mu\mu$ $K^+ \to \pi \ell \ell^-$ $K_s - K_L$ mass difference in gauge theories Produced on λ quarks (Ref. 1, 2, 3)	$\nu_\mu + \lambda \to C + \mu^-$ $\to \mu + \nu + \lambda$ $\to$ hadrons (λ quarks) (Magnitude suppressed by the number of λ quarks in the nucleon	Noisy dimuon events with strange particles in final state Violation of $\Delta S = \Delta Q$ Rule Breakdown of charge symmetry and threshold effect
(V + A) Quarks (FANCY or FICTITIOUS)	No reason (Ref. 4)	$\nu_\mu + F \to F' + \mu^-$ F, F' coupled by $(V \underline{+} A)$ $\frac{dN^{\nu_\mu}}{dy} \sim (1 - y)^2$; $\frac{dN^{\bar{\nu}_\mu}}{dy} \sim$ const	Peaking at large y in $\bar{\nu}$ reaction and small y in ν_μ reaction Apparent charge symmetry violation
Higgs Particle	Common to most Gauge Theories (Ref. 2 and 5)		Noisy dimuon events May break μ-e universality
Vector Gluon	Breaks Bjorken scaling ($e^+e^- \to$ hadrons?) (Ref. 6)	Produced at low Q^2 and high E_ν	Structure in $\frac{dN}{dw^2}$ at low Q^2

virtually all possible particles is allowed. Only particles which have new quantum numbers which are conserved by the weak interaction cannot be produced singly. For example, leptons with a new lepton number must be pair produced and, therefore, neutrino beams are no more useful than perhaps proton or photon beams. Another possibly useful property of neutrino collisions is the relative ease with which high momentum transfer reations are initialed because of the short range nature of the weak interaction. This might be useful if particles were "tightly bound" inside the hadron. It is useful to separate the new kinds of particles into the traditional classification of hadrons, leptons and semi-weakly interacting particles such as the W boson.

In Table 1 is given a selected list of new hadrons and a brief explanation for their suspected existence. The particle for which the greatest enthusiasm has been shown is one that carries a new quantum number called charm.[1,2] In fact this could refer to a class of particles in some models. Strictly speaking the existential motivation is for a new quantum number similar to strangeness that is useful in certain models to explain the suppression of $\Delta S = 1$ neutral currents. The expected properties of these particles has recently been reviewed by M. K. Gaillard.[3,4] Other possible new hadrons are listed in Table 1.[5,6,7]

The oldest speculated particle (or groups of particles) is the intermediate vector boson which transmits the weak interaction.[8] The expected properties of this particle are by now well known. The greatest unknown in searching for this particle is the mass; speculation on the mass range from a few GeV to greater than 37 GeV, however.[9] Some gauge theories make very specific predictions.[10] Another unknown property is the branching ratio into the leptonic final state. The recent CEA-SPEAR measurements of[11]

$$e^+e^- \rightarrow \text{hadrons} \tag{1}$$

indicate a surprisingly strong coupling of high mass virtual photons with hadrons. This effect would also enhance W decay into hadrons relative to the leptonic final state.

The charged intermediate vector boson has recently acquired a neutral partner with the discovery of neutral currents.[12] The experimental signature for the production of a W^o is straight forward but there have been no detailed estimates of the production cross section as of yet.[13] These calculations are necessary to access the feasibility of a unique search for the W^o using neutrino beam.

TABLE 2

EXPERIMENTAL SIGNATURE FOR NEW WEAK PARTICLE PRODUCTION

	PARTICLE	DECAY MODE	PRODUCTION PROCESS	SIGNATURE	BACKGROUND PROCESS
Ia	W^+	$W^+ \to \mu^+ \nu_\mu$	$\nu_\mu + Z \to \mu^- W^+ z$	Quiet dimuon production	$\nu_\mu + N \to \mu^- + X$ $\to \pi \to \mu$
Ib	W^+	$W^+ \to$ hadrons	$\nu_\mu + Z \to \mu^- W^+ z$	Characteristic Energy dependent structure in low x, high y inelastic structure functions	$\nu_\mu + N \to \mu^- + X$
IIa	W°	$W° \to \mu^+ \mu^-$	$\nu_\mu + Z \to \nu_\mu W° Z$	Quiet dimuon events	$\nu_\mu + N \to \nu_\mu + X$
IIIa	M^+	$M^+ \to \mu^+ \nu_\mu \nu_\mu$	$\nu_\mu + N \to M^+ + X$	Apparent nonconservation of lepton number (μ^+ produced in ν_μ beam) or peak at high y in dN/dy for μ^+ events	($\bar{\nu}$ Contamination in ν_μ beam) $\nu_\mu + N \to \mu^- + X$ (μ^- undetected) $\to \pi^+ \to \mu^+$
IIIb	M^+	$M^+ \to \nu_\mu + X^+$	$\nu_\mu + N \to M^+ + X$	Muonless events with characteristic energy threshold	$\nu_\mu + N \to \nu_\mu + X$
IVa	M°	$M° \to \nu_\mu + X$	$\nu_\mu + N \to M° + X$	Muonless events with characteristic energy threshold	$\nu_\mu + N \to \nu_\mu + X$
IVb	M°	$M° \to \mu^+ \mu^- \nu_\mu$	$\nu_\mu + N \to M° + X$	Noisy dimuon events	$\nu_\mu + N \to \mu^- + X$ $\to \pi^+ \to \mu^+$

New weakly interacting particles have also been suggested from time to time.[14] Heavy leptons with the same lepton numbers as the electron and muon fit naturally into gauge theories providing a mechanism by which divergent diagrams can be cancelled.[15,16] It is frequently said jokingly that gauge theories need "either neutral currents or heavy leptons and preferably both." In Table 2 is summarized the properties of the gauge theory heavy leptons. Heavy neutral leptons also play a role in some theories.[16] Such particles can be produced by neutral current interactions making the search a search feasible with neutrino beams. We can make crude estimates of the cross section for these particles using the relation

$$\sigma(\nu_\mu + N \rightarrow M^\circ + x) = R^\nu \sigma(\nu_\mu + N \rightarrow M^+ + x)$$

where R^ν is the measured ratio of muonless to muon event rate.[14,17]

Other varieties of heavy leptons could exist, for example, with new lepton numbers.[14] These particles would be produced in pairs and neutrino beams would not be any better suited to carry out a definitive search than other techniques. However, the charged leptons are expected to be pair produced by electrons, photons or protons and would be produced in the proton beam dump that is associated with the neutrino beam. Furthermore, the charged leptons could decay into lighter (neutrino like) leptons if they exist giving a new component of penetrating particles to the neutrino beam. The neutral lepton would interact in the same detectors used to study neutrino interactions frequently giving a muonless signature. However the observed large rates for neutral current events makes the search for such particles in beam dump experiments difficult but not impossible. e^+e^- colliding beam experiments are perhaps the best way to search for this variety of heavy lepton.[14]

2. KINEMATICS AND THE EXPERIMENTAL SIGNATURE FOR NEW PARTICLE PRODUCTION

We imagine that the neutrino collision can be broken up into two vertices, the leptonic and the hadronic. New particle production can occur at either vertex. Figure 1 shows this schematically. The Bjorken x and y variables are defined at the production process by

$$y = E_x/E_\nu ; \; x = Q^2/2mE_x$$

KINEMATICS OF PRODUCTION

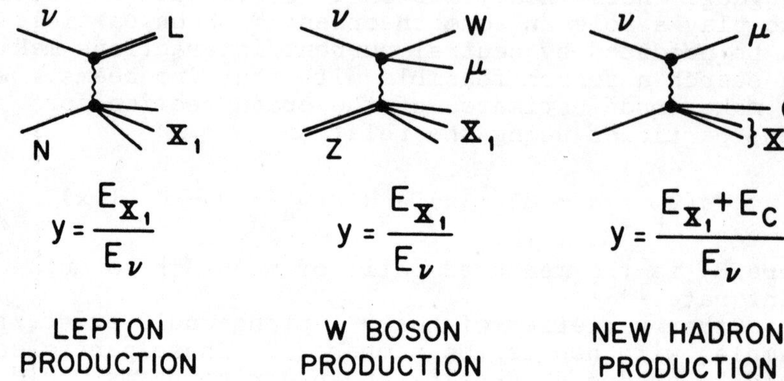

KINEMATICS OF DECAY

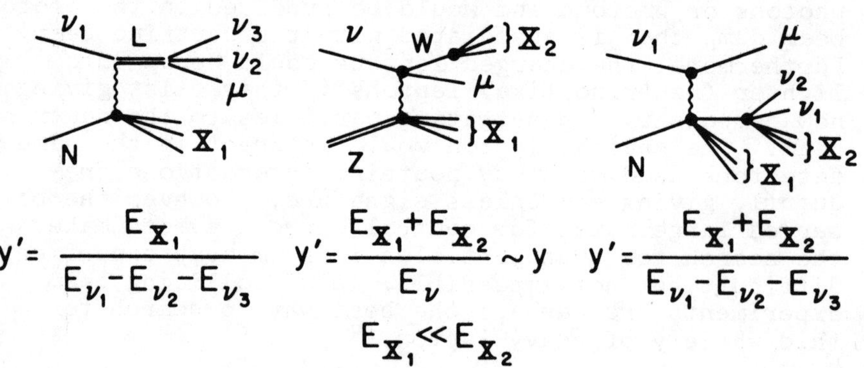

FIG. 1

If a short lived particle is produced and decays some of the energy from this particle may escape detection in the form of neutrino energy. The apparent total energy in the collision becomes

$$E' = E_\nu - E_{(\text{outgoing neutrinos})}$$

and the y becomes

$$y' = E_x/E'.$$

E' and y' are the measurable variables for experiments that use a broad band neutrino beam.

The kinematics of new particle production at the hadronic vertex are controlled by two variables

$$v = Q^2/s \text{ and } W^2/s,$$

where W^2 is the hadronic invariant mass squared. In Fig. 2 we show the contours of these variables on the (x, y) plot. The production of a particle of mass M_H requires that

$$\frac{W^2}{s} > \frac{(M_H)^2}{s}.$$

There may also be a form factor depending on Q^2 for the production of the new particle. For example, the particle could produced only when

$$Q^2 \sim M_H^2$$

or may be produced at any Q^2.[18] Figure 2 shows the allowed kinematic region on the (x, y) plot as the neutrino energy crosses the threshold for producing the particle. The observation of such characteristic structure and the attendent scaling breaking would be a signal for new particle production. Figure 3 shows the schematic (x, y) distributions for new particle production as threshold is crossed. It is possible that at high energy scaling will be reestablished again and the structure functions might again return to those expected on the basis of the simple parton model. Some specific examples of this type have been calculated by Barger for charmed particle production.[19] In Fig. 4 we show for illustration energy dependence of the slope parameter ($\sigma_T = \alpha^\nu E_\nu$) for a typical case. Note the vast energy range required to reestablish neutrino scaling. Figure 5 shows the distortion of the

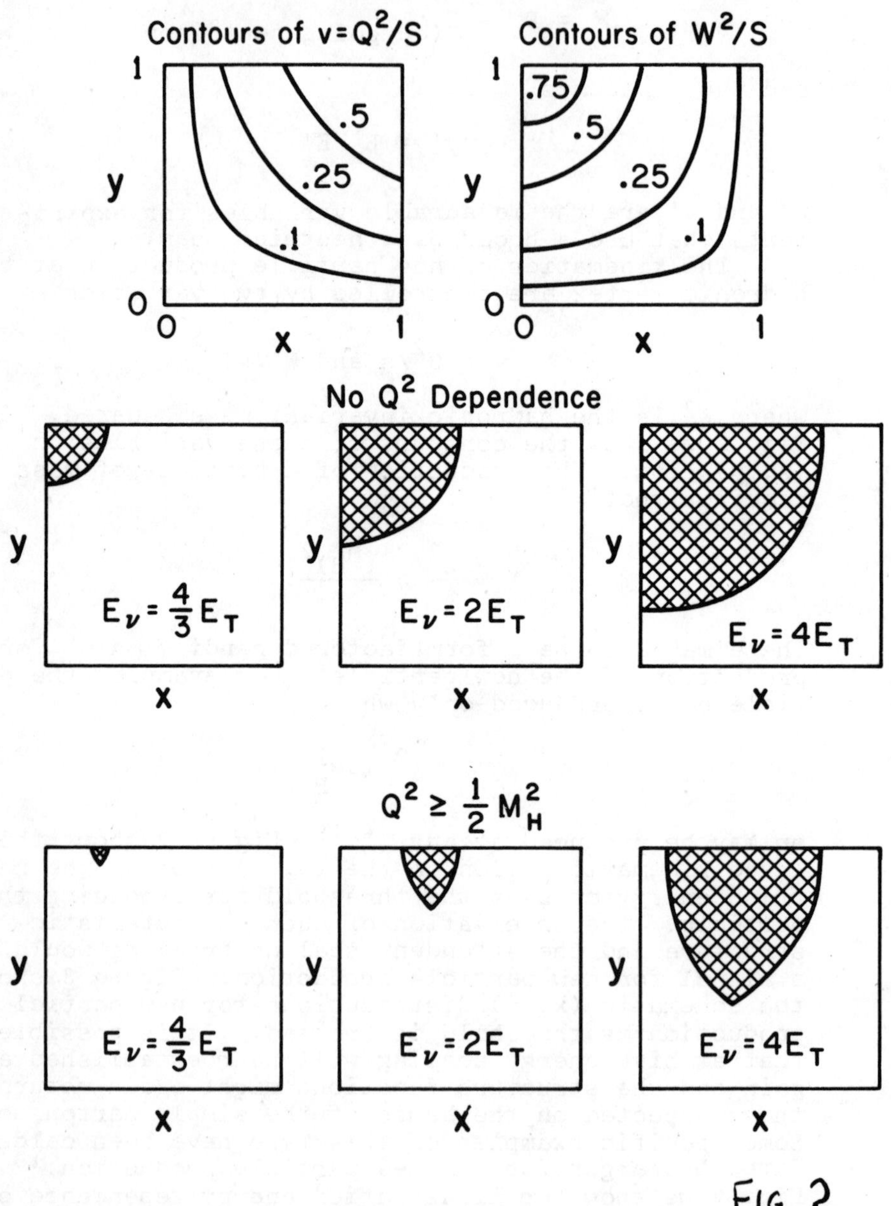
PRODUCTION OF NEW HADRONIC STATES
IN DEEP INELASTIC COLLISIONS
Contours of v=Q²/S
Contours of W²/S
.5
.25
.1
.75
.5
.25
.1
y
x
0
1
No Q² Dependence
E_ν = 4/3 E_T
E_ν = 2E_T
E_ν = 4E_T
Q² ≥ 1/2 M_H²
E_ν = 4/3 E_T
E_ν = 2E_T
E_ν = 4E_T
FIG 2

DISTORTION OF THE (x,y) PLOT BY NEW HADRON PRODUCTION

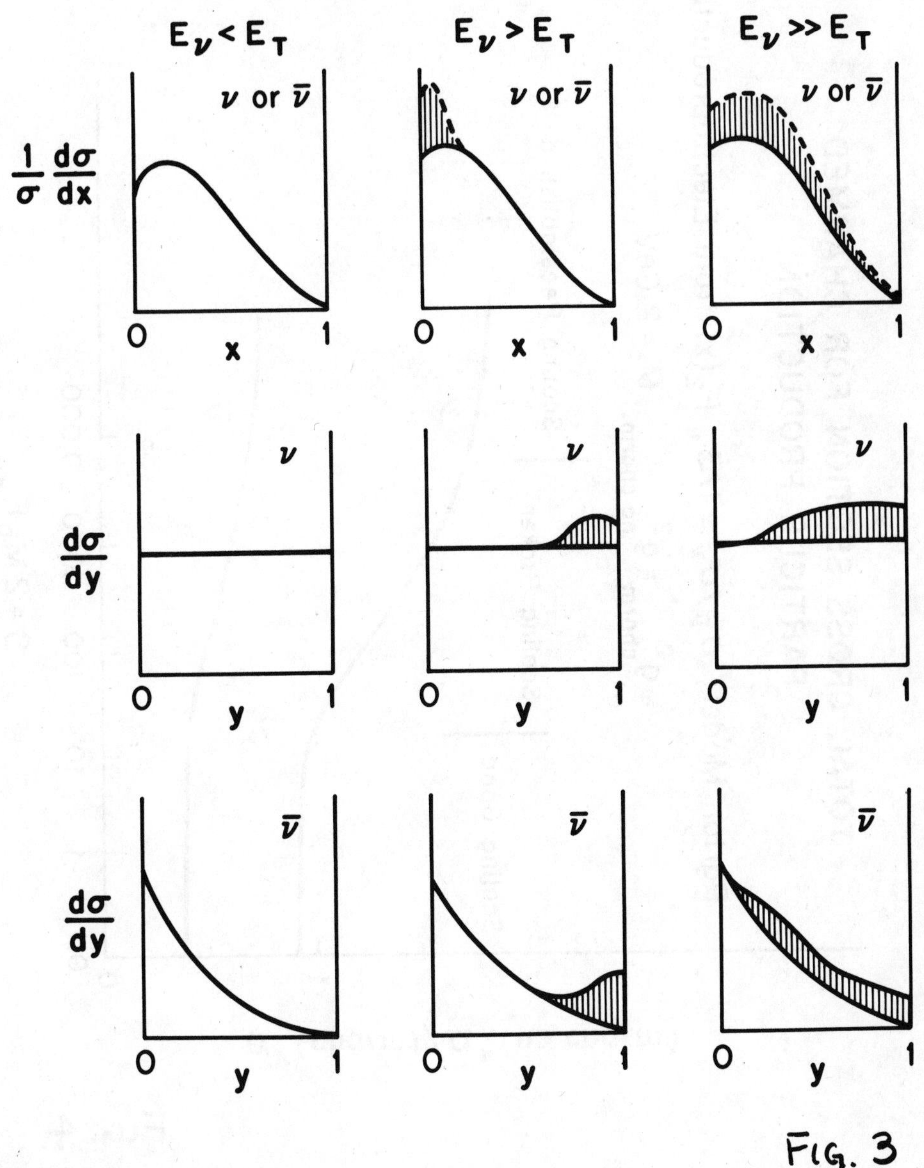

FIG. 3

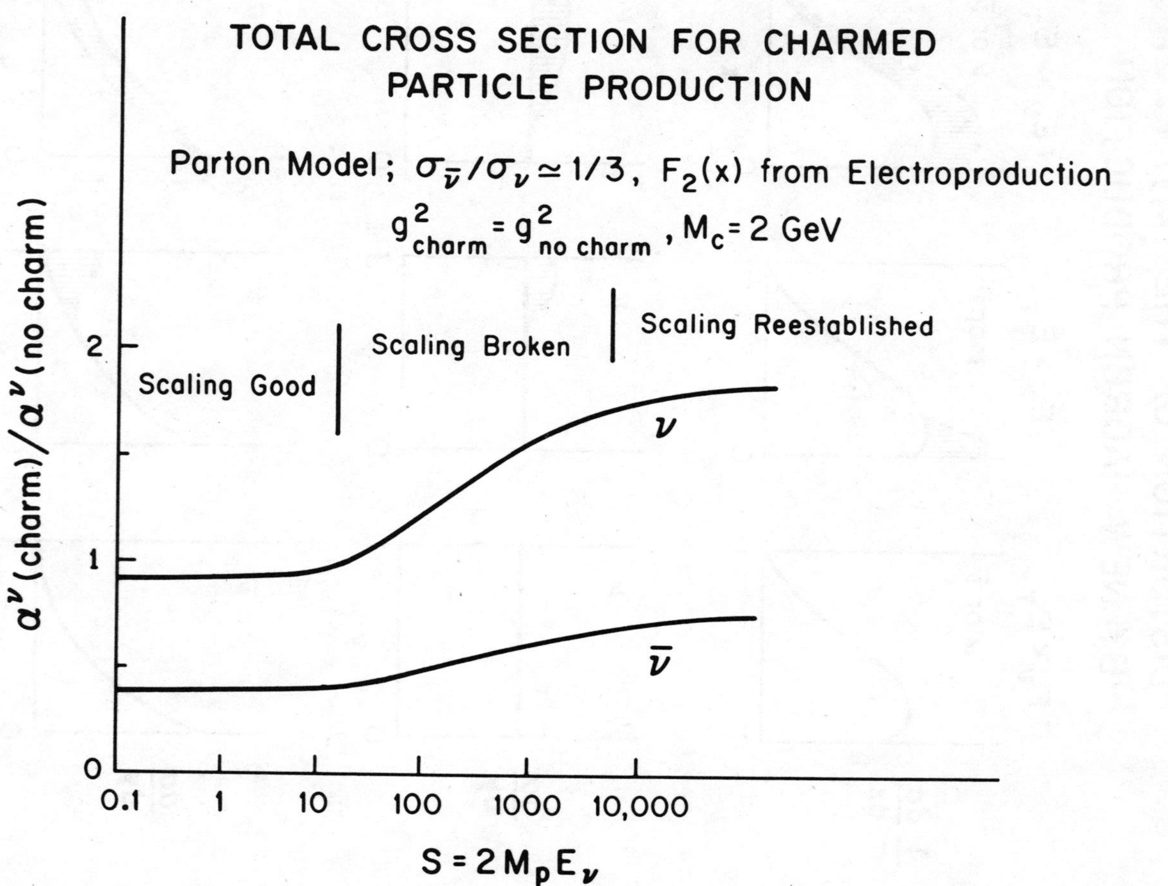
TOTAL CROSS SECTION FOR CHARMED
PARTICLE PRODUCTION
Parton Model; $\sigma_{\bar{\nu}}/\sigma_{\nu} \simeq 1/3$, $F_2(x)$ from Electroproduction
$g^2_{charm} = g^2_{no\ charm}$, $M_c = 2$ GeV
Scaling Good
Scaling Broken
Scaling Reestablished
ν
$\bar{\nu}$
$\sigma^{\nu}(\text{charm})/\sigma^{\nu}(\text{no charm})$
0
1
2
0.1
1
10
100
1000
10,000
$S = 2 M_p E_{\nu}$

FIG. 4

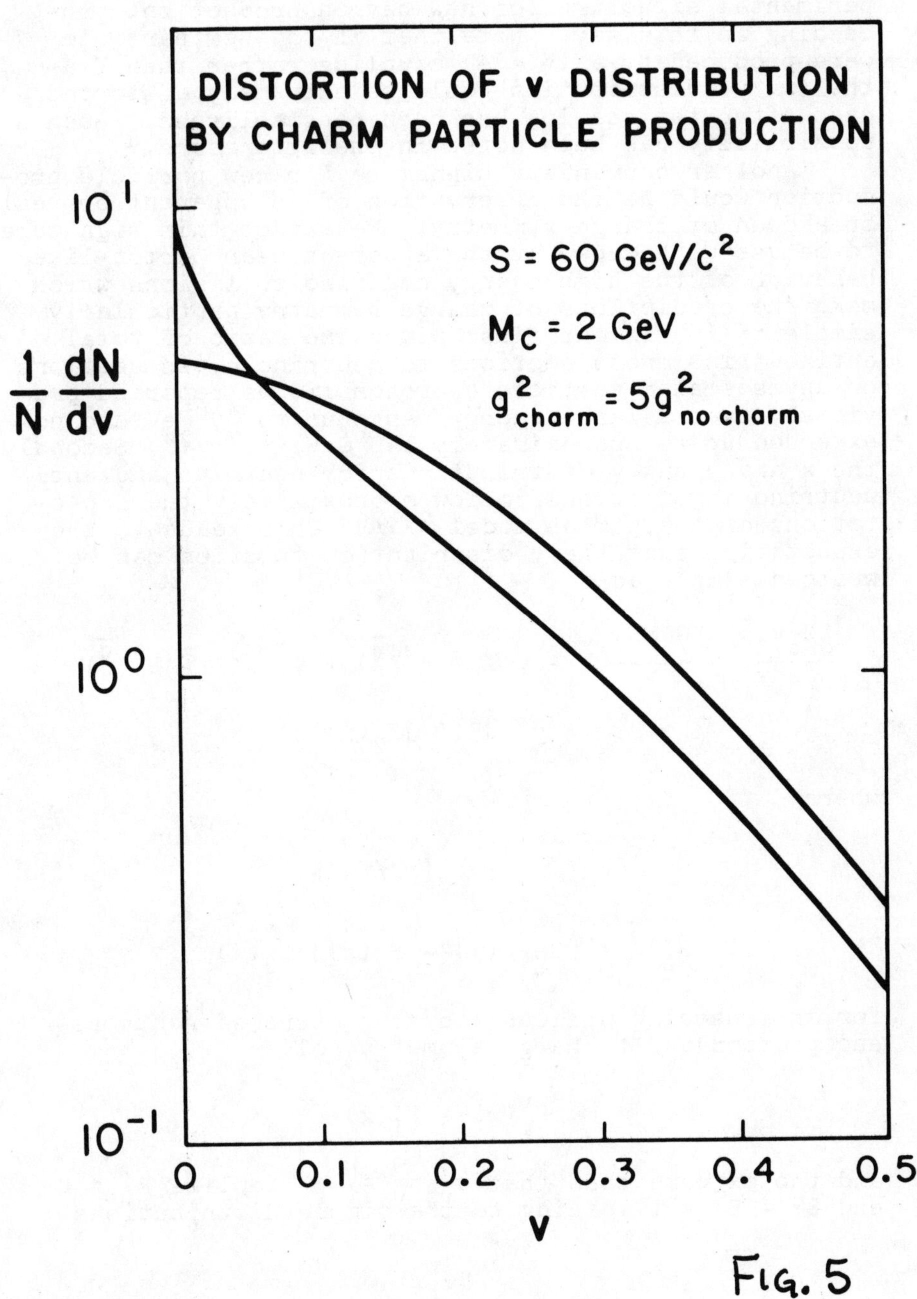
DISTORTION OF v DISTRIBUTION
BY CHARM PARTICLE PRODUCTION
10^1
10^0
10^{-1}
$\frac{1}{N}\frac{dN}{dv}$
$S = 60$ GeV/c^2
$M_C = 2$ GeV
$g^2_{charm} = 5g^2_{no\ charm}$
0
0.1
0.2
0.3
0.4
0.5
v

FIG. 5

1/ N dN/dv distribution for a specific case (mass $\sim$ 2 GeV, E_ν = 30 GeV).[19] Table 1 lists the resulting experimental signature for new hadron production, proceeding in this way. Note that if the new particle were produced by a (V + A) coupling rather than V-A the (x, y) distribution could be extremely distorted even after the scaling has been reestablished. Such a possiblility has been noted in the literature.[5]

Another convenient signature for new particle production would be the observation of an apparent or real breakdown of charge symmetry. We expect this signature to be useful because of the apparent near parton-like behavior of the high energy neutrino collisions which make the predictions of charge symmetry particularly simple.[20,21] In the first place the ratio of total antineutrino cross sections to neutrino cross sections on an isoscalar particle approach 1/3 as reported previously over a large energy range up to 70 GeV and now extended up to approximately 140 GeV.[22,23,24] Secondly the x and y and v distributions for neutrino and antineutrino interactions follow approximately the expectations of the parton model.[25,26] This leads to the expectation that the y distribution function can be written simply as[27]

$$\frac{d^2\sigma^{\nu,\bar{\nu}}}{dxdy} = \frac{G^2ME_\nu}{\pi}[1 - y(1 \mp B^{\nu,\bar{\nu}}) + y^2/2(1 \mp B^{\nu,\bar{\nu}}) + y^2/2R_L^{\nu,\bar{\nu}}]F_2(x)^{\nu,\bar{\nu}}$$

where

$$B^{\nu,\bar{\nu}} = -xF_3^{\nu,\bar{\nu}}(x)/F_2^{\nu,\bar{\nu}}(x) \qquad (3)$$

$$R_L^{\nu,\bar{\nu}} = [2xF_1^{\nu,\bar{\nu}}(x) - F_2^{\nu,\bar{\nu}}(x)]/F_2^{\nu,\bar{\nu}}(x) \qquad (3)$$

for an isoscalar nucleus (equal numbers of neutrons and protons). If charge symmetry holds

$$F_i^\nu(x) = F_i^{\bar{\nu}}(x)$$

and the extreme case that $\sigma_{\bar{\nu}}/\sigma_\nu$ = 1/3 implies $R_L \simeq 0$ and $B^\nu = B^{\bar{\nu}} = 1$ leading to the simple distributions

$$\frac{d\sigma^\nu}{dxdy} = \frac{G^2ME_\nu}{\pi}F_2^\nu(x) \qquad (4)$$

$$\frac{d\sigma^{\bar{\nu}}}{dxdy} = \frac{G^2ME_\nu}{\pi}[1 - y]^2F_2^\nu(x) \tag{5}$$

If we assume, for simplicity, a small deviation from charge symmetry in the $F_3(x)$ form factor such that

$$B^\nu = 1 + \varepsilon \tag{6}$$

$$B^{\bar{\nu}} = 1 - \varepsilon \tag{7}$$

leading to distributions of the form

$$\frac{d\sigma^\nu}{dxdy} = \frac{G^2ME_\nu}{\pi}[1 + \varepsilon y[1 - y/2]]F_2(x) \tag{8}$$

$$\frac{d\sigma^{\bar{\nu}}}{dxdy} = \frac{G^2ME_\nu}{\pi}[[1 - y]^2 + \varepsilon y[1 - y/2]]F_2(x) \tag{9}$$

and for $y \to 1$, $d\sigma^{\bar{\nu}}/dy$ is particularly sensitive to the violation of charge symmetry. The resulting ratio of cross sections becomes

$$\frac{\sigma_{\bar{\nu}}}{\sigma_\nu} = \frac{1/3[1 + \varepsilon]}{[1 + 1/3\varepsilon]}$$

and is obviously less sensitive to the violation of charge symmetry. Of course the violation of charge symmetry might occur in all the form factors and a more complicated behavior would be obtained. More general parameterizations for the breakdown of charge symmetry appear in the literature.[28] We note also that the production of strange particles is expected to violate charge symmetry at the level of a few percent.

If massive new leptonic particles were produced charge symmetry could still be valid but additional form factors such as W_4 and W_5 would be important near threshold and the simple formulas given above would no longer be valid. This could result in an apparent violation of charge symmetry. At higher energies the simple parton model predictions might become valid again in this case. However, other kinds of new particle production might violate charge symmetry but ε could be very small and experimentally the effects of these particles would be hard to see in neutrino interactions but easier in the antineutrino case. Note that if there were a small admixture of antipartons in the nucleon (ε') the distribution functions would instead become

$$\frac{d\sigma^{\nu}}{dxdy} = \frac{G^2ME_{\nu}}{\pi}[1 + \varepsilon'(1 - y]^2 F_2(x) \tag{10}$$

$$\frac{d\sigma^{\bar{\nu}}}{dxdy} = \frac{GME_{\nu}}{\pi}[[1 - y]^2 + \varepsilon']F_2(x) \tag{11}$$

and the charge symmetry concerning distribution eq. (11) would be hard to distinguish from the charge symmetry violating distribution eq. (9).

Another test for new hadronic particle production is through the test of various sum rules.[29] The Adler and Gross-Llewellyn Smith sum rules provide a sensitive test of the application of presently accepted concepts of current algebra and the more speculative constituent models of elementary particles.[29] A particularly simple formulation of these sum rules occur if moments of the x or v distributions are used.[30] We assume that Bjorken scaling holds and that the Callan-Gross relation holds (or the spin 1/2 parton model which is equivalent). For isoscalar target a generalized relation is obtained for the moments in v

$$\langle v^n\rangle_{\nu} - \frac{\sigma_{\bar{\nu}}}{\sigma_{\nu}}\langle v^n\rangle_{\bar{\nu}} = -[\frac{2}{n+2} - \frac{1}{n+3}]\frac{\int x^{n+1} F_3(x)dx}{\int F_2(x)dx}$$

for n = -1 (the Gross-Llewellyn Smith sum rule)

$$\langle v^{-1}\rangle_{\nu} - \frac{\sigma_{\bar{\nu}}}{\sigma_{\nu}}\langle v^{-1}\rangle_{\bar{\nu}} = 3$$

for consituents with baryon number 1 and

$$\langle v^{-1}\rangle_{\nu} - \frac{\sigma_{\bar{\nu}}}{\sigma_{\nu}}\langle v^{-1}\rangle_{\bar{\nu}} = 9$$

for baryon number 1/3 constituent. Several authors have generalized this sum rule to include various other constituents such as charmed particles.[2,31] Note that the n = 0 moments give

$$1 - \sigma_{\bar{\nu}}/\sigma_{\nu} = -\tfrac{2}{3}\int xF_3dx/\int F_2dx$$

(=2/3 on the parton model)

and the n = 1 moment is given by

$$\langle v\rangle_\nu - \frac{\sigma_{\bar{\nu}}}{\sigma_\nu}\langle v\rangle_{\bar{\nu}} = \frac{10}{24}\frac{\int xF_2^\nu dx}{\int F_2^\nu dx}$$

Using CVC we may relate

$$\frac{\int xF_2^\nu dx}{\int F_2^\nu dx} = \frac{\int xF_2^{ed} dx}{\int F_2^{ed} dx}$$

and evaluating the righthand side from SLAC data gives[30]

$$\langle v\rangle_\nu - \frac{\sigma_{\bar{\nu}}}{\sigma_\nu}\langle v\rangle_{\bar{\nu}} \simeq 0.1$$

Presumably similar moment sum rules using a different scaling variable can also be derived for the neutral current channel. The evaluation of such sum rules will only provide a sensitive test for the presence of new particle production if tested well above threshold. A similar moment analysis for the Adler sum rule is easily formulated.[30]

An indirect search for the W^+ or W° bosons can be made by searching for effects of the propagator in the Q^2, x or v distributions in charged current or neutral current interactions.[32] This technique will only work easily if the simple Bjorken scaling continues to hold at high energies and high Q^2. The modified distribution for neutrino interaction is

$$\frac{d\sigma^\nu}{dxdy} = \frac{G^2ME_\nu}{\pi}\frac{F_2(x)}{[1 + \frac{vS}{M_W^2}]^2} \qquad (12)$$

where $v = xy = Q^2/s$ and $s = 2mE_\nu$.[33] In the limit that $M_W \to \infty$, we obtain the relation

$$\langle v\rangle = \tfrac{1}{2}\int xF_2(x)dx/\int F_2(x)dx = \tfrac{1}{2}\langle x\rangle \qquad (13)$$

and $\langle Q^2\rangle = \langle v\rangle S$; the mean Q^2 depends linearily on E_ν with a slope given by the first moment in x.[34] The large v behavior is very sensitive to the effects of a W propagator and generally allows for a search up to much higher masses than the direct production.[33]

The kinematic signature for W production and subsequent decay through the hadronic channel is the enhancement of a small region of the (x, y) plot at large y and small x as shown in Fig. 6a.[35,36] At energies well above threshold the cross section for W production exceeds the ordinary deep inelastic scattering and thus the bulk of the events will cluster in the small region shown in Fig. 6a. Diffractive production of new states would follow the structure shown in Fig. 6b.

The production and decay of a heavy lepton M^+ is illustrated in Fig. 6c as a function of E_ν. (See Fig. 1 also.) The projected distributions on the y and y' axis are shown in Fig. 7 illustrating the functional dependence that results when the neutrino energy considerably exceeds the threshold energy E_T. Note that an exceedingly sharp peak at $y' \sim 1$ develops as $E_\nu > E_T$ giving a unique signature. The expected cross section for M^+ production for various mass values as calculated by Albright and Jarlskog is shown in Fig. 8.[37]

In the preceeding examples new particle production has resulted in a distortion of the (x, y) distribution with a characteristic energy threshold behavior. Another signature for new particle production is the apparent nonconservation of lepton number such as in the production of an M^+ by a neutrino beam resulting in a μ^+ or e^+ in the final state, i.e.

$$\begin{array}{l} \nu_\mu + N \rightarrow M^+ + X \\ \qquad\qquad \hookrightarrow \mu^+ + \nu_\mu + \nu_\mu \end{array} \tag{12}$$

The violation of other selection rules such as $\Delta S = \Delta Q$ or $\Delta I = 1$ can also be the signature for new particle production.

The production of events without charged leptons in the final state or apparent neutral currents would be a very clean signature for new lepton production if neutral current reactions did not already exist for neutrino beams. However, new lepton production will additionally have a threshold dependence which may still provide a clean experimental signature. The cleanliness of this signature will depend on the behavior of the neutral current y distribution if broad band neutrino beams are used. Figure 9 shows the expected threshold dependence for the ratio

$$\begin{array}{l} R^\nu_{M^+} = \sigma[\nu_\mu + N \rightarrow M^+ + X]/\sigma(\nu_\mu + N \rightarrow \mu^- + X) \\ \qquad\qquad\qquad\qquad \hookrightarrow \nu_\mu + X' \end{array} \tag{13}$$

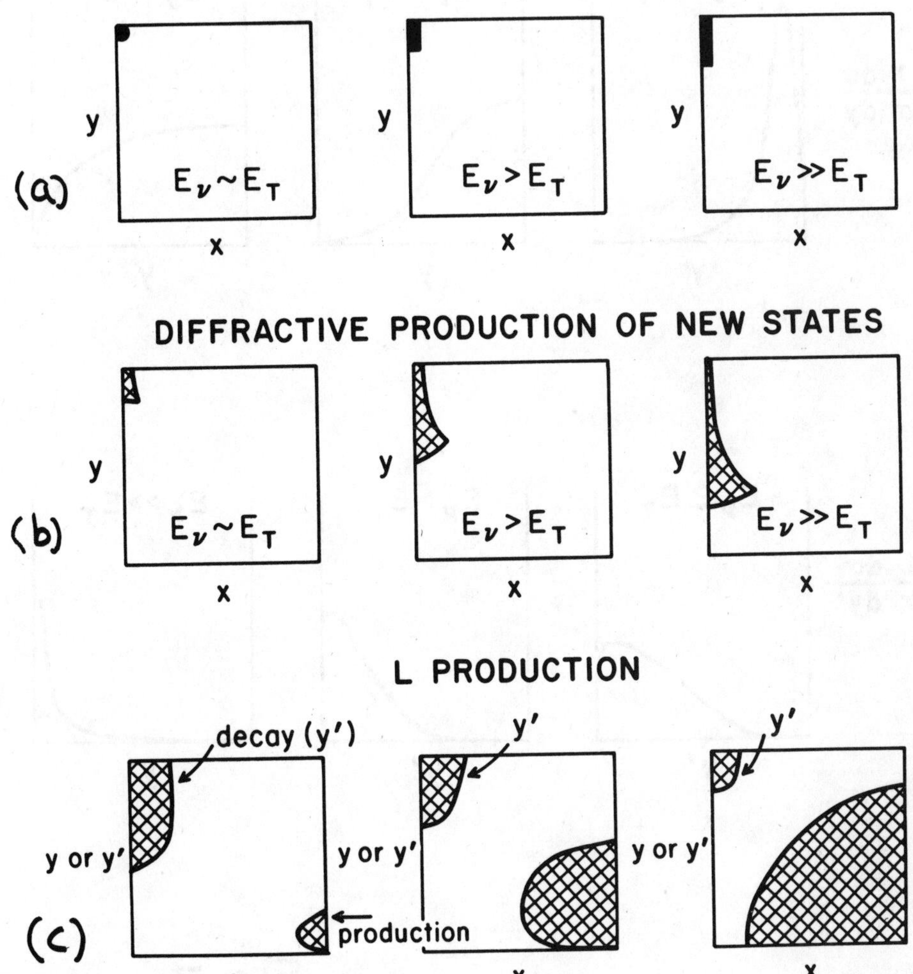
PRODUCTION OF NEW STATES AT LOW Q2
PRODUCTION OF W BOSON
(a)
y
x
Eν~ET
Eν>ET
Eν>>ET
DIFFRACTIVE PRODUCTION OF NEW STATES
(b)
L PRODUCTION
decay (y')
y'
y or y'
production
(c)

FIG. 6

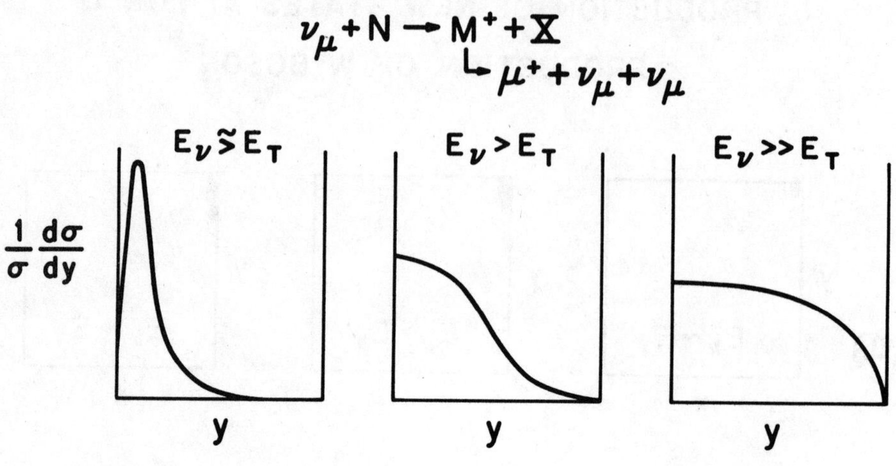
$\nu_\mu + N \rightarrow M^+ + X$
$\hookrightarrow \mu^+ + \nu_\mu + \nu_\mu$
$E_\nu \gtrsim E_T$
$E_\nu > E_T$
$E_\nu \gg E_T$
$\frac{1}{\sigma}\frac{d\sigma}{dy}$
y
y
y

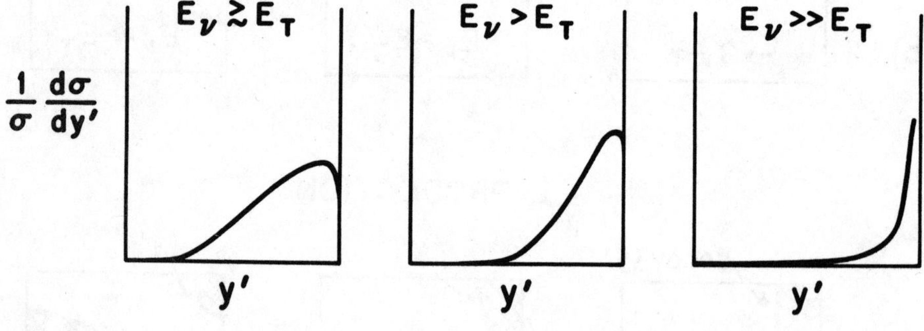
$E_\nu \gtrsim E_T$
$E_\nu > E_T$
$E_\nu \gg E_T$
$\frac{1}{\sigma}\frac{d\sigma}{dy'}$
y'
y'
y'

FIG. 7

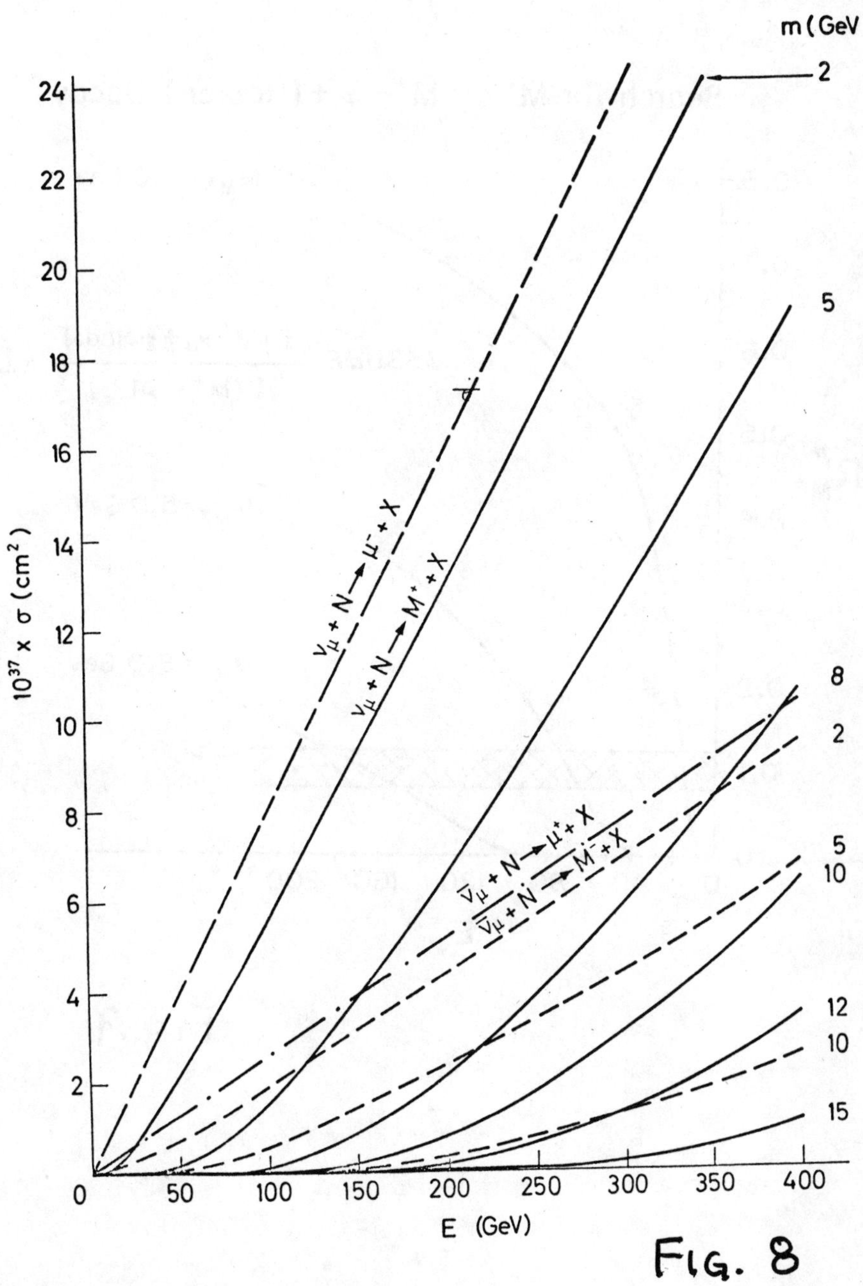
m(GeV
2
5
8
2
5
10
12
10
15
24
22
20
18
16
14
12
10
8
6
4
2
0
10^{37} x σ (cm^2)
$\nu_\mu + N \rightarrow \mu^- + X$
$\nu_\mu + N \rightarrow M^+ + X$
$\bar{\nu}_\mu + N \rightarrow \mu^+ + X$
$\bar{\nu}_\mu + N \rightarrow M^- + X$
50
100
150
200
250
300
350
400
E (GeV)

FIG. 8

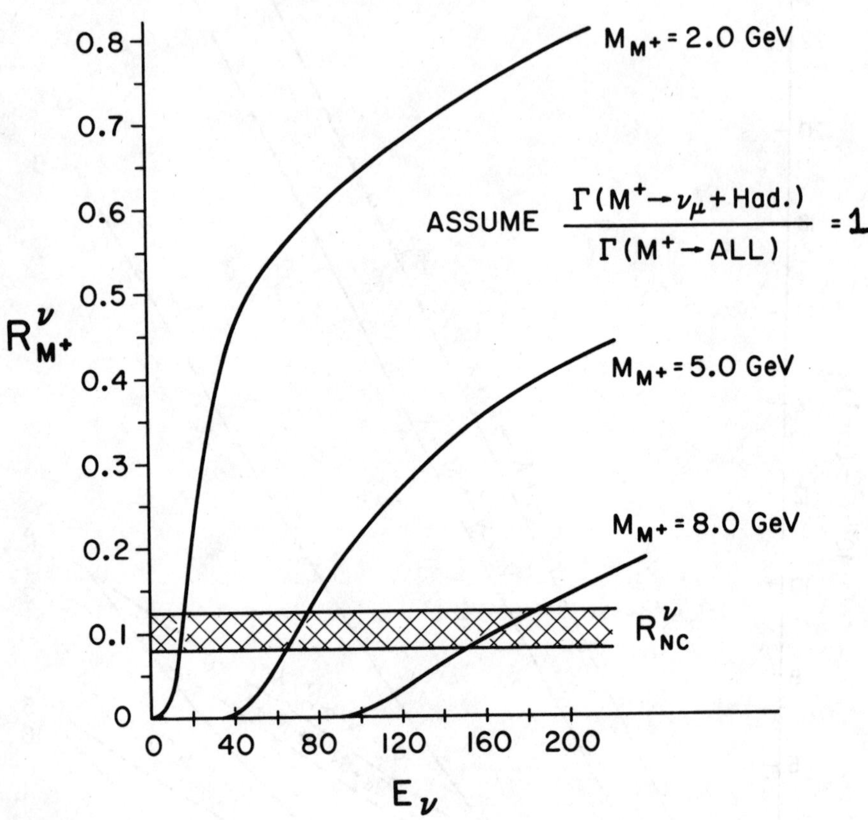

Search for M⁺ by M⁺→ν+(Hadron) Decay
0.8
0.7
0.6
0.5
0.4
0.3
0.2
0.1
0
$R^{\nu}_{M^+}$
M_{M^+} = 2.0 GeV
ASSUME $\frac{\Gamma(M^+ \to \nu_\mu + \text{Had.})}{\Gamma(M^+ \to \text{ALL})} = 1$
M_{M^+} = 5.0 GeV
M_{M^+} = 8.0 GeV
R^{ν}_{NC}
0
40
80
120
160
200
E_ν

FIG. 9

assuming that

$$\frac{\Gamma(M^+ \rightarrow \nu_\mu + X')}{\Gamma(M^+ \rightarrow \text{all})} \sim 1 \qquad (14)$$

and as a function of incident neutrino energy.[37] We note that the production of an M° and decay into the $M^\circ \rightarrow \nu_\mu + X$ channel would also produce a threshold effect but reduced in magnitude by approximating the ratio of neutral current to charged current cross sections.

Finally the production of two charged leptons in the final state is a traditional signature for W production but can be extended to the production of other short lived particles. The signature for W production and decay by the leptonic mode is the production of a "quiet" dimuon event, i.e. a dimuon with very little hadronic energy in the final state.[38] The production of other new short lived particles that have leptonic or semileptonic decay modes can result in "noisy" events with additional energy going into hadrons in the final state.

Table 2 lists the experimental signature for several kinds of new weakly interacting particles. In the case of the $W^\mp$ boson, M^+ lepton and M° lepton two different decay modes are indicated and the resulting signature. It is expected that if the particle is searched for using both decay modes, the resulting search will be essentially independent of the theoretical estimates for the decay branching fraction.[36] Of course, the search will be dependent on the theoretical estimates of the production cross section. However, in many cases this is expected to be fairly reliably estimated.

In Table 3 new particle production is characterized by the number of muons in the final state and the experimental signature within an event class. For the selected set of particles illustrated here it appears that the experimental signature is sufficiently unique to unravel these particles. Of course, nature may be more clever and confuse the issue.

3. SEARCH FOR NEW PARTICLES IN THE MUONLESS CHANNEL

The search for new particles in this channel for inclusive experiments relies on the observation of threshold effects as a function of E'. It is possible that a neutral current interaction with a specific (x, y) distribution could give rise to similar structure in absence of new particle production with a

TABLE 3

EVENT CLASS	PARTICLE	SIMPLEST EXPERIMENTAL SIGNATURE
0μ	$\nu_\mu + N \to M^+ + X$ $\to \nu_\mu + X$	Threshold Effect
	$\nu_\mu + N \to M^o + X$ $\to \nu_\mu + X$	Threshold Effect
1μ	$\nu_\mu + N \to M^+ + X$ $\to \mu^+ + \nu_\mu + \nu_\mu$	Apparent nonconservation of lepton number Structure in (E', y')
	$\nu_\mu + N \to \mu^- + X + C$ $\bar{\nu}_\mu + N \to \mu^+ + X + C$ (C = new hadron)	Energy dependent structure in (x, y) plot, total cross section rises faster than linear Apparent or real violation or charge symmetry
	$\nu_\mu + Z \to W^+ + \mu^- + Z$ $\to X$	Sharp energy dependent structure at very small x, large y
2μ (Quiet)	$\nu_\mu + Z \to W^+ + \mu^- + Z$ $\to \mu^+\nu_\mu$	Dimuon with very low momentum μ^- and higher momentum μ^+
	$\nu_\mu + Z \to W^o + \nu_\mu + Z$ $\to \mu^+\mu^-$	Symmetric μ^+, μ^- pair with characteristic invariant mass
2μ (Noisy)	$\nu_\mu + Z \to \mu^- + X + C$ $\to \mu^+ + \nu + X$ (C = hadron with semileptonic decay)	Low energy μ^-, Characteristic E' dependence
	$\nu_\mu + N \to M^o + X$ $\to \mu^+\mu^-\nu_\mu$	Low energy μ^+

broad band neutrino beam. We, therefore, first digress to the question of the y dependence of the neutral current interaction. For the V, A interaction, assuming charge symmetry and scaling the general y distribution can be written as[39]

$$\frac{d\sigma^{\nu}}{dy} = f_L^{NC} + (1 - y)^2 f_R^{NC} + 2(1 - y) f_s^{NC}$$

$$\frac{d\sigma^{\bar{\nu}}}{dy} = f_R^{NC} + (1 - y)^2 f_L^{NC} + 2(1 - y) f_s^{NC}$$

where the f functions are all positive definite. Sakurai has indicated that the expected distribution for a scalar interaction behaves as[40]

$$\frac{d\sigma^{\bar{\nu}}}{dy} = \frac{d\sigma^{\nu}}{dy} \sim y^2 .$$

In the same formulation the y distribution for the charged current is given by

$$\frac{d\sigma^{\nu}}{dy} = f_L^{CC} + (1 - y)^2 f_R^{CC} + 2(1 - y) f_s^{CC}$$

$$\frac{d\sigma^{\bar{\nu}}}{dy} = f_R^{CC} + (1 - y)^2 f_L^{CC} + 2(1 - y) f_s^{CC}$$

The simple parton model gives

$$f_R^{CC} = f_s^{CC} = 0$$

and

$$\frac{d\sigma^{\nu}}{dy} \propto \text{const.}; \quad \frac{d\sigma^{\bar{\nu}}}{dy} \propto (1 - y)^2$$

In the case of the neutral current there are four simple possibilities.[41]

I $\quad f_L^{NC} = 0 = f_s^{NC} \qquad \frac{d\sigma^{\nu}}{dy} \propto (1 - y)^2 \qquad R^{\bar{\nu}} = 9R^{\nu}$

$\quad f_R^{NC} = 1 \qquad \frac{d\sigma^{\bar{\nu}}}{dy} \propto \text{const.}$

(V + A interaction)

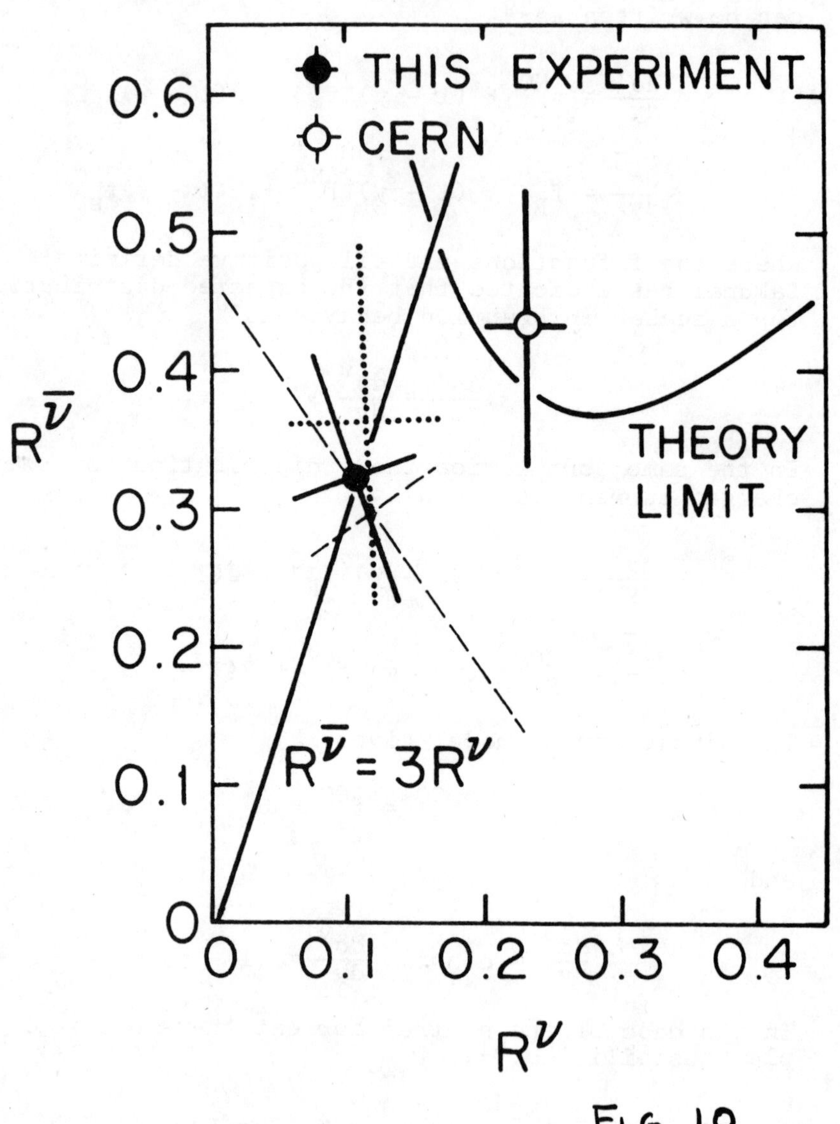
THIS EXPERIMENT
CERN
THEORY
LIMIT
$R^{\bar{\nu}} = 3R^{\nu}$
$R^{\bar{\nu}}$
R^{ν}
0.6
0.5
0.4
0.3
0.2
0.1
0
0
0.1
0.2
0.3
0.4

Fig. 10

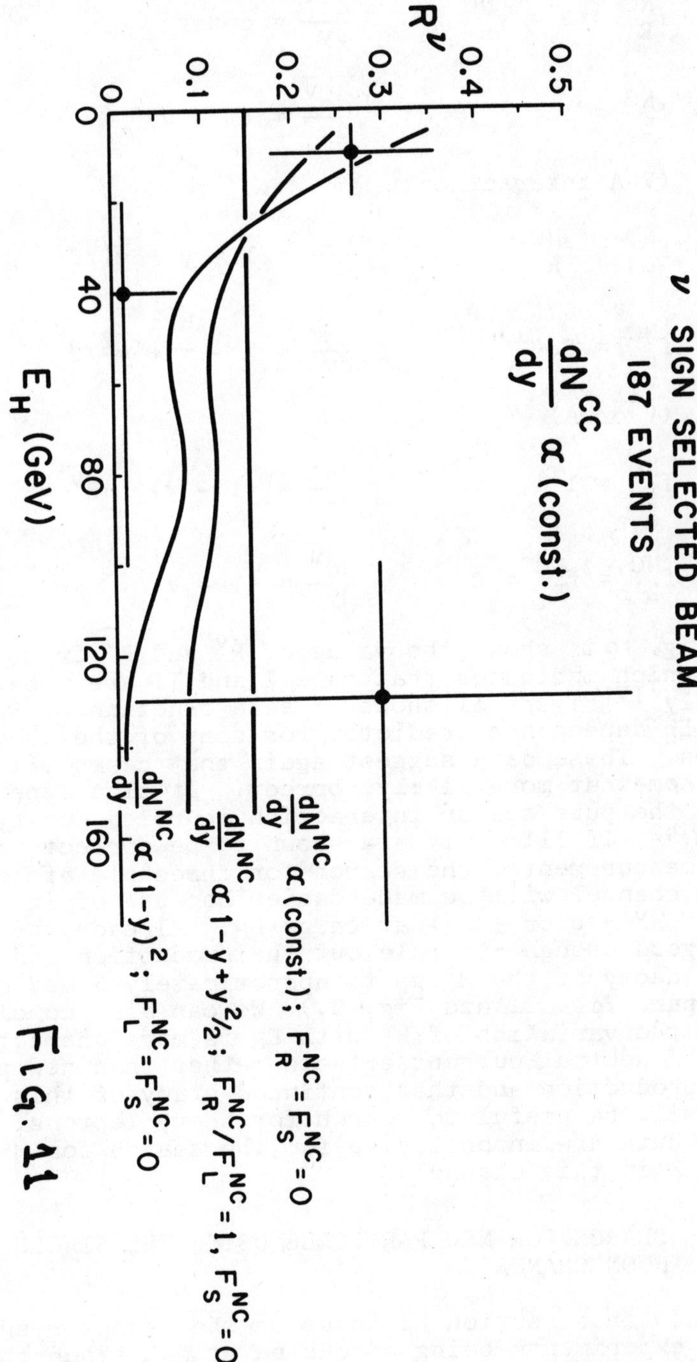
ν SIGN SELECTED BEAM
187 EVENTS
$\frac{dN^{CC}}{dy} \propto$ (const.)
R^{ν}
0.5
0.4
0.3
0.2
0.1
0
0
40
80
120
160
E_H (GeV)
$\frac{dN^{NC}}{dy} \propto$ (const.); $F_R^{NC} = F_S^{NC} = 0$
$\frac{dN^{NC}}{dy} \propto 1-y+y^2/2$; $F_R^{NC}/F_L^{NC} = 1$, $F_S^{NC} = 0$
$\frac{dN^{NC}}{dy} \propto (1-y)^2$; $F_L^{NC} = F_S^{NC} = 0$

FIG. 11

II $f_R^{NC} = 0 = f_s^{NC}$ $\frac{d\sigma^\nu}{dy} \propto \text{const.}$ $R^\nu = R^{\bar{\nu}}$

$f_L^{NC} = 1$ $\frac{d\sigma^{\bar{\nu}}}{dy} \propto (1 - y)^2$

(V-A interaction)

III $f_L^{NC} = f_R^{NC}$ $\frac{d\sigma^\nu}{dy} \propto 1 - y + y^2/2$ $R^{\bar{\nu}} = 3R^\nu$

$f_s^{NC} = 0$ $\frac{d\sigma^{\bar{\nu}}}{dy} \propto 1 - y + y^2/2$

(V or A)

IV $f_s^{NC} = 1$ $\frac{d\sigma^\nu}{dy} \propto (1 - y)$ $R^{\bar{\nu}} = 3R^\nu$

$f_L^{NC} = f_R^{NC} = 0$ $\frac{d\sigma^{\bar{\nu}}}{dy} \propto (1 - y)$

In Fig. 10 is shown the value of R^ν and $R^{\bar{\nu}}$ from the NAL E1A which indicates that case I and II are somewhat unlikely.[17] Figure 11 shows R^ν as a function of E_H and the E_H dependence predicted for some of the above cases. These data suggest again that cases III or IV are somewhat more likely correct. It also appears that the pure scalar interaction is ruled out by these data.[40] If III or IV are shown to be correct with better measurements, the search for threshold affects in this channel will be made easier because of in this limit $R^\nu \to 0$ or $\frac{1}{2}\langle R^\nu\rangle$ at large E_H. Already these data are good enough to rule out the production and muonless decay of the M^+ up to approximately 5 GeV mass. (Compare Fig. 11 and Fig. 9.) We can also conclude that the variation of R^ν with E_H is more characteristic of neutral current effects rather than new particle production and that continued study of this channel will be useful to search for heavy leptons. Present data are inconclusive for the search for $M^o \to \nu_\mu + X$ in this channel.

4. SEARCH FOR NEW PARTICLES USING THE SINGLE MUON CHANNEL

In this section we focus on the recent results from experiments being conducted at NAL since the results of earlier experiments are confined to lower energies and are well known. At NAL there are two presently operating experiments, E21 and E1A.[42] In both cases the general features of the detectors are similar. They consist of a Target-Calorimeter followed by toroidal iron core magnets. In detail the detectors are very different in cross sectional area, target material and the muon angular acceptance of the magnet. For the E21 detector, the target consist of slabs of iron and the E1A target contains mainly carbon. Both detectors have been exposed to the dichromatic beam but only the E1A detector has run extensively with a broad band horn or unfocussed beam. Figures 12 and 13 show the schematical details of these detectors. These experiments use a variety of techniques to "calibrate" the energy response of the calorimeter. Figure 14 shows the calorimeter response of the E1A detector to a monocromatic pion beam of 25 GeV incident on the detector.

The first question that must be addressed in studying the single muon channel at high energies is the extent to which Bjorken scaling continues to hold. Although this is interesting in its own right, a unique signature for new particle production requires an understanding of the background processes, which approximate Bjorken scaling allows. In Fig. 15 is shown a comparison of the ν distribution for neutrino scattering obtained at CERN and NAL.[43] At large ν the distributions are consistent as would be expected if scaling holds up to neutrino energy of $\sim$ 50 GeV. There is a possible deviation at the lower ν values with the high energy data cross section at small ν being the larger, however, the CERN data is not fully in the scaling energy region. Another test of scaling is the linear rise of the total cross section. Figure 16 shows the relative energy dependence of the total neutrino cross section obtained in NAL E1A. An arbitrary scale is used here since the slope of the cross section with neutrino energy has only been crudely determined at an average energy of $\sim$ 30 GeV.[23] The cross section appears to rise at least as fast as linear up to energies of approximately 200 GeV. Note that a substantial non-linear increase of the cross section at the higher energies is not ruled out.

In Fig. 17 is shown the dependence $\langle Q^2 \rangle$ on E_ν from E1A and a comparison with the expectations of scaling using $F_2(x)$ from the SLAC data.[44,45] Up to 100 GeV there is good agreement with the scaling predictions. Above 100 GeV the slope of $\langle Q^2 \rangle$ versus E_ν seems to be in better agreement with a sharper $F_2(x)$ than obtained in the SLAC experiments.[45] If the x distribution that is directly measured in E1A is used to obtain $F_2(x)$ (by assuming $dN/dx \propto F_2(x)$) the predicted slope is in bet-

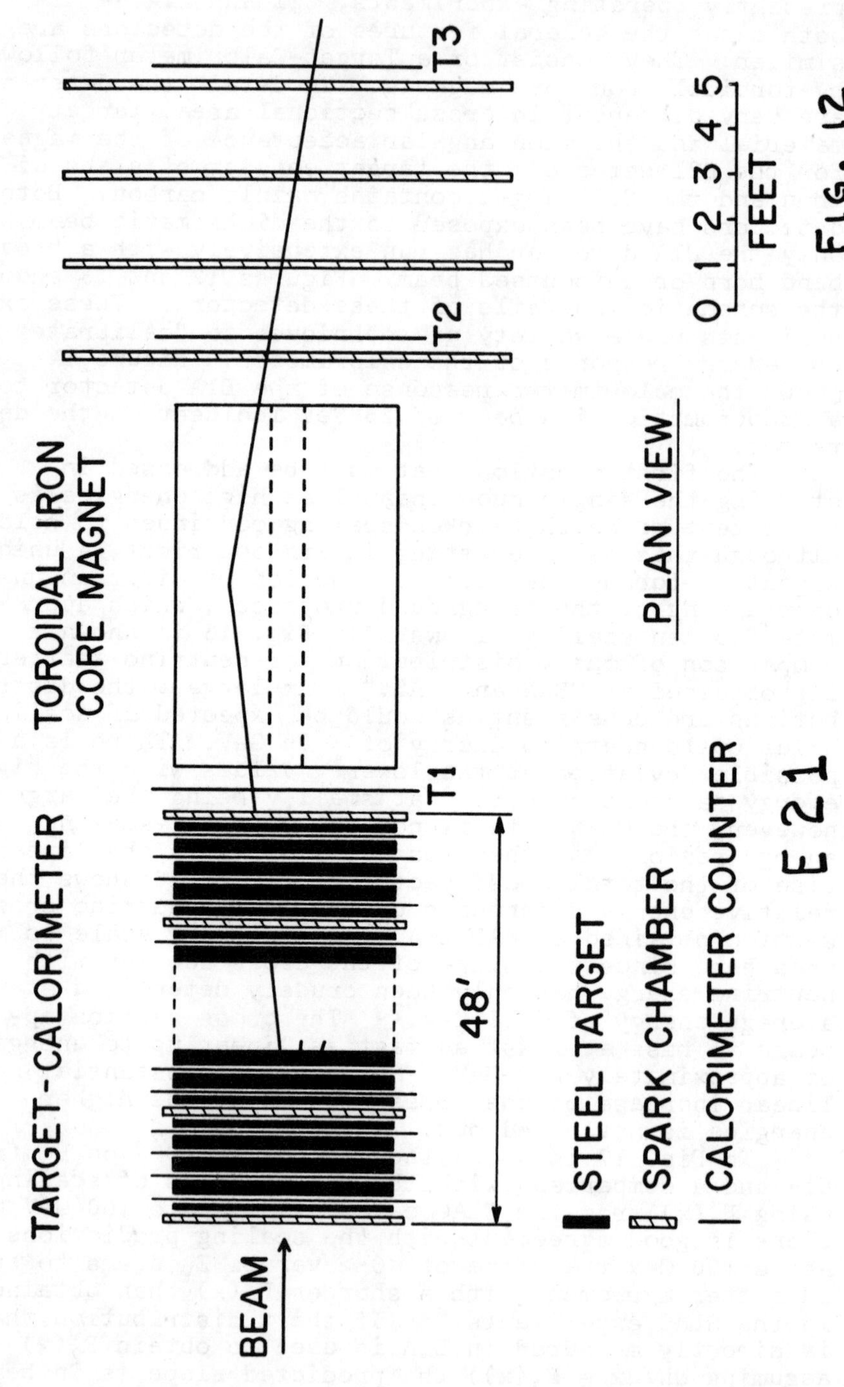
TARGET-CALORIMETER
TOROIDAL IRON CORE MAGNET
BEAM
48'
T1
T2
T3
STEEL TARGET
SPARK CHAMBER
CALORIMETER COUNTER
PLAN VIEW
0 1 2 3 4 5
FEET
E 21
FIG. 12

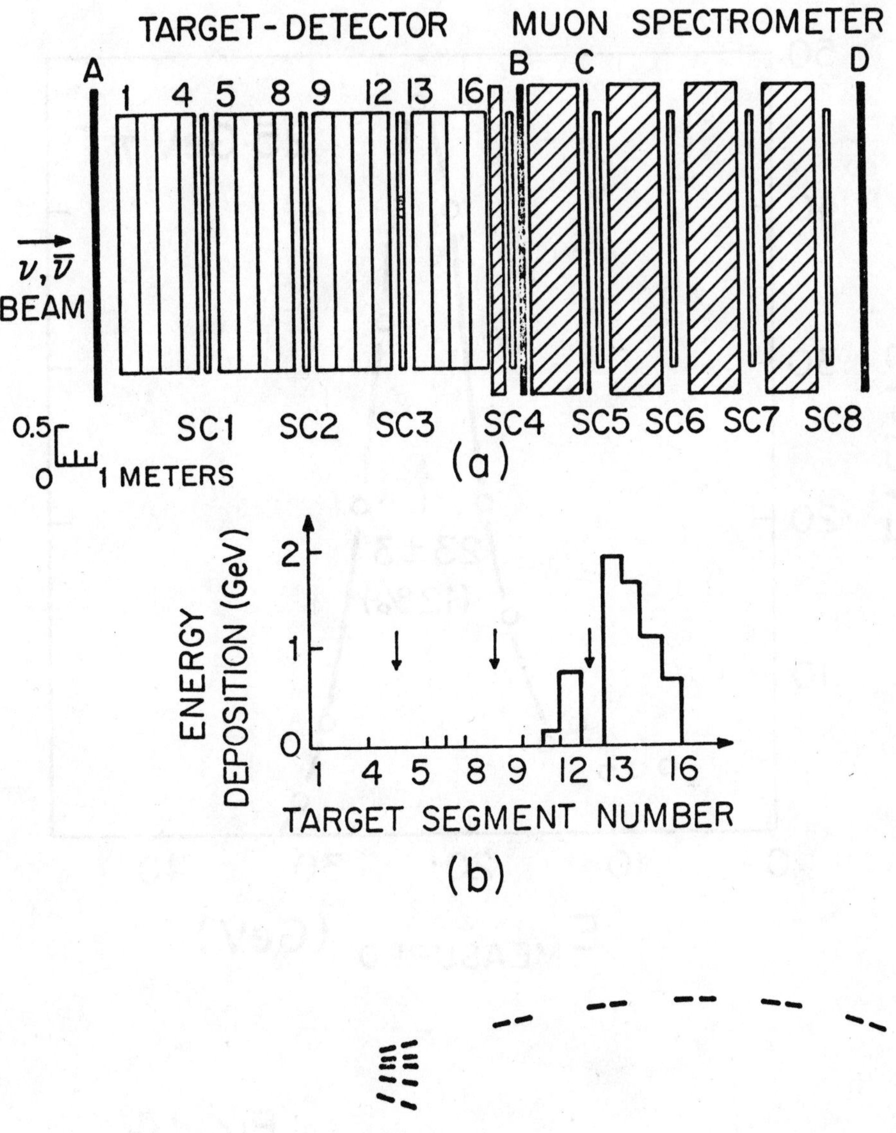

SC1 SC2 SC3 SC4 SC5 SC6 SC7 SC8

(c)

E1A

FIG. 13

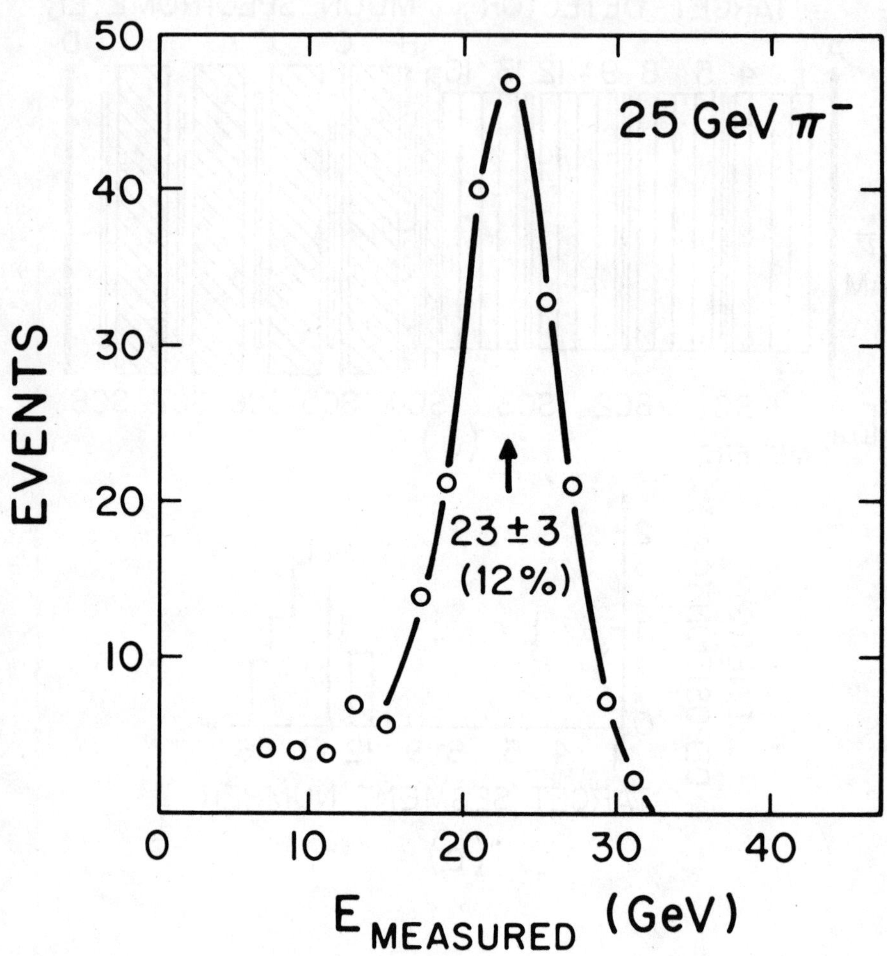

FIG. 14

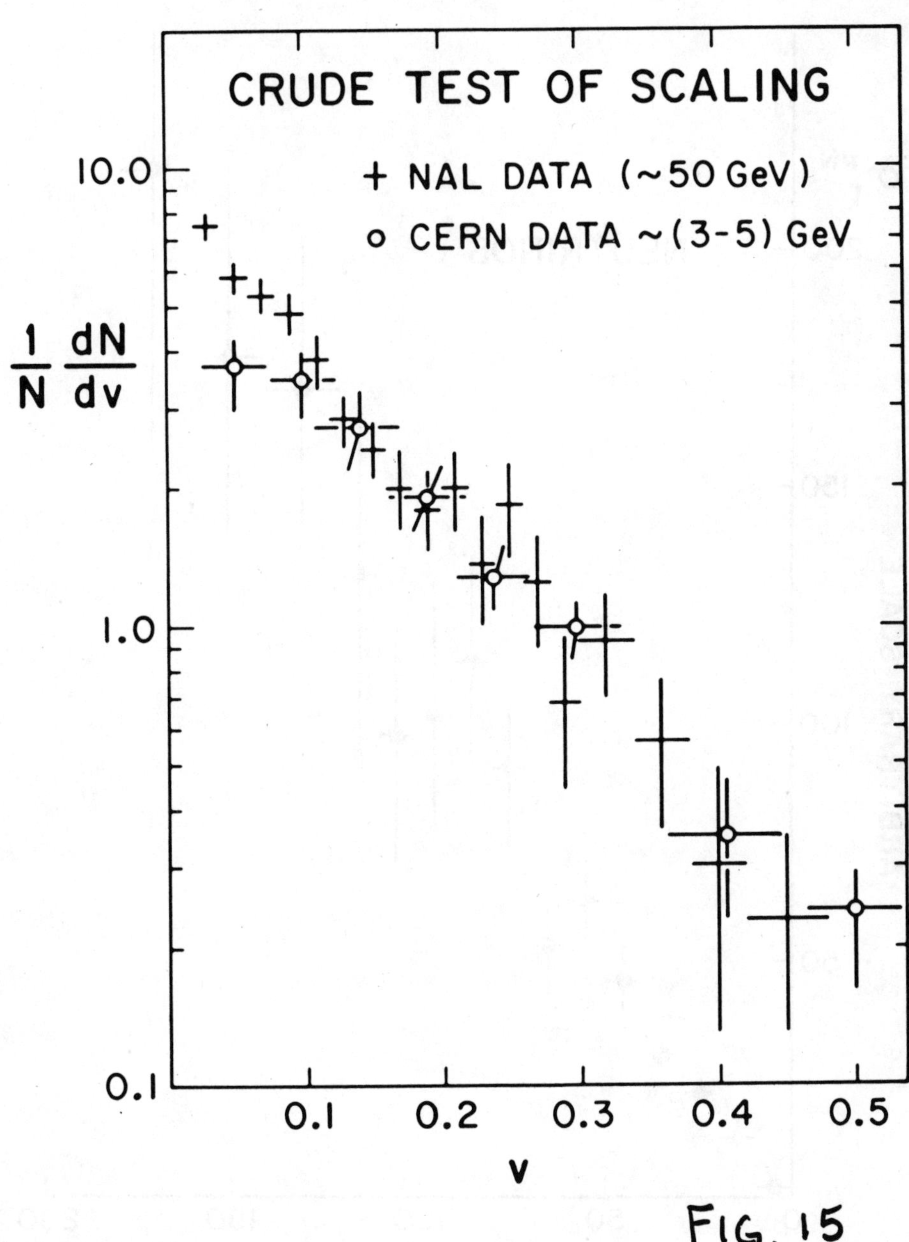
CRUDE TEST OF SCALING
+ NAL DATA (~50 GeV)
o CERN DATA ~(3-5) GeV
10.0
1.0
0.1
1/N dN/dv
0.1
0.2
0.3
0.4
0.5
v

FIG. 15

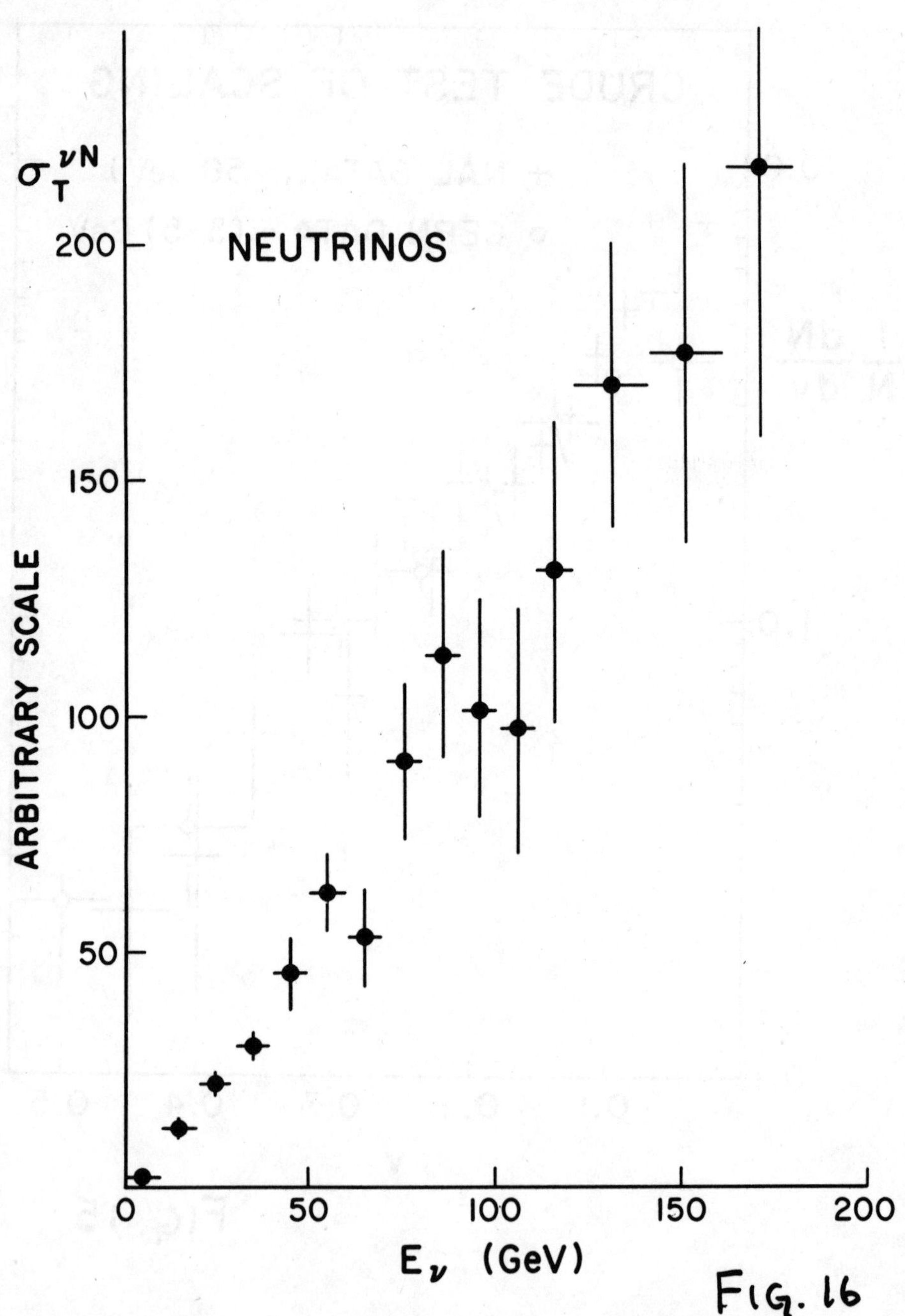
$\sigma_T^{\nu N}$
ARBITRARY SCALE
NEUTRINOS
200
150
100
50
0
50
100
150
200
E_ν (GeV)

FIG. 16

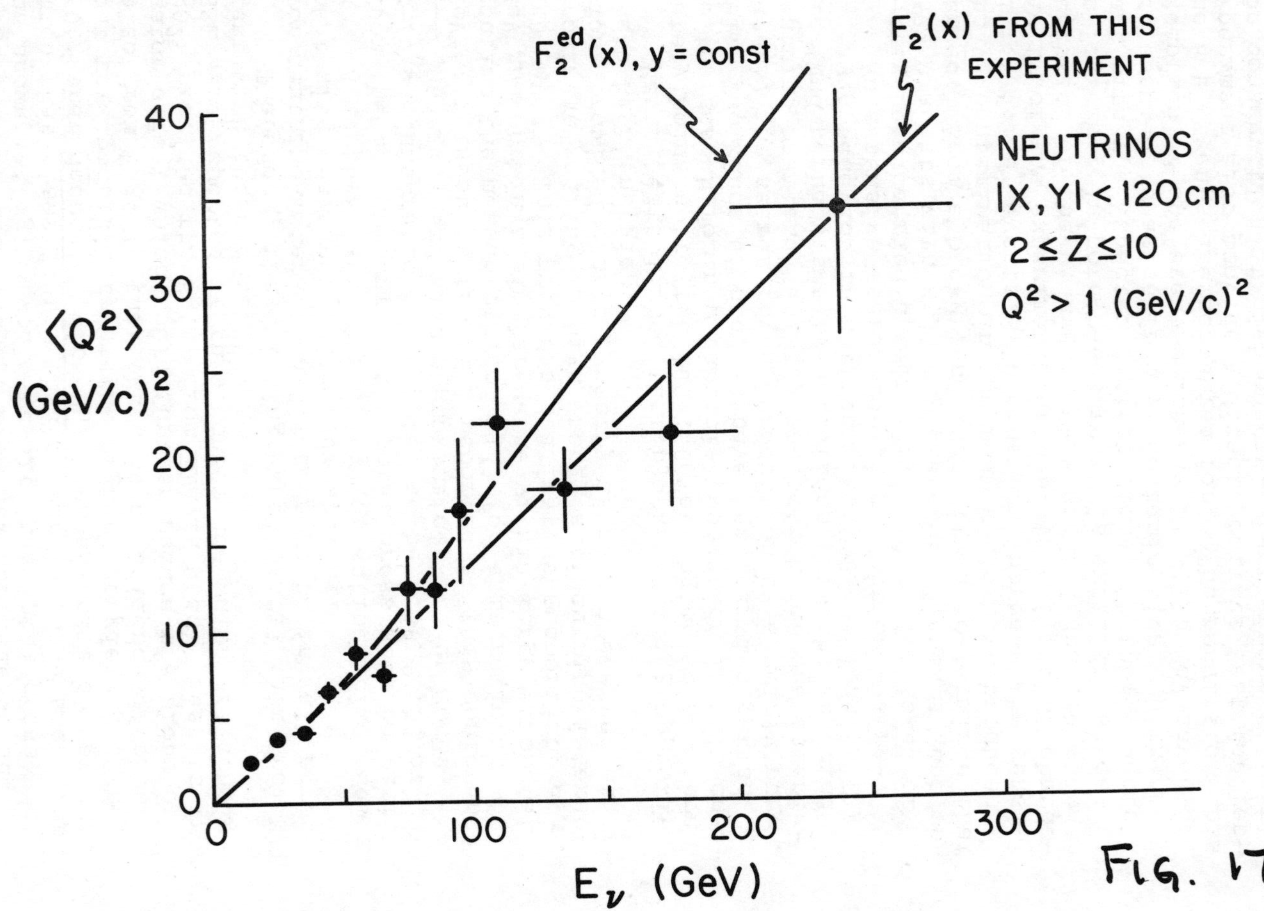

⟨Q²⟩
(GeV/c)²
0
10
20
30
40
0
100
200
300
E_ν (GeV)
F₂^ed (x), y = const
F₂ (x) FROM THIS
EXPERIMENT
NEUTRINOS
|X,Y| < 120 cm
2 ≤ Z ≤ 10
Q² > 1 (GeV/c)²

Fig. 17

ter agreement. It is interesting to note the sharpening of the x distribution from new particle production as shown in Fig. 3 would result in a total cross section that rises faster than linearity and a slope for $\langle Q^2 \rangle$ that decreases with E_ν. Present data do not rule out this possibility at the higher energies. We see also from Fig. 17 that there is no evidence of propagator effect which would further reduce the slope of $\langle Q^2 \rangle$ provided the $F_2(x)$ from E1A is used. Results from E21 were reported at this conference which are consistent with these data.[44]

Therefore, the existing experimental data from NAL for neutrino interactions is seen to be in fair agreement with the expectation from scaling.

A further test of scaling for antineutrinos comes from the measurement of the total cross section ratio $\sigma_{\bar\nu}/\sigma_\nu$. This ratio is reported to be 0.38 ± .03 at CERN energies and also approximately 1/3 up to 70 GeV[22,23] Results from E21 reported at this conference confirm the measurement of ∿1/3 for this ratio at ∿50 GeV and obtain a similar value within large errors at 140 GeV.[44] This provides qualitative support for approximate scaling in high energy antineutrino interactions provided charge symmetry is also preserved at these high energies.

Having established that Bjorken scaling approximately holds up to the 100-150 GeV region we now turn to the search for new particle production in this channel. Two kinds of studies have been carried out so far (1) the search for μ^+ production by a nearly pure neutrino beam and[46,47] (2) the search for characteristic (and perhaps small) deviations from the expected $(1 - y)^2$ behavior in antineutrino interactions in E1A.

An extensive search for μ^+ production has recently been carried out in E21.[47] Out of a total of 1522 ν events with a μ^- signature, 8 μ^+ events were observed. From a background study it was determined that the $\bar\nu_\mu$ contamination in the beam would have produced 11 ± 5.1 μ^+ events. In order to increase the sensitivity of the experiment further, the y' and E' characteristics of the μ^+ events were studied and are shown in Fig. 18. In both cases there is excellent agreement with the expected distributions, if these events are assumed to arise from $\bar\nu_\mu$ contamination. With the exception of one event which has E' ∿ 110 GeV seven of the 8 μ^+ events are accounted for. The results of this experiment also provide a direct positive test for lepton conservation at high energy. Furthermore a limit can be placed on the mass and M^+ lepton using production calculations already discussed in Section 2. The resulting mass limit is sensitive to the branching ratio for $M^+ \to \mu^+\nu_\mu\nu_\mu$ and Fig. 19 shows the 90% confi-

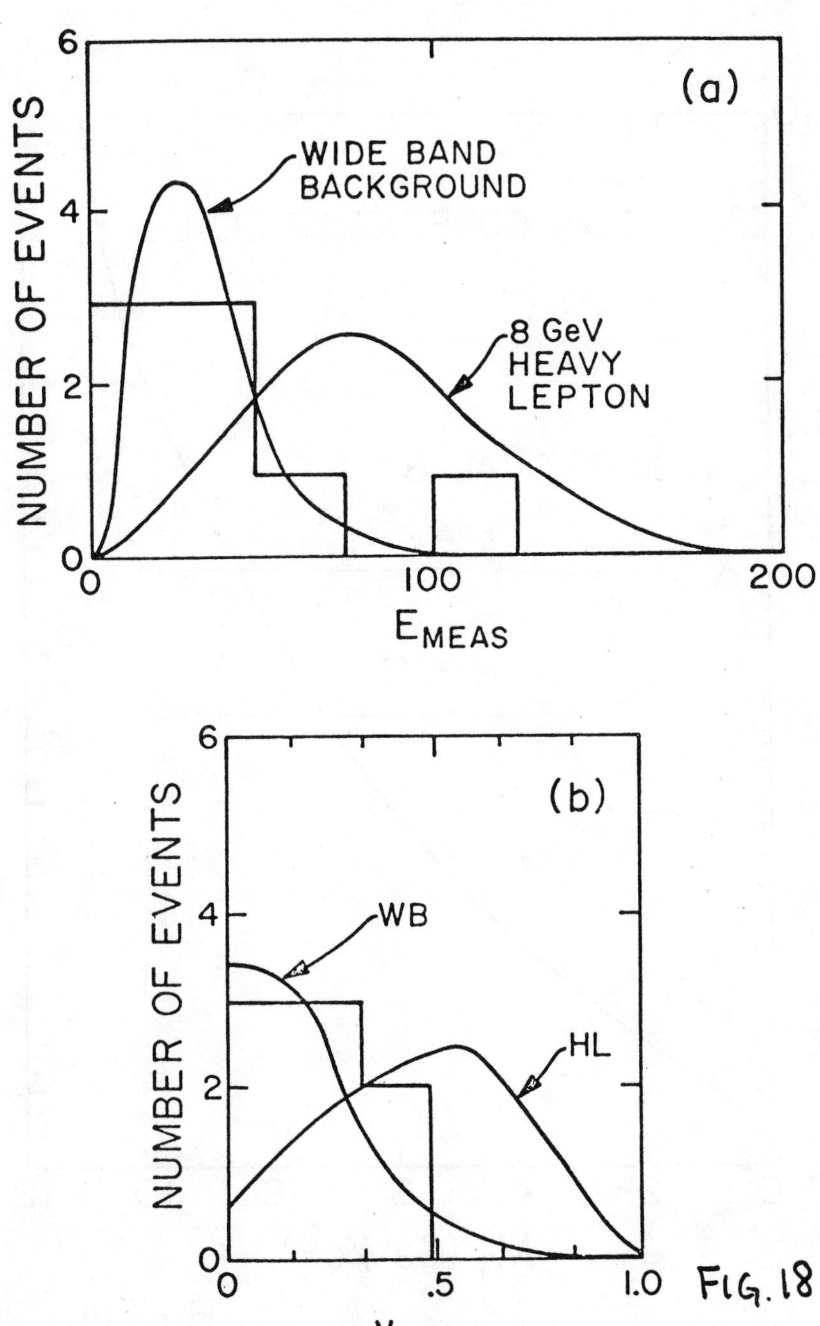
6
4
2
0
NUMBER OF EVENTS
(a)
WIDE BAND
BACKGROUND
8 GeV
HEAVY
LEPTON
0
100
200
EMEAS
6
4
2
0
NUMBER OF EVENTS
(b)
WB
HL
0
.5
1.0
yMEAS
FIG. 18

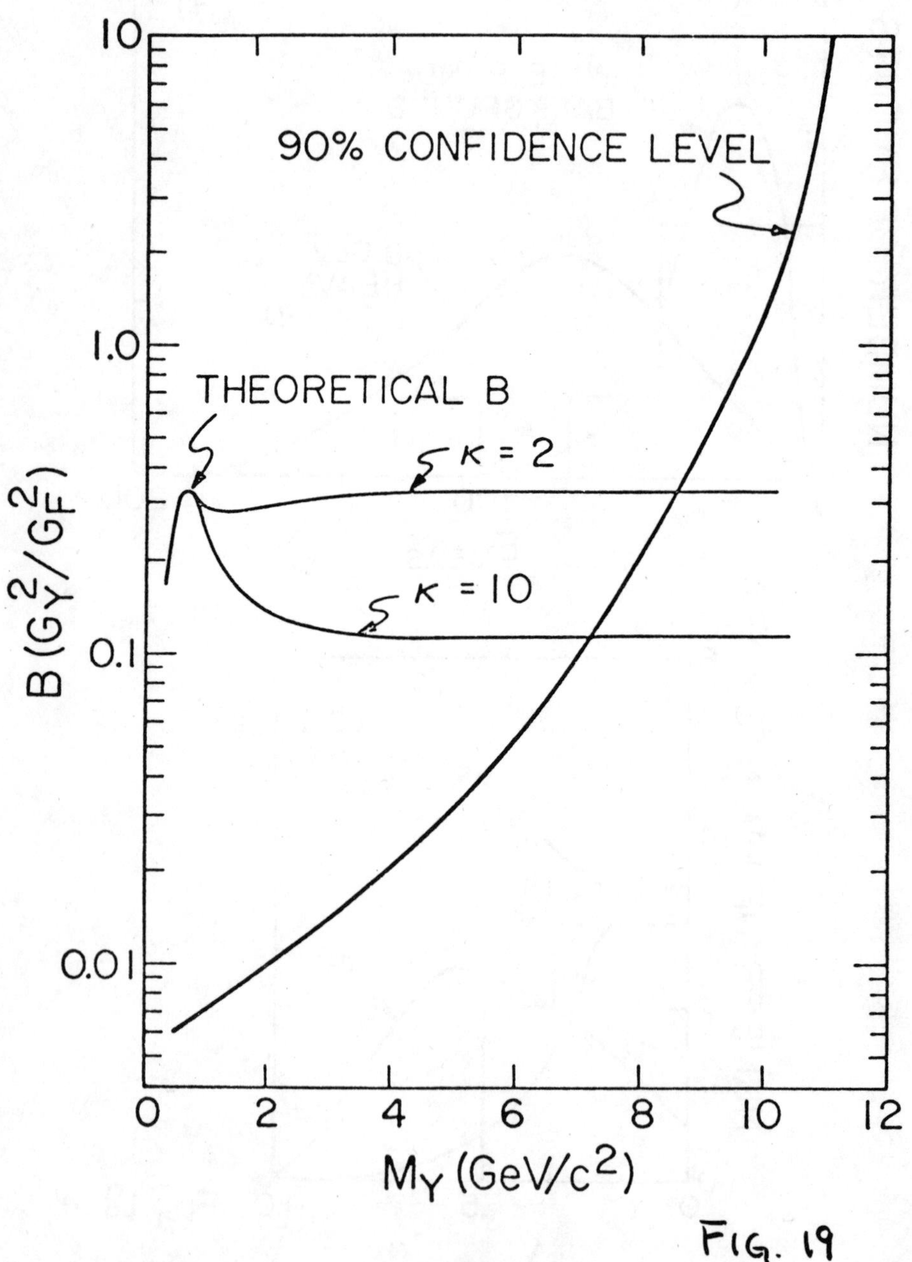
10
1.0
0.1
0.01
90% CONFIDENCE LEVEL
THEORETICAL B
κ = 2
κ = 10
B (G$_Y^2$/G$_F^2$)
0
2
4
6
8
10
12
M$_Y$ (GeV/c^2)

FIG. 19

dence limit on the mass as a function of the branching ratio into this channel. Also shown are the branching ratios expected for various theoretical models that rely on current algebra and the asymptotic value for the cross section ratio

$$K = \frac{\sigma(e^+e^- \to \text{hadrons})}{\sigma(e^+e^- \to \mu^+\mu^-)}.$$

The detailed study and understanding of the x and y distributions in high energy neutrino collisions requires an understanding of the resolution functions for the measured variables. For deep inelastic scattering using a non monochromatic beam only three variables are measured directly θ_μ, P_μ and E_H. The errors in the scaling variables are given in terms of the errors in these variables as

$$\left(\frac{\delta v}{v}\right)^2 = 4\left(\frac{\delta\theta_\mu}{\theta_\mu}\right)^2 + \left(\frac{\delta P_\mu}{P_\mu}\right)^2$$

$$\left(\frac{\delta x}{x}\right)^2 = \left(\frac{\delta v}{v}\right)^2 + (1 - y)^2\left(\frac{\delta E_H}{E_H}\right)^2$$

$$\left(\frac{\delta y}{y}\right)^2 = (1 - y)^2\left(\frac{\delta E_H}{E_H}\right)^2 + (1 - y)^2\left(\frac{\delta P_\mu}{P_\mu}\right)^2$$

There is an interesting property of these error formulas. If y is large the error in x and y is almost independent of $(\delta E_H/E_H)$. In this case the error in x is directly related to the error in v. Typically in experiment E1A,

$$\left(\frac{\delta\theta_\mu}{\theta_\mu}\right)^2 \sim 10^{-2},\ \left(\frac{\delta P_\mu}{P_\mu}\right) \sim 10^{-2}$$

and $(\delta E_H/E_H)^2 \sim 4 \times 10^{-2}$ and therefore for $y \sim 1/2$; $(\delta x/x)^2$ depends appreciably on $(\delta\theta_\mu/\theta_\mu)^2$ and less so on $(\delta P_\mu/P_\mu)^2$ and $(\delta E_H/E_H)^2$. At small y, $(\delta x/x)$ receives equal contribution from $\delta\theta_\mu/\theta_\mu$ and $\delta E_H/E_H$.

We now turn to a discussion of the x and y distribution. In Figs. 20 and 21 are shown the x distributions obtained for ν_μ and $\bar{\nu}_\mu$ interactions in E1A. These distributions are also compared with the $F_2(x)$ obtained from SLAC data for various neutrino energy cuts. No striking variation of these distributions

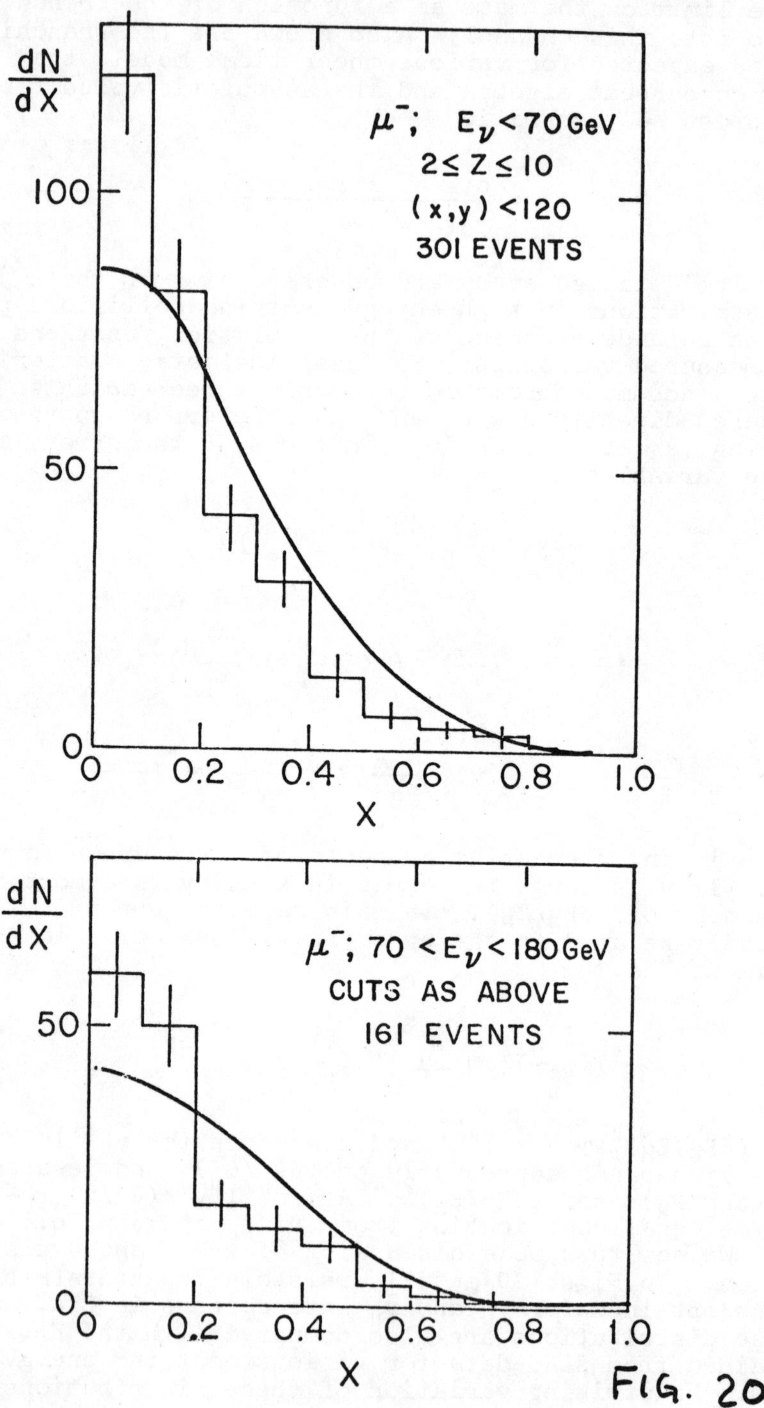
dN/dX
μ^-; E_ν < 70 GeV
2 ≤ Z ≤ 10
(x,y) < 120
301 EVENTS
100
50
0
0 0.2 0.4 0.6 0.8 1.0
X
dN/dX
μ^-; 70 < E_ν < 180 GeV
CUTS AS ABOVE
161 EVENTS
50
0
0 0.2 0.4 0.6 0.8 1.0
X

FIG. 20

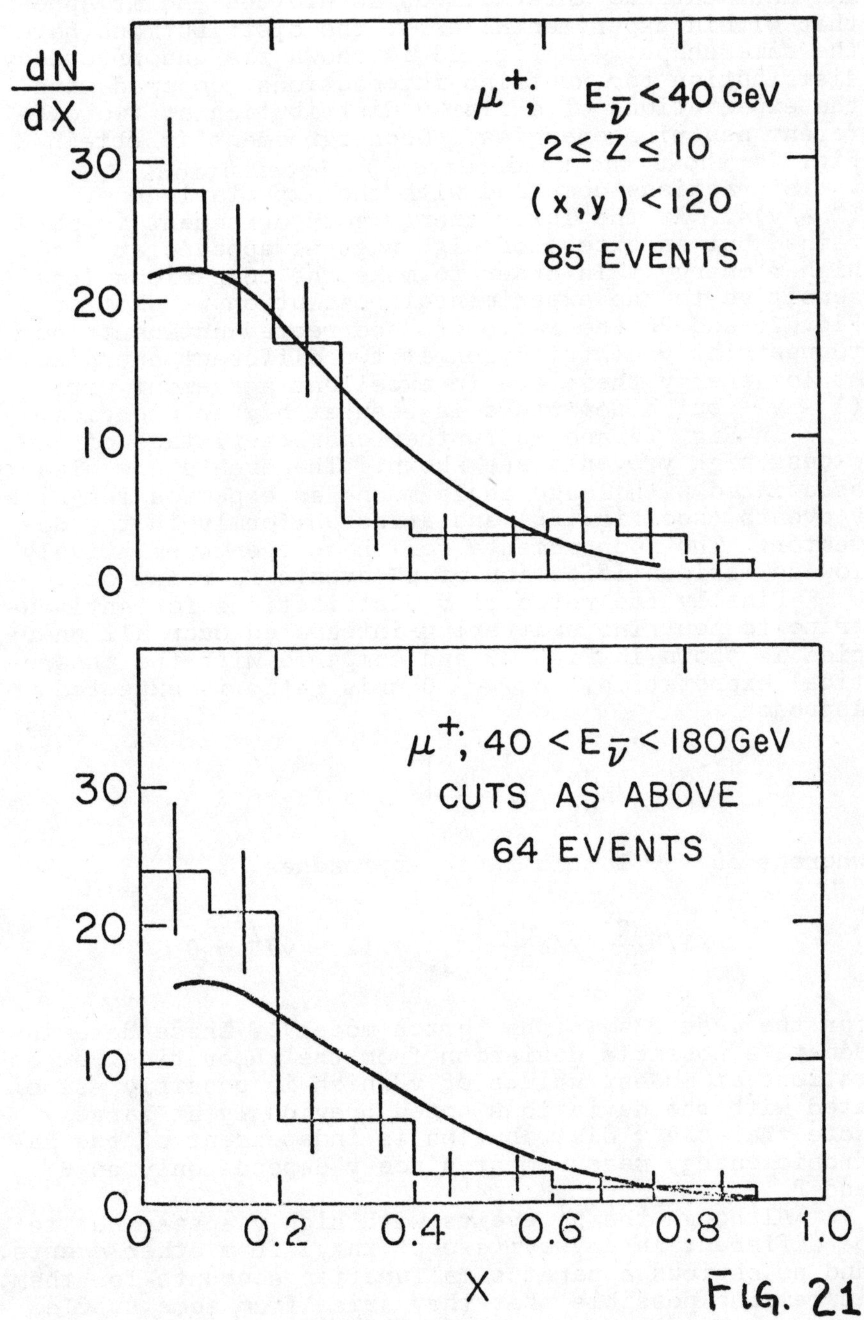
dN/dX
μ^+; $E_{\bar{\nu}}$ < 40 GeV
2 ≤ Z ≤ 10
(x, y) < 120
85 EVENTS
30
20
10
0
μ^+; 40 < $E_{\bar{\nu}}$ < 180 GeV
CUTS AS ABOVE
64 EVENTS
30
20
10
0
0
0.2
0.4
0.6
0.8
1.0
X

FIG. 21

with neutrino energy is observed providing additional evidence for the validity of scaling. In Fig. 22 the ratio of distributions as a function of x for neutrino and antineutrino interactions is plotted and it appears that within experimental error the distributions have the same shape. In Fig. 23 is shown the uncorrected y distribution for neutrino interactions compared with the expectations of a flat y distribution at two different neutrino energies. Good agreement is obtained. Fig. 24 shows the uncorrected y distributions for $\bar{\nu}_\mu$ interactions compared with the expectations of $(1 - y)^2$. At the lower energy good agreement is obtained but an excess of high y event appears at the higher energy. In order to make the comparison less sensitive to the experimental resolution we plot in Fig. 25 and 26 the ratio of uncorrected antineutrino to neutrino y distribution at two different energies. At low energy these are in excellent agreement with $(1 - y)^2$ but a departure is seen at higher energies.

In Fig. 27 and 28 further characteristics of the excess high y events are shown. The events are clearly associated with large angle muons as expected for high y events (see Fig. 28) and arise uniformly in the detector. The muon momenta for these events relatively low so misidentification of μ^- events is ruled out.

Finally the ratio of v distributions for antineutrino to neutrino scattering integrated over all energies is shown in Fig. 29 and compared with the theoretical expectation. As $v \to 0$ this ratio is expected to approach

$$1/N\frac{dN^{\bar{\nu}}}{dv}/1/N\frac{dN^{\nu}}{dv}\Bigg]_{v\to 0} \to \sigma_\nu/\sigma_{\bar{\nu}} \simeq 3$$

whereas as $v \to 1$ this ratio approaches

$$1/N\frac{dN^{\bar{\nu}}}{dv}/1/N\frac{dN^{\nu}}{dv}\Bigg]_{v\to 1} \to (1 - v)^2 \to 0$$

for the case B = 1 (the parton model). These data indicate a possible deviation from the theoretical expectations at modest values of v which is possibly associated with the deviations noted previously at large y. Note that the v distribution is independent of the hadronic energy measurement since v depends only on θ_μ and P_μ.

Although the μ^+ events with high y appear not to be different in anyway (except in y) from other events, and no obvious apparatus malfunction accounts for them, it remains possible that they arise from some subtle

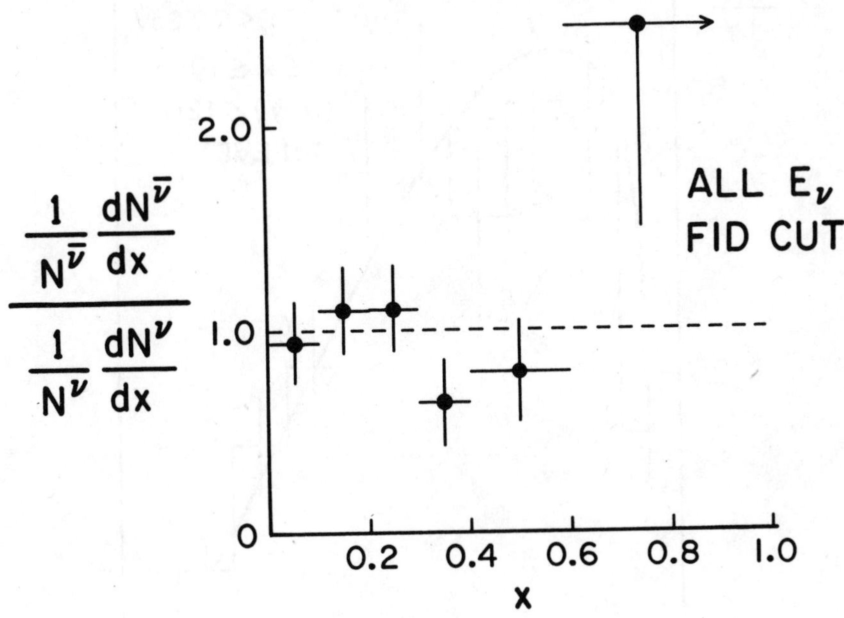
$\frac{\frac{1}{N^{\bar{\nu}}}\frac{dN^{\bar{\nu}}}{dx}}{\frac{1}{N^{\nu}}\frac{dN^{\nu}}{dx}}$
2.0
1.0
0
0.2
0.4
0.6
0.8
1.0
x
ALL E_ν
FID CUT

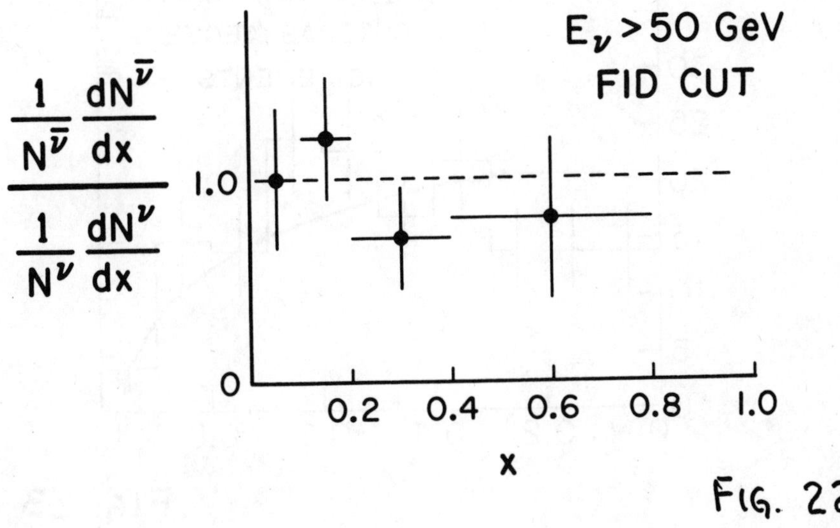
$\frac{\frac{1}{N^{\bar{\nu}}}\frac{dN^{\bar{\nu}}}{dx}}{\frac{1}{N^{\nu}}\frac{dN^{\nu}}{dx}}$
1.0
0
0.2
0.4
0.6
0.8
1.0
x
E_ν > 50 GeV
FID CUT

FIG. 22

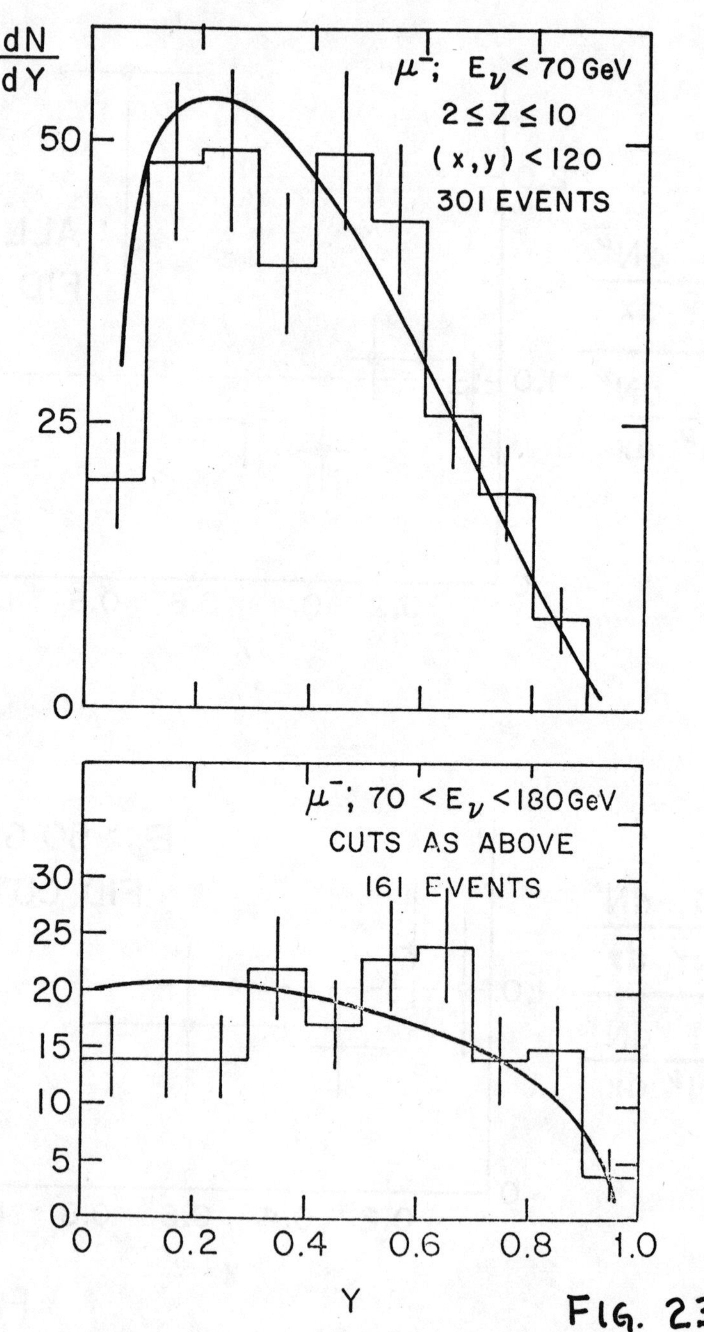
dN/dY
μ^-; E_ν < 70 GeV
2 ≤ Z ≤ 10
(x,y) < 120
301 EVENTS
50
25
0
μ^-; 70 < E_ν < 180 GeV
CUTS AS ABOVE
161 EVENTS
30
25
20
15
10
5
0
0 0.2 0.4 0.6 0.8 1.0
Y

FIG. 23

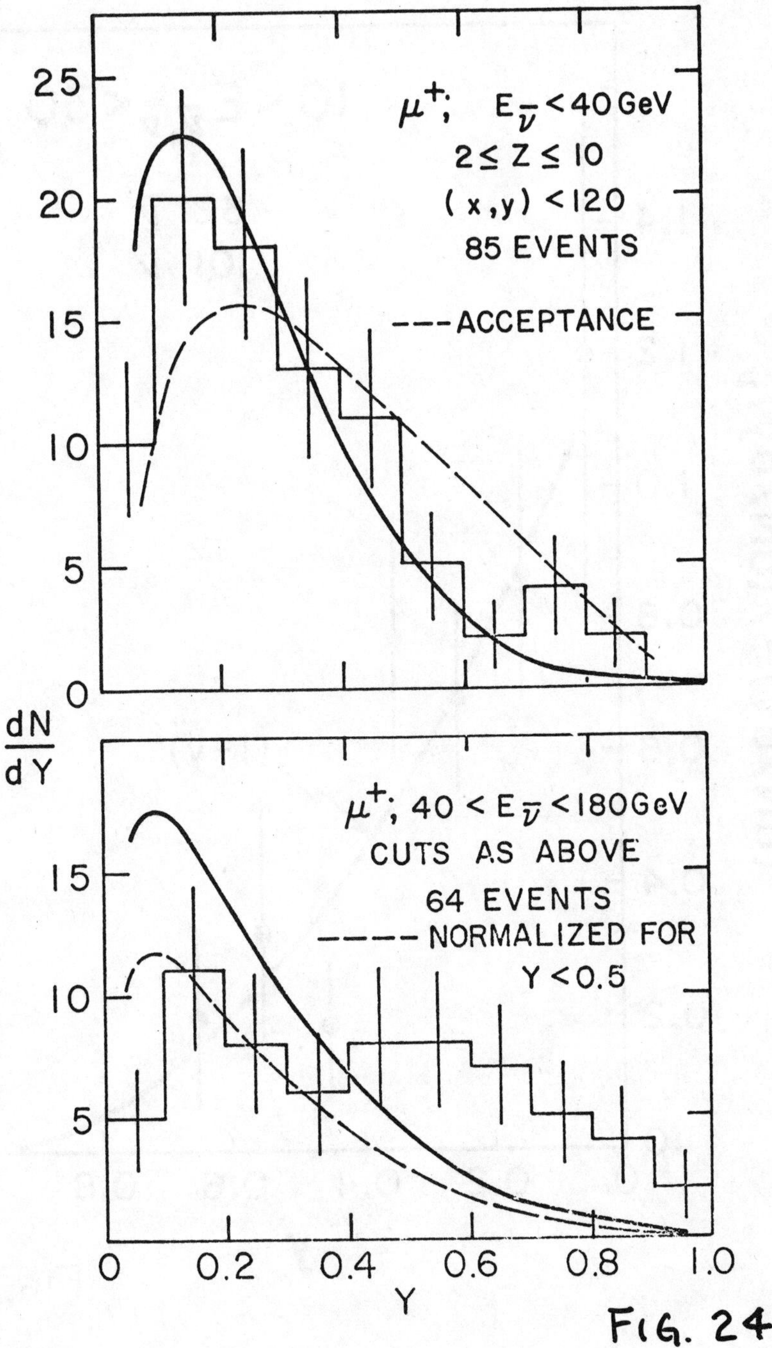
25
20
15
10
5
0
μ^+; $E_{\bar{\nu}}$ < 40 GeV
2 ≤ Z ≤ 10
(x , y) < 120
85 EVENTS
---ACCEPTANCE
dN/dY
15
10
5
μ^+; 40 < $E_{\bar{\nu}}$ < 180 GeV
CUTS AS ABOVE
64 EVENTS
----NORMALIZED FOR
Y < 0.5
0
0.2
0.4
0.6
0.8
1.0
Y

FIG. 24

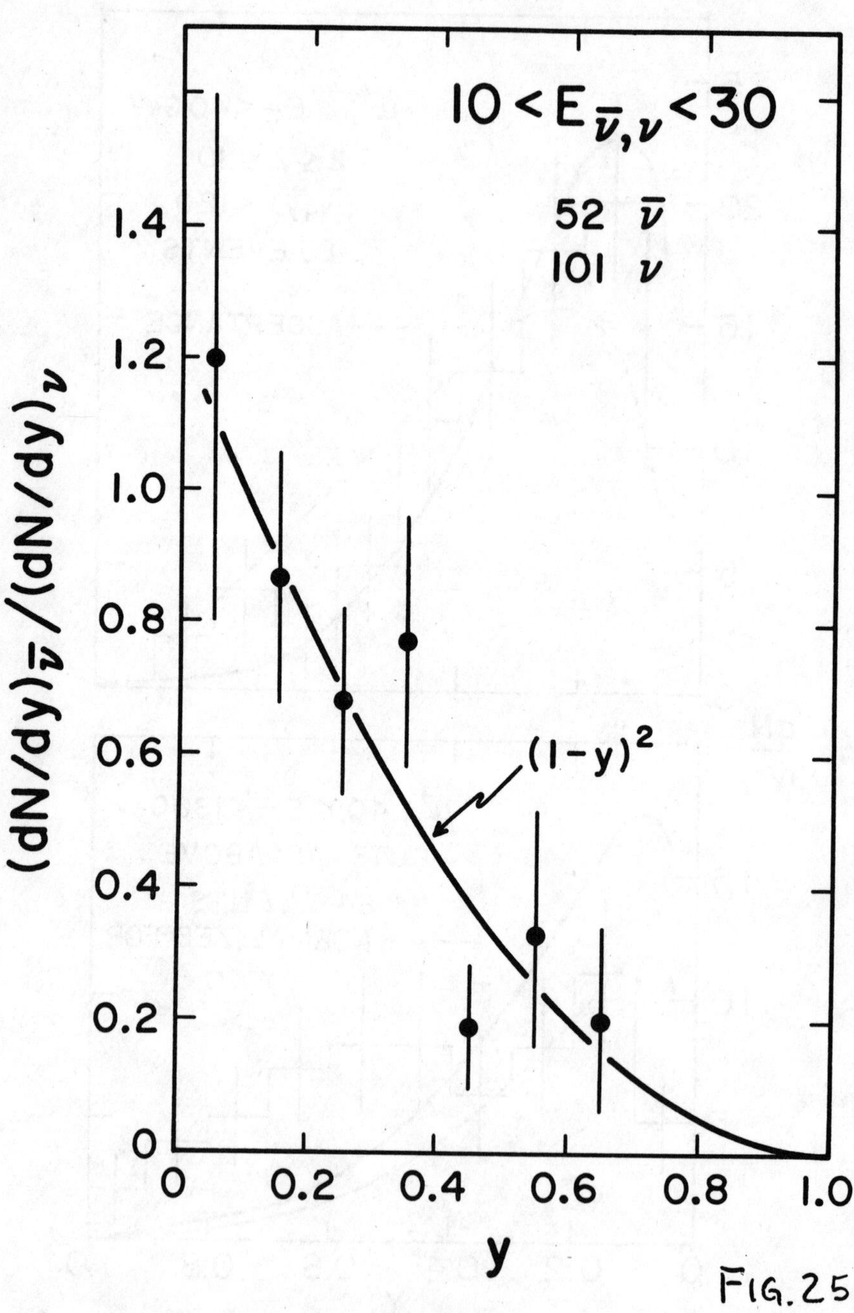

$10 < E_{\bar{\nu},\nu} < 30$
52 $\bar{\nu}$
101 ν
$(1-y)^2$
$(dN/dy)_{\bar{\nu}}/(dN/dy)_{\nu}$
1.4
1.2
1.0
0.8
0.6
0.4
0.2
0
0
0.2
0.4
0.6
0.8
1.0
y

FIG. 25

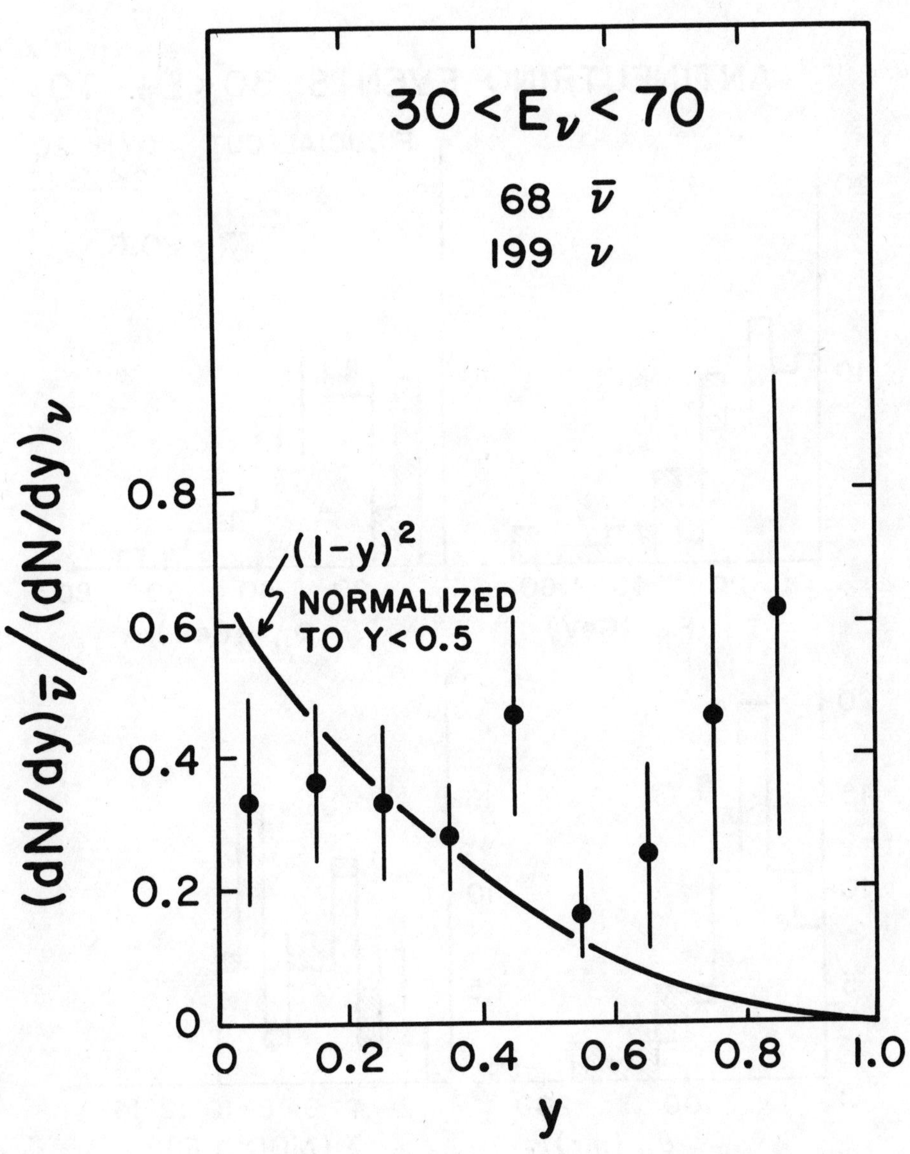

30 < E_ν < 70
68 ν̄
199 ν
(1-y)^2
NORMALIZED
TO Y<0.5
(dN/dy)ν̄/(dN/dy)ν
0.8
0.6
0.4
0.2
0
0
0.2
0.4
0.6
0.8
1.0
y

FIG. 26

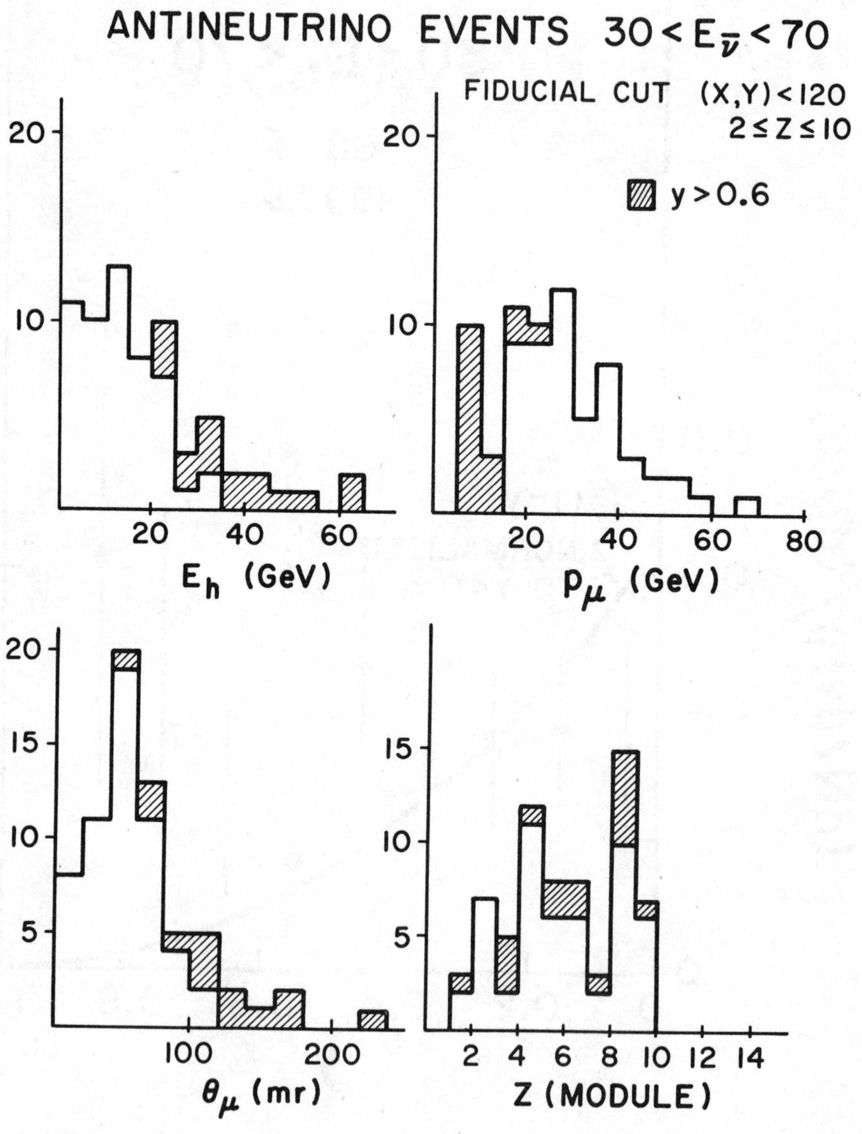
ANTINEUTRINO EVENTS 30 < E$_{\bar{\nu}}$ < 70
FIDUCIAL CUT (X,Y) < 120
2 ≤ Z ≤ 10
y > 0.6
20
10
20
40
60
E$_h$ (GeV)
20
10
20
40
60
80
p$_\mu$ (GeV)
20
15
10
5
100
200
θ$_\mu$ (mr)
15
10
5
2
4
6
8
10
12
14
Z (MODULE)

FIG. 27

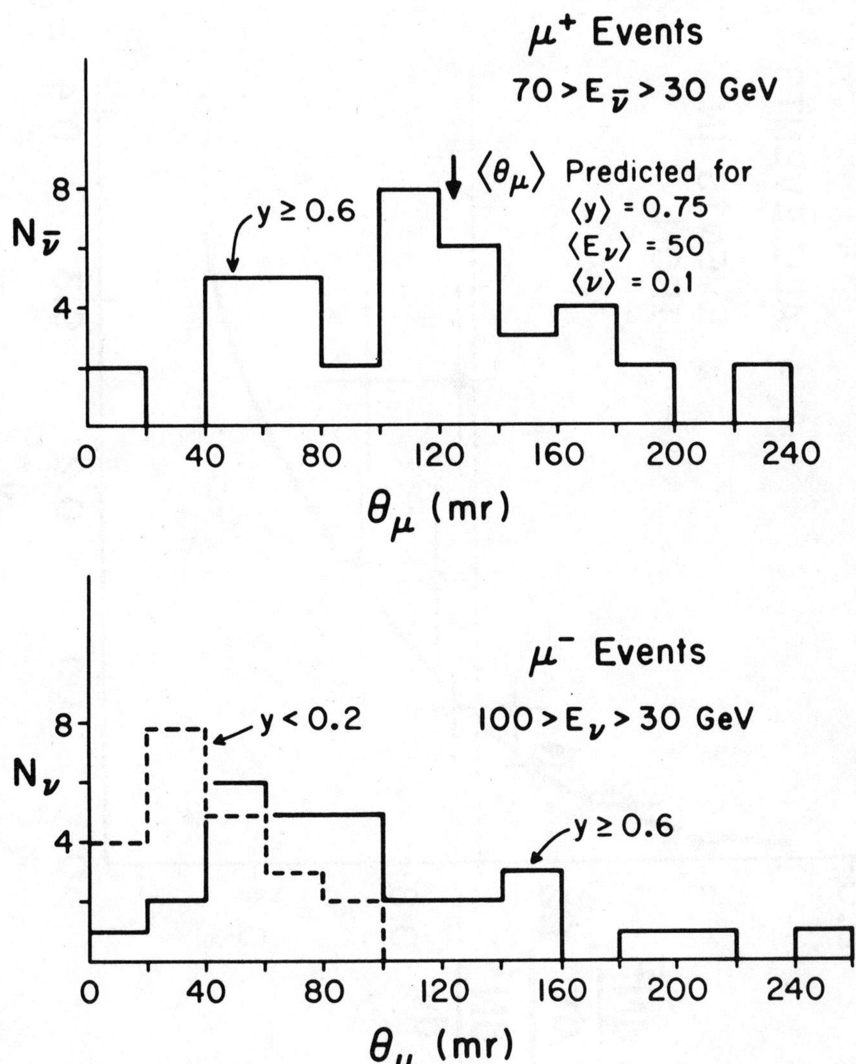
μ^+ Events
$70 > E_{\bar{\nu}} > 30$ GeV
$\langle\theta_\mu\rangle$ Predicted for
$\langle y\rangle = 0.75$
$\langle E_\nu\rangle = 50$
$\langle \nu\rangle = 0.1$
$y \geq 0.6$
$N_{\bar{\nu}}$
8
4
0
40
80
120
160
200
240
θ_μ (mr)
μ^- Events
$100 > E_\nu > 30$ GeV
$y < 0.2$
$y \geq 0.6$
N_ν
8
4
0
40
80
120
160
200
240
θ_μ (mr)

FIG. 28

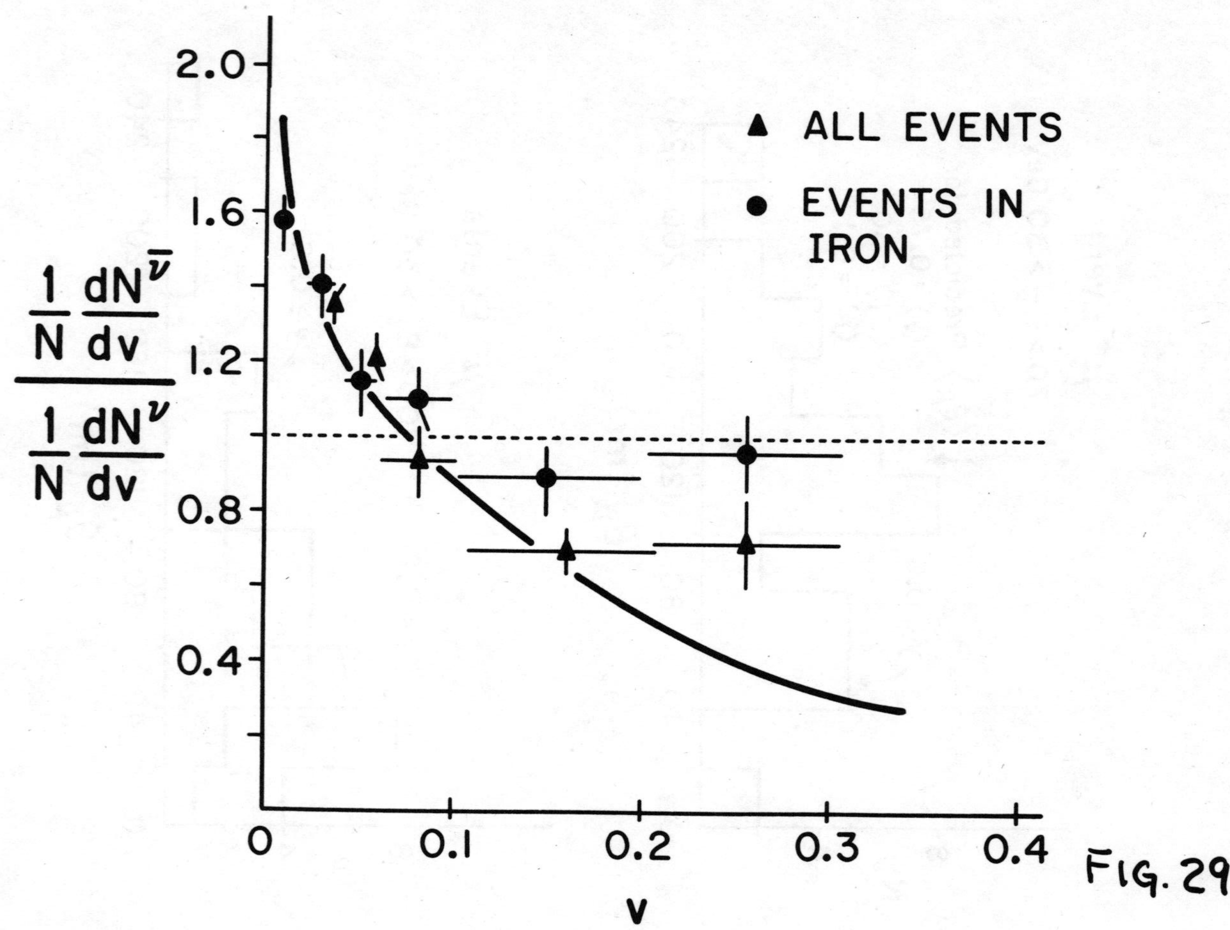

▲ ALL EVENTS
● EVENTS IN IRON
$\frac{1}{N}\frac{dN^{\bar{\nu}}}{dv} \Big/ \frac{1}{N}\frac{dN^{\nu}}{dv}$
2.0
1.6
1.2
0.8
0.4
0
0.1
0.2
0.3
0.4
v

FIG. 29

aberration in experimental method that has been overlooked. If, however, they prove genuine and are reproduced in subsequent experiments, they are indications of a significant effective deviation from scale invariance or charge symmetry or both and could arise from some form of new particle production as for example outlined in Sections 1 and 2.

5. DIMUON EVENT SIGNATURE

Since the beginning of accelerator produced neutrino beam experiments, an intensive search has been carried out for dimuon final states. This search is now being continued to higher energies by using the new generation of calorimeter - spectrometer detectors. There are some definite advantages for the new detectors when bonified dimuon events are observed since the accompanying hadronic energy can now be measured. This allows for an extremely clean experimental signature for the quiet dimuon events and thus a very sensitive search for the W boson.

As with the earlier experiments there are important backgrounds for dimuon production that arise from $\pi \to \mu$ decay and $K \to \mu$ decay, i.e.

$$\nu_\mu + N \to \mu^- + \pi^\pm + X$$
$$\to \mu^\pm + \nu_\mu$$

Furthermore, the production of vector mesons can result in dimuon events or trimuon events such as

$$\nu_\mu + N \to \nu_\mu + (\rho,\ \phi) + X$$
$$\to \mu^+ + \mu^-$$

and

$$\nu_\mu + N \to \mu^- + (\rho,\ \phi) + X$$
$$\to \mu^+ + \mu^-$$

The background for quiet dimuons has been estimated to be very small and approximately 10^{-5} per event for a noisy dimuon event is more difficult to determine but two methods have been used in E1A.

(1) Estimate the number of charged pions that accompany a deep inelastic collision from vari-

ous models that also are in agreement with electroproduction data. Let the $\pi \to \mu$ decay in an absorbing medium and estimate the rate. One prediction from this calcuation is that the overwhelming majority of dimuon events will have one very low energy muon.

(2) Expose the detector to a high energy hadron beam that interacts in the first interaction length and measure the probability that a $\pi \to \mu$ decay occurs.

Both methods of estimating the probability per event for a $\pi \to \mu$ (or $K \to \mu$) decay in E1A give an estimate of 10^{-3} to 10^{-4} for the rate per event. In addition the majority of these events are expected to have low energy muons of momentum below 5 GeV. Thus it appears that the search for quiet dimuons in E1A can be carried to the level of $\sim 10^{-4} - 10^{-5}$ and for noisy dimuons from 10^{-3} to 10^{-4} per neutrino event. The background from vector meson production is not expected to exceed this rate.

No quiet dimuon events have so far been observed in E1A and the limit of the production cross section

$$\frac{\sigma(\nu_\mu + N \to \mu^+\mu^-\nu_\mu + N)}{\sigma(\nu_\mu + N \to \mu^- + X)} \leq 10^{-3}$$

has been obtained. Two good candidates for noisy dimuon events have been obtained. The dimuons from these events extrapolate satisfactorily to a vertex and the two muons traverse separate scintillation counters and are therefore known to be in-time events. A picture of one of these events is shown in Fig. 30. Table 4 lists the parameters of the muons and the accompanying hadronic energy. The production cross section ratio for the noisy dimuon events to all deep inelastic events based on these events is approximately 5×10^{-3}. These events are somewhat unlikely to arise from background. However, a better background study as well as better statistics are needed in order to reach a definite conclusion regarding noisy dimuon production.

The limit on quiet dimuon events can be used to set a limit on the W^+ mass and the limit on noisy dimuon event rate can be used to limit the M^o mass as reported in Table 5.

6. SUMMARY

High energy neutrino collisions provides a rich field to search for new particle production and to test several important theories that seek to explain hadrons

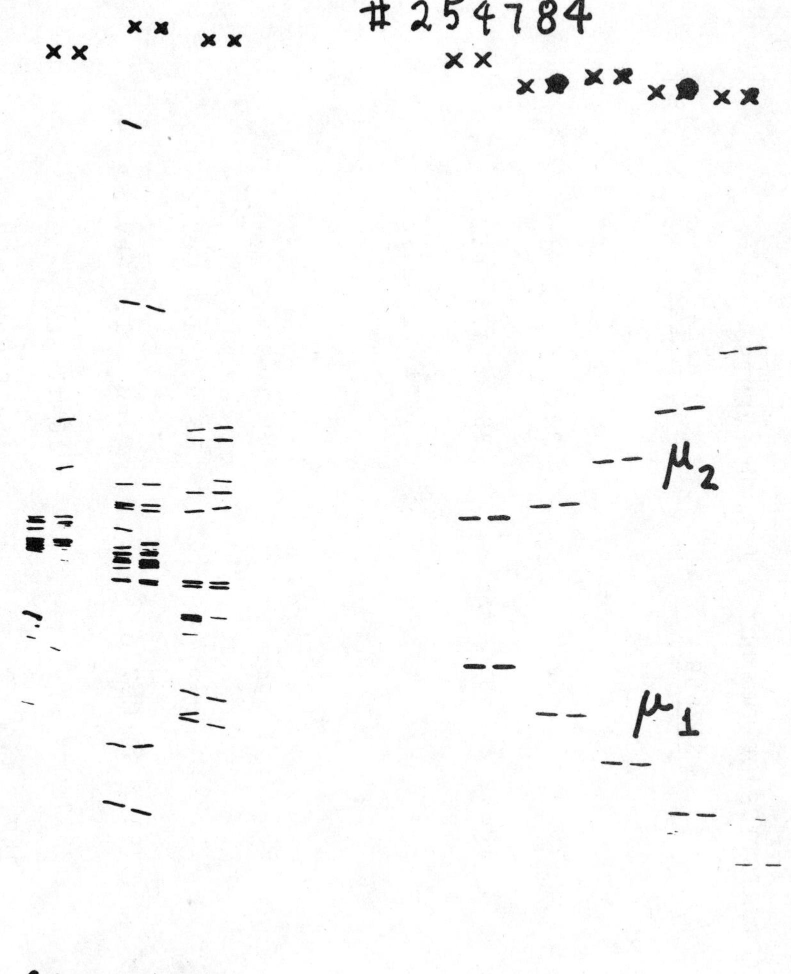

FIG 30

TABLE 4

CHARACTERISTICS OF TWO NOISY DIMUON EVENTS

EVENT NUMBER	253659	254784
Charge of First Muon	-	-
θ_μ^1	29.4 ± 2.6 mr	32.7 ± 1.2 mr
P_μ^1	106.9 ± 5.8 GeV/c	36. ± 0.7 GeV/c
$P_\perp^1$	3.1 ± 0.2 GeV/c	1.2 ± .1 GeV/c
Charge of Second Muon	+	+
θ_μ^2	55.5 ± 1.5 mr	45.1 ± 3.9 mr
P_μ^2	16.7 ± .2 GeV/c	13.9 ± 1.0 GeV/c
$P_\perp^2$	.93 ± .1 GeV/c	0.63 ± 1.0 GeV/c
E_{Hadron}	23.7 GeV	104.7 GeV
Total Visible Energy	147.3 GeV	154.6 GeV

in terms of constituents and to unify weak and electromagnetic interactions. Most of the new particles that have been suggested in this contex have a unique experimental signature and therefore rather definitive searches for these particles can be made. In many cases the production cross sections can be reliably estimated and the dependence of the search on the detailed calculation of the branching ratios into specific channel can be minimized by search for a given new particle using several decay channels. An example of this technique occurs for W production where both the hadronic and leptonic decays can be used to provide an inclusive search.

The search for new particle production is made somewhat easier at least in the neutrino energy range up to 100 GeV because the background from deep inelastic scattering process appear to follow approximately the predictions of the simple parton model. This allows sensitive tests of charge symmetry and various current algebra sum rules. In addition, present experimental data are in agreement with the predictions of scaling for neutrino collisions up to the highest energies studied of $\sim$ 250 GeV. This leads to the hope that the search for non-locality due to W exchange can be extended up to very high masses perhaps above 50 GeV.

Furthermore the study of antineutrino interactions at large y or large v appears to be extremely sensitive to new particle production because of the expected large suppression of the background cross section.

The present experimental data for antineutrino interactions is an approximate agreement with the expectations derived from a SLAC data and the simple parton model. The ratio of antineutrino to neutrino cross section appears to be in the vicinity of (0.3-0.5) up to 140 GeV. Preliminary evidence exists that the antineutrino y distribution is distorted for $y > 0.5$ and that this distortion increases with energy. If this distortion is confirmed then a small but significant deviation from Bjorken scaling is indicated. This breakdown of scaling could be due to new particle production and several models accounting for such a distortion were given in the introduction.

With the discovery of muonless events in neutrino collisions and their probable interpretation as due to neutral currents a new category of reactions is added to the roster for new particle production. For example non-locality due to W^o exchange could be in principle, identified in the neutral current channel and heavy neutral leptons can also be produced in this channel.

The experimental search for dimuon events produced by high energy neutrinos is reaching a significant level.

In the "quiet" channel the abscence of events already allows a restrictive mass limit for the W boson. The search for dimuon events with appreciable accompanying hadronic energy is in an early stage. Two candidates for dimuon events were reported but with the background for these events is not yet reliably estimated. The crude present estimates for the background rate indicate that these events are unlikely to be due to background but only perhaps at the 10% level. Already the low rate for such events can be used to put mass limits on M° production. The continued study of this channel is clearly extremely desirable.

In table 5 we summarize the mass limits that have been placed on some new particles in experiments so far completed at NAL. We may anticipate that limits on charm particle production will be available as more information becomes available concerning strangeness changing interactions. If the distortion of the antineutrino y distribution discussed in Section 4 were attributed to charm particles production the charm particle mass could well be 5-8 GeV since the distortion occurs at higher antineutrino energies. However theoretical models that seek to explain the suppression of the $\Delta S = 1$ neutral currents probably restrict the mass to be less than $\sim$10 GeV.[48] Likewise the existence of noisy dimuon events could result from charm particle production and the present upper limit on such dimuons is consistent with almost any mass value.[4]

I wish to thank J. Bjorken, V. Barger, E. Paschos and my colleagues in E1A for many helpful discussions.

TABLE 5

MASS LIMITS ON NEW PARTICLES

PARTICLE	DECAY MODE OR SIGNATURE	MASS LIMIT	REMARKS AND REFERENCE
W^+	$W^+ \rightarrow \mu^+ \nu_\mu$	$\geq$ 8.0 GeV	Assume B = 0.25
	$W^+ \rightarrow$ hadrons	$\geq$ 5 GeV	$B \rightarrow 1$
	Propagator Modification	$\geq$ 10 GeV	Assumes Bjorken scaling
W^o	$W^o \rightarrow \mu^+ \mu^-$	Very Low	The cross section for $\nu_\mu + Z \rightarrow W^o \nu_\mu Z$ is very small
M^+	$M^+ \rightarrow \mu^+ \nu_\mu \nu_\mu$	$\gtrsim$ 8.4 GeV	Assume B = 0.3
	$M^+ \rightarrow \nu_\mu$ + hadrons	$\geq$ 5 GeV	Assume $B \rightarrow 1$
M^o	$M^o \rightarrow \mu^+ \mu^- \nu_\mu$	$\geq$ 2 GeV	Assume B = 0.5 and $\frac{\sigma(\nu_\mu + N \rightarrow M^o + X)}{\sigma(\nu_\mu + N \rightarrow M^+ + X)} \propto R^\nu \sim 0.1$
	$M^o \rightarrow \nu_\mu$ + hadrons	No limit	Threshold effect in R^ν
Charm	$C \rightarrow \mu^+ + \nu_\mu$ + (hadrons)	No limit	Depends on number of λ quarks in nucleon
	$C \rightarrow$ hadrons (Breaking of charge symmetry or y distribution structure)	No limit	

References

*Supported in part by the U.S. Atomic Energy Commission under contract AT(11-1)-881-407

1. J. D. Bjorken and S. L. Glashow, Phys. Letters 11, 255 (1964), S. L. Glashow, J. Iliopoulas and L. Maiani, Phys. Rev. D2, 1285 (1970).
2. A more complete set of references included in M.A. B. Big and A. Sirlin, Gauge Theories of Weak Interactions, to be published in Annual Reviews of Nuclear Science, Vol. 24.
3. G. Snow, Nucl. Phys. B35, 107 (1971).
4. M. K. Gailliard, Notes on Charmed Particle Searches in Neutrino Experiments, National Accelerator Lab Report FN-259, G. Altarelli et al., Phys. Letters B448, 435 (1974).
5. A. DeRujula, H. George, S. L. Glashow and H. R. Quinn, Fact and Fancy in Neutrino Physics, Harvard Preprint (1973).
6. See for example, C. H. Llewellyn Smith, An Introduction to Renormalizable Models of Weak Interactions and Their Experimental Consequences, CERN Report Th1710 (1973) or B. W. Lee, Proceedings of the XVI Conference on High Energy Physics, NAL (1972).
7. M. Chanowitz and S. Drell, Phys. Rev. Letters 30,, 807 (1973).
8. H. Yukawa, Proceedings of the Physics-Math Society of Japan (3) 17, 48 (1935).
9. A complete list of references to this "old fashioned" approach to the W mass is given in D. Cline, Prospects for the Detection of Higher Order Weak Interactions and Study of Weak Interactions at High Energy, Wisconsin Preprint and published in the Proceedings of the 1972 Coral Gables Conference.
10. S. Weinberg, Phys. Rev. Letters 27, 1688 (1971) and A. Salem in Elementary Particle Physics (ed. N. Svartholm, Almquist and Wikselles, Stockholm (1968), p. 367.
11. A. Lutke et al., Phys. Rev. Letters, 30, 1189 (1973), B. Richter, invited talk at the Irvine Conference, University of California, Irvine, December (1973).
12. F. J. Hasert et al., Phys. Letters 46B, 138 (1973); A. Benvenuti, et al., Phys. Rev. Letters 32, 800 (1974); B. Aubert et al., Further Observation of Muonless Neutrino Induced Inelastic Interactions, Harvard-Pennsylvania-Wisconsin NAL Preprint (1974).

13. The W° could in principle be produced in the Coulomb field by the interaction with the quadrupole moment.
14. See for early review, M. Perl, SLAC PUB 1002 (1972).
15. H. George and S. L. Glashow, Phys. Rev. Letters 28, 1494 (1972).
16. J. D. Bjorken and C. H. Llewellyn Smith, Phys. Rev. D7, 887 (1973).
17. B. Aubert et al., Measurement of the Rates of Muonless Deep Inelastic Neutrino and Antineutrino Interactions, Harvard-Pennsylvania-Wisconsin-NAL Preprint (1974).
18. This was suggested to me by J. D. Bjorken.
19. V. Barger, private communication.
20. See for example A. De Rujula and S. L. Glashow, Phys. Letters 46, B337 (1973).
21. P. Musset, CERN-TC-L (Int. 73-1).
22. T. Eichten et al., Phys. Letters B46, 274 (1973), B46, 281 (1973).
23. A. Benvenuti et al., Phys. Rev. Letters 30, 1084 (1973) and 32, 125 (1974).
24. See the report by F. Sculle et al., contribution to this conference.
25. D. Reeder, contribution to the Bonn Conference (1973) (See P. Mussets Report at this Conference).
26. B. C. Barrish et al., Phys. Rev. Letters 31, 180 (1973).
27. J. D. Bjorken, Phys. Rev. 779, 1547 (1969); J. D. Bjorken and E. A. Paschos, Phys. Rev. D1, 3151 (1970).
28. See for example, Fayyazuddin and Riazuddin, CERN Preprint TH-1807.
29. For a recent review see S. L. Adler, Accelerator Neutrino Physics, Present and Future - A Review for Theorists and Experimentalists, NAL Conference-74/39 THY.
30. D. Cline and E. A. Paschos, Phys. Rev. D8, 984 (1973).
31. See for example A. De Rujula and S. L. Gashow, Phys. Letters 46B, 337 (1973).
32. Or using a scaling variable in the neutral currents that is better suited to the experimental measurements that can be performed. For example, the variable $u = E_H \theta_H^2/2mp$ where θ_H is the hadronic angle and E_H the energy which is analogous to the muon variable $v = xy$.
33. J. D. Bjorken, D. Cline and A. K. Mann, Phys. Rev. D8, 3207 (1973).
34. G. Myatt and D. H. Perkins, Phys. Letters 34B, 542 (1971).

35. R. W. Brown, R. H. Hobbs and J. Smith, Phys. Rev. D4, 794 (1971).
36. D. Cline, A. K. Mann and C. Rubbia, Detection of the Weak Intermediate Boson Through Its Hadronic Decay Modes, Phys. Rev. Letters 25, 1309 (1970).
37. These results are taken from C H. Albright and C. Jarlskog, Neutrino Production of M^{+h} and E^{+} Heavy Leptons (I); CERN Preprint Th-1828, References to other calculations are contained in this work.
38. The dimuons produced through W production and leptonic decay will have characteristic m^{+} and μ^{-} momenta and angles. See R. W. Brown and J. Smith, Phys. Rev. D3, 207 (1971) and references therein.
39. We thank V. Barger for a discussion on this point. A more complete formulation has been given by G. Rajasekaran and K. V. L. Sarma, TATA Institute, Preprint TIFR/TH/74-10. See also S. Pakvasa and S. F. Tuan, University of Hawaii, Preprint UG-511-171-73.
40. J. S. Sakurai, Remarks on Neutral Current Interactions, reported at this conference.
41. R^{ν} and $R^{\bar{\nu}}$ are the standard ratio of neutral current to charged current cross sections integrated over all y.
42. E21 is a CAL Tech-NAL Collaboration and E1A is a Harvard-Pennsylvania-Wisconsin-NAL Collaboration.
43. The CERN data was obtained from C. Franzinetti, private communication, and the NAL data is from E1A.
44. For early data from E21 in this regard see reference 26 and the report of F. Sculli to this conference.
45. G. Miller et al., Phys. Rev. D5, 528 (1972).
46. B. C. Barrish et al., Phys. Rev. Letters 31, 410 (1973).
47. B. C. Barrish et al., Search for a Gauge Theory Heavy Lepton, CAL Tech Preprint (1974).
48. See for example M. K. Gailliard and B. W. Lee, Rare Decay Modes of the K-Meson in Gauge Theories, NAL Preprint (1974).

SOLAR NEUTRINOS*

Roger K. Ulrich
University of California, Los Angeles, Calif. 90024 and
California Institute of Technology, Pasadena, Calif. 91109

ABSTRACT

The current status of the experimental efforts by Davis to detect neutrinos from the sun is reviewed from the point of view of the theory of solar models. The essential physical input for a standard model of the sun is presented. Ten different hypothetical explanations of the discrepancy between the theoretically expected ^{37}Ar production rate and the observed production rate are discussed. None of these hypotheses are entirely satisfactory. One, the low Z model, is marginally attractive but it requires the sun to have a very unusual chemical composition. The errors in the fluxes from the standard model and the low Z model resulting from errors in the input data lead to ^{37}Ar production rates of 5.6 ± 1.8 and $1.4 \pm 0.35 \times 10^{-36}$ atoms/target atom/sec, respectively.

INTRODUCTION

The experimental efforts by Ray Davis and his co-workers[1] to detect neutrinos from the sun through the reaction $^{37}Cl + \nu_{\odot} \rightarrow {}^{37}Ar + e^-$ have thus far only succeeded in setting an upper limit on the rate of production of ^{37}Ar. This upper limit is well below the rate of ^{37}Ar production expected from models of the solar interior and the conflict presents a serious challenge to the theory of stellar interiors. In this discussion I will adopt the point of view that an error in the theory of stellar interiors is the cause of the conflict between the Davis upper limit on the rate of production of ^{37}Ar of 1 SNU (1 SNU = 10^{-36} product atoms/target atom/sec) and the theoretical rate of production calculated from standard solar models of 6 SNU.[2] While changes in the theory of the weak interaction have been proposed,[3] I cannot test these ideas and therefore will not discuss them. Instead the discussion here centers on the definition and critical analysis of the standard solar model and various alternate solar models. The standard solar model involves the simplest set of assumptions that can be made about the

*Supported in part by the NSF [GP-36687X] and [GP-28027].

sun's structure and evolution. Of prime importance among these assumptions are 1) the sun currently produces its energy from nuclear reactions; 2) the sun was chemically homogeneous before nuclear reactions began with a composition the same as the present solar surface; 3) the core of the sun does not rotate rapidly enough to require the addition of a centripetal acceleration term to the equation of hydrostatic equilibrium. Other assumptions of lesser importance in the construction of a standard solar model are listed in Table I and will be discussed in a later section of this paper. Much of the remaining discussion in this section and in the next section depends in a crucial way on the validity of assumption 1) even though other of the standard model assumptions may break-down. Unless otherwise stated, all models are assumed to obey assumption 1).

Table I Input Physics for a Standard Solar Model

I. Conservation Equations
   A. Mass
   B. Momentum (hydrostatic equilibrium)
   C. Energy
   D. Nuclear transformations

II. Supplementary Physics
   A. Inverse square law of gravity
   B. Equations of State
   C. Opacity function
   D. Nuclear reaction cross sections
   E. Convection theory

III. Initial Condition of Uniform Composition

IV. Observational Constraints
   A. Age = 4.7×10^{9} yrs
   B. L = 3.81×10^{33} ergs/sec
   C. $G M_{\odot} = 6.67 \times 10^{-8} \times 1.98 \times 10^{33}$
   D. Composition of the solar surface
   E. $R_{\odot} \cdot 6.96 \times 10^{10}$ cm
   F. Laboratory measurements of opacity and nuclear cross sections
   G. Solar oblateness

Before beginning the main discussion I would like to bring out an important feature of the Davis experiment which may not be widely recognized. The production of ^{37}Ar from ^{37}Cl occurs primarily as a result of the capture of neutrinos from three reactions in the solar interior:

$$^{8}B(e^{+}\nu)\ ^{8}Be^{*} \tag{1}$$

$$^{7}Be(e^{-},\nu)\ ^{7}Li \tag{2}$$

$$^{1}H(e^{-}p,\nu)\ ^{2}D \tag{3}$$

The neutrinos from each of these reactions differ greatly from one another both in their probability of capture by ^{37}Cl and in their sensitivity to the details of solar models. At one extreme reaction (1) produces energetic neutrinos which are easily captured by ^{37}Cl, but it does not occur frequently in the solar interior and the number of these neutrinos depends sensitively on the details of the solar model. At the other extreme are the neutrinos produced by reaction (3). The energy of these neutrinos is only about a factor of two above the threshold for capture and the cross section is low. Although these neutrinos are also relatively rare, their rate of production is a small but slowly changing density and temperature dependent factor times the rate of the $^{1}H(p,e^{+}\nu)^{2}D$ reaction. This latter reaction determines the rate of energy production in the sun and is subject to the integral constraint that nuclear reactions must provide the solar luminosity. Thus the number of ^{37}Ar atoms produced by reaction (3) is virtually independent of the details of the solar model. Reaction (2) is intermediate between reactions (1) and (3), both with respect to ease of detection and sensitivity to model details.

The relative contribution of each of these reactions to the ^{37}Ar production rate depends on the total rate under consideration. At a level of 6 SNU as expected from standard models most of the ^{37}Ar production is due to ^{8}B neutrinos. It is easy to make a small change in the input physics for the model and change in the production rate by a factor of up to two. However, as modifications are introduced which bring the theoretical production rate down to about 1.5 SNU, the primary contribution begins to come from reactions (2) and (3). Below 1.5 SNU level further reduction in the production rate is very difficult to achieve. Because of this character of the solar models and the ^{37}Cl detection reactions, there are a number of hypotheses which can account for a production rate of 1.5 SNU but few if any, that can account for a production rate below 1 SNU. The 1.5 SNU level is therefore a rather important benchmark for the Davis experiment and it appears that this benchmark has probably already been passed (see the concluding section). There are two additional benchmarks which still lie well within the experimental error. These are at the 0.5 SNU level where it is virtually impossible for nuclear reactions to be currently producing the solar luminosity and at the 0.25 SNU level where the entire counting rate is due to reaction (3). The last benchmark has long been recognized, while the importance of the 0.5 SNU level has only recently been emphasized.[4] The argument establishing the importance of the 0.5 SNU counting rate is given below.

NUCLEAR ENERGY MINIMUM COUNTING RATE

The range of possible variation in the temperature distribution in the sun is severely limited by the constraint that nuclear reactions produce the solar luminosity. The various nuclear reactions leading to energy production in the sun each depend on different powers of the temperature; however, except for reaction (3) they all depend on the same power of the density. Since reaction (3) is not important for energy production, the ratios of the rates of production of neutrinos from reactions other than reaction (3) to the rate of production of energy are dependent on temperature alone. Specification of the luminosity at each point throughout the model as a function of the temperature at that point is therefore sufficient to determine the fluxes of the various neutrinos at the earth ϕ_i. The functions $d\phi_i/dL$ are given in Fig. 1 as functions of temperature. Some samples of L(T), the luminosity as a function of temperature, are given in Fig. 2 for a standard solar model (line 1), the low Z model proposed by Bahcall and Ulrich[5] and discussed below (line 2) and an arbitrary model (line 3) not derived from the equations of stellar structure but simply extrapolated from lines 1 and 2. The arbitrary model yields a production rate of 0.5 SNU. Variations in the form of L(T) do not change the production rate by more than 0.05 SNU. Only by lowering the overall temperature level of line 3 could the production rate begin to approach the ultimate 0.25 SNU level. Such modifications run into serious difficulties with the amount of mass required to produce the solar luminosity. The mass requirement depends on density and thus is more closely related to a specific solar model than is $d\phi_i/dL$. Nonetheless, all models of the sun available to me produce 95% of their energy in a mass less than 0.37 solar masses. If the average density in the arbitrary model 3 is the same as in standard models, the mass required to produce 95% of the luminosity is 0.5 solar masses. This problem quickly becomes prohibitive as the average temperature is decreased. A more complete discussion of these ideas can be found in Ref. 4.

THE STANDARD SOLAR MODEL

When a standard solar model is calculated, a set of assumptions is made. Some of these assumptions are based on fundamental laws of nature (e.g., conservation of energy, etc.) while others are based on a more complex theory (e.g., the opacity function, etc.). In order to obtain a non-standard model for the sun, at least one of the basic assumptions must be either abandoned or reinterpreted. The task for the remainder of this paper is therefore to list and examine each of the assumptions underlying the standard solar model. Probably the greatest hazard in this undertaking comes from the possibility that an important assumption has been left off the list and therefore will not be scrutinized.

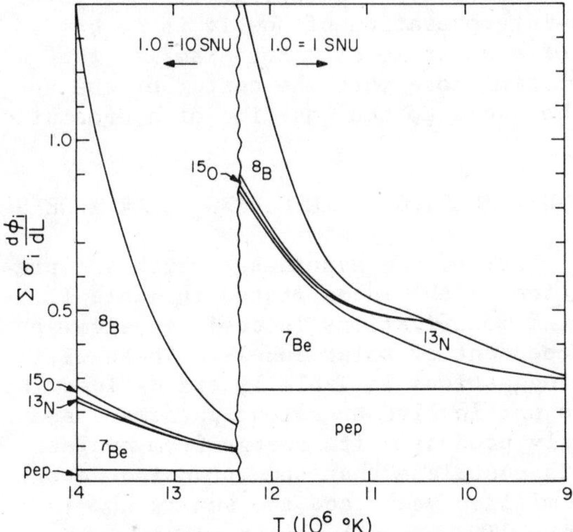

Fig. 1. The ratio of the increment in counting rate to the increment in luminosity versus temperature. The matter undergoing the nuclear reactions has a chemical composition which results from 4.7×10^9 years of exposure to a gradually increasing temperature.

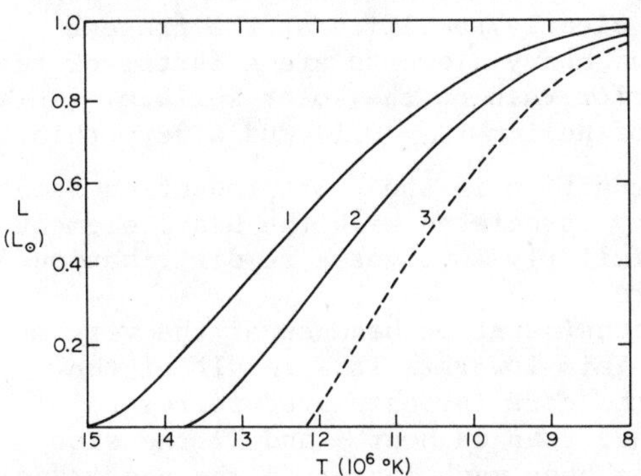

Fig. 2. Luminosity versus temperature. The lines labeled 1, 2, and 3 are respectively the current standard model, a model with minimum opacity, and a hypothetical model.

Table I gives the basic list of assumptions used to calculate a standard solar model. The assumptions are gathered into four groupings: the first group contains basic conservation laws, the second group contains derived physical laws and data, the third group contains only the assumption of the initial distribution of chemical composition and the final group contains the data necessary to distinguish a model of the sun from a model of an arbitrary star. Each of these assumptions must be examined critically to determine if some physical model might lead to a re-interpretation of how it is to be applied to the construction of a solar model. For example, the introduction of a rapidly rotating core into the center of the sun requires an additional term be added to the equation of hydrostatic equilibrium.

SUGGESTED EXPLANATIONS OF THE DISCREPANCY INVOLVING SOLAR MODELS

Table II gives a partial list of the hypotheses which are presently known to me. In addition to the ideas listed in Table II, there are a number of suggested modifications in basic physical processes which are largely independent of solar models. These will not be discussed here. The suggestions in Table II are divided into two groups: group A models do not involve transient phenomena and require the sun to be currently producing its energy from nuclear reactions while group B models postulate that some transient event occurred within the last ten million years and the sun is now in a period of thermal readjustment. The group A models are subject to the nuclear reaction minimum counting rate while the group B models are not. Although the group B models are less appealing than group A models because they require the sun to be in a special phase of evolution, they do, at least in principle, permit lower theoretical counting rates to be achieved. In fact however, only one of the suggested models in group B has succeeded in reducing the theoretical counting rate.

Several models in group A involve altering the distribution of elements within the sun. In model A1 the interior of the sun is mixed continuously and the central hydrogen abundance remains high. The cause for the mixing is not clearly specified at the present time. Model A2 proposes that the heavy elements are a factor of ten less abundant in the solar interior than on the solar surface. This model requires a low abundance of helium ($Y = 0.1$) and a very thin surface convection zone. The reduction in the ^{37}Ar production rate is due to the decrease in opacity associated with the heavy element abundance decrease. Model A5 similarly involves a re-distribution of the hydrogen in the solar core.

Model A6 deserves special consideration because of the very low counting rate it achieves. This low rate is a result of the combination of two factors: 1) the core is postulated to rotate very rapidly with a period of less than an hour — and 2) the size of the core is postulated to have been much larger in the past than it is at present. The first factor introduces a centripetal acceleration equal to as much as 1/2 the acceleration of gravity while the second factor reduces the early luminosity of the sun and makes the sun effectively younger. The centripetal acceleration reduces the pressure and temperature gradients and produces a model which resembles models of lower mass stars. In spite of the low production rates these rapidly rotating models achieve, there are a number of problems which probably make such models unacceptable as the cause for Davis' experimental result. The problems are 1) The simultaneous attainment of a counting rate as low as 1 SNU and an oblateness less than the measured[13] oblateness of 5×10^{-5} requires a careful adjustment of core size and rotation rate. It now seems likely[14] that the actual oblateness of the sun is less than 5×10^{-5} so that the constraint is more severe. Furthermore, calculations by Rood and Ulrich[11] yield a theoretical oblateness 20 times larger than that given in Ref. 9. 2) The process of spin down involves the transport of angular momentum out of the solar core. These is a very good possibility that the agent transporting the angular momentum is matter circulating between the rotating core and the solar surface. Such circulation currents would almost certainly transport matter from the solar surface to high temperature regions where nuclear reactions occur. The abundances of such light isotopes as ^{3}He, ^{7}Li, ^{9}Be, ^{10}B, and ^{13}C would be altered more than is permitted by the observed abundances. 3) The luminosity and surface temperature of the models during the early stages of evolution are substantially altered from normal values. The displacement is roughly paralled to the normal main sequence so that the effect is not as severe as the calculated initial solar luminosity of 0.03 $L_{\odot}$ would suggest. However, a spread in the lower main sequence of about one magnitude is indicated. Such a large spread is definitely not compatible with the observed color-magnitude diagram of the Praesape cluster.[15]

The perturbed models in group B attempt to circumvent the constraint imposed by the solar luminosity through the introduction of a transitory perturbation. If the perturbation starts from an essentially standard solar model, then it must reduce the central temperature by 20% in a time of less than 10^{6} yrs. Such a large and rapid perturbation is difficult to introduce as a result of a consistent physical model. It has been shown[10] following a suggestion by Fowler[16] that rapid mixing of ^{3}He into the solar core can introduce a sufficiently large perturbation that Davis's results could be easily explained. The essential problem with this idea is the absence of a mechanism to cause the mixing. In particular it is essential for the mixing to occur on a time scale shorter than 5×10^{5} yrs in order to prevent the ^{3}He abundance from coming to an equilibrium value consistent with the local temperature. One proposed mechanism[10] might conceivably cause mixing, but the time scale associated with this process is probably in excess of 10^{7} yrs. The possible pulsational instability involved with this mixing process is driven by ^{3}He burning and hence mixing cannot proceed more rapidly than the average ^{3}He time scale.

Table II Proposed Solar Models

	Hypothesis	References	$\Sigma\sigma\Phi$ (SNU)
A.	Non-perturbed models		
1.	Continuous mixing	5,6	1.5
2.	Low interior abundance of heavy elements abundance	5	1.4
3.	Large core magnetic field	5,7,4	> 7
4.	$G = 6.0 \times 10^{-8}$, $M_\odot = 2.2 \times 10^{33}$	4	4.5
5.	Pure He core	8	3-4
6.	Rapid core rotation	9	~ 1.0
B.	Perturbed models		
1.	$0.75\ M_\odot$ mixed in 5×10^5 yrs	10	0.25
2.	Jerky spin down		?
3.	Rapid core spin up	11	13
4.	End of hydrogen burning	12	10

The second and third models of group B rely on an interaction between rapid rotation and the transport of angular momentum. In model B2 it is supposed that a rapidly rotating core abruptly decreases its rotation rate. As the associated centripetal acceleration decreases, the core must quickly contract. Unfortunately, the consequence of this contraction is a rapid, adiabatic compression which heats the core matter so that the sense of the effect is wrong to resolve the neutrino problem. Model B3 attempts to abruptly cause mixing through an increase in the core rotation rate. According to Sakurai,[17] Eddington-Sweet circulation currents induced by rapid rotation cause angular momentum to be transported into the solar core. As the angular velocity and centripetal acceleration increase, the pressure gradient must drop. The temperature gradient, however, is fixed by the condition that the diffusion of radiation must carry the luminosity of the sun out of the core.

Thus the adiabatic temperature gradient, which is closely coupled to the pressure gradient, is decreased while the radiative gradient is not. Convective instability can result and possibly mix ^{3}He into the solar core. Preliminary investigations[11] have not succeeded in finding this instability.

The final model on the list of hypotheses known to me was suggested to J. N. Bahcall and myself by M. Schwarzschild and is based on the supposition that the sun is now at the end of its main sequence life. During the transition from burning hydrogen in the center to burning hydrogen in a shell source, stellar models sometimes undergo a brief period of gravitational contraction when the nuclear luminosity is less than the total luminosity. Standard evolutionary calculations indicate that the sun does not undergo any rapid readjustment at the end of the main sequence. Figure 3 shows the behavior of the neutrino fluxes when a standard model leaves the main sequence. No major decrease in any of the fluxes is seen and it is apparent that some further modification of the solar model is needed. In contrast more massive stars do undergo a brief period of gravitational contraction when leaving the main sequence. This contraction is caused by the cessation of convection in their cores. A convective core in a solar model could be caused by a non-uniform distribution of heavy elements. I have increased the combined carbon and nitrogen abundance from 0.005 by mass to 0.088 by mass in the inner 0.1 $M_\odot$ of a model of the sun. I took the initial hydrogen abundance by mass to be 0.710 in the core and 0.90 in the region outside the core. The low envelope abundance was needed to achieve the proper luminosity at the end of the main sequence. A small convective core formed initially with a mass of 0.04 $M_\odot$ and

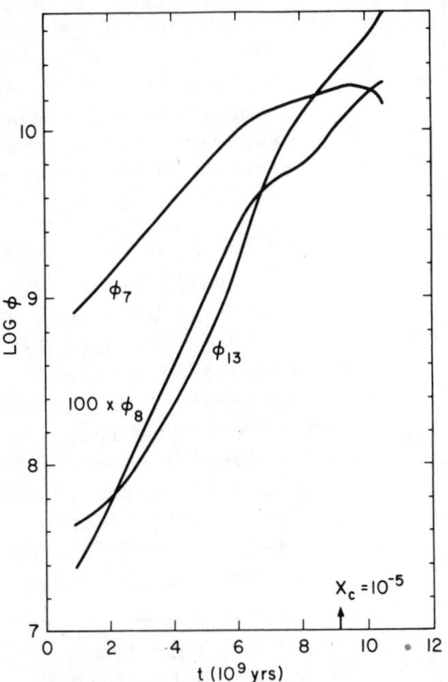

Fig. 3. The behavior of the neutrino fluxes in a normal solar model as hydrogen is exhausted in the center. This model does not have a convective core. The luminosity at the time of core hydrogen exhaustion was substantially greater than the present solar luminosity. The qualitative behavior of the fluxes should not be effected by the luminosity.

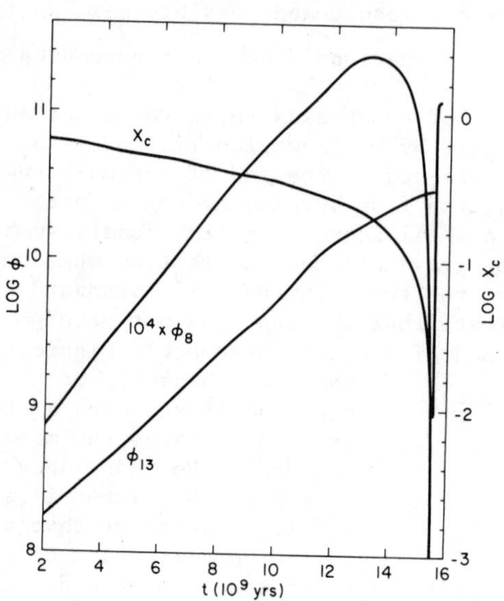

Fig. 4. The behavior of the neutrino fluxes as hydrogen is exhausted in a solar model with a core containing an overabundance of CNO nuclei. Also shown is X_c, the central hydrogen abundance. This model contains a convective core. The luminosity at the time of hydrogen exhaustion was 10% greater than the present solar luminosity. The excessive age of 15.5 billion years could be reduced by starting with $X_c = 0.30$ instead of 0.70.

grew later to 0.08 $M_\odot$. Figure 4 shows the behavior of ϕ_8, ϕ_{13}, and X_c with time. An initial central hydrogen abundance of 0.30 by mass would be required in order for the sun to exhaust its central hydrogen in 4.7×10^9 yrs. Such a low central hydrogen abundance might result from chemical fractionation early in the history of the solar system. Figure 5 shows the behavior of the neutrino fluxes as the sun readjusts the location of the energy producing regions. The minimum in the flux of ^{13}N and ^{15}O neutrinos is very high. At all times the CN cycle produces 1/3 or more of the solar luminosity and the minimum counting rate achieved is about 8 SNU. The difficulty with the model is the length of time required for the convective core to shrink $- 5 \times 10^7$ yrs. During this time the outer parts of the model heat up and continue to provide most of the solar luminosity. Gravitational contraction never supplies more than 10% of the luminosity. Clearly this model is not a possible explanation for Davis' results.

ERROR ANALYSIS OF THE STANDARD MODEL AND THE LOW Z MODEL

The significance of the discrepancy between the results reported by Davis and the theoretical expectations from various solar models cannot be appreciated without an analysis of the errors in the solar model. I will address this discussion primarily to errors

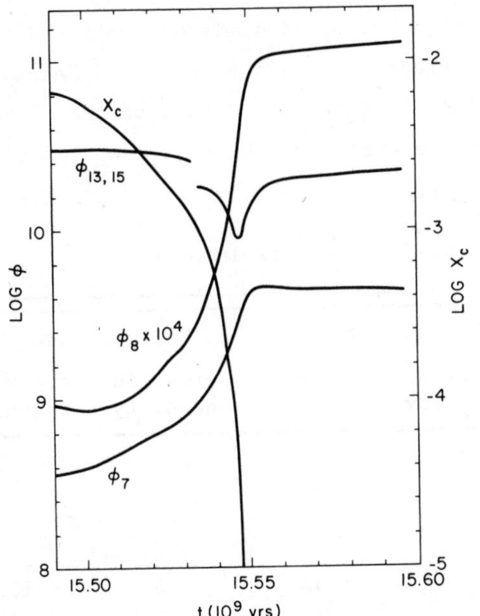

Fig. 5. Details of Fig. 4 at the time of core hydrogen exhaustion.

in the measured parameters given in Section IV of Table I. In addition there may be errors of ommission in the basic physics of Section II of Table I. The size of these latter errors cannot be reliably estimated and I have used the magnitude of the most recent corrections to the various physical relations as a guide. In general the response to the predicted fluxes tends to be uniform in the logarithm of the fluxes. As the more sensitive ^8B flux drops out of $\Sigma\sigma\phi$, the predicted production rate tends to become more reliable. This was the point of the earlier discussion of the benchmark at 1.5 SNU.

The quantities subject to error which have a bearing on the neutrino fluxes are listed in Table III together with an estimate of their standard error. The error of the solar constant was reviewed recently by Labs and Neckel.[18] They concluded that the formal error from twelve indepent measurements is 0.3%; however, they caution that systematic are probably on the order of 1%. The errors in the nuclear cross sections are taken from Kavanagh[19] except for the weak interactions. The 5% error in $S_{1,1}$ is from Bahcall and May[20] plus 1% for the error in the meson exchange correction.[21]

The errors in the opacity are assumed to be a result of five statistically independent errors — each extending over a range of 30 to 40% in temperature. They are mostly a result of complicated physical processes which cannot presently be treated properly. Some of the error for the standard model is also due to the uncertainty in the abundances of important heavy elements such as iron. The errors in the equation of state are assumed to result from two independent errors — one near $T = 8 \times 10^6\ ^oK$ and the second in some other portion of the model. This error is hardest to estimate of any listed. For the equation of state the distribution of the error through the model is crucial; if it is uniformly spread, it has little effect on the counting rate while, if it is concentrated near 8×10^6 °K, it can have a big effect. Finally Bahcall and Sears[22] give an error of 10% in the cross section for capture of the ^8B neutrino by ^{37}Cl. The effects of the errors on the neutrino fluxes were found from the calculations reported by Bahcall et al.[23] and Ulrich.[4]

Table III Error Analysis (all errors in percent)

Quantity	Standard Model		Low Z Model	
	error in quantity	error in $\Sigma\sigma\phi$	error in quantity	error in $\Sigma\sigma\phi$
1. Luminosity	1	6	1	4
2. Nuclear cross sections				
$S_{1,1}$	5	13	5	10
$S_{3,3}$	11	3	11	1.5
$S_{3,4}$	17	14	17	11
$S_{1,7}$	9	7	9	4
"$S_{e,7}$"	10	8	10	5
3. Opacity between T_1 and T_2 (10^6 °K)				
10 — 15	10	13	5	7
7 — 10	15	9	5	3
5 — 7	15	9	5	3
3.5 — 5	15	6	5	3
4. Equation of state				
near $T = 8 \times 10^6$ °K	1	8	1	8
Elsewhere	1	8	1	8
5. ν capture cross sections	10	10	10	10
RMS error in $\Sigma\sigma\phi$		33		24

The ^{37}Ar production rates for the standard model and the low Z model are 5.6 ± 1.8 SNU and 1.4 ± 0.35 SNU, respectively. Taking Davis's result[1] without run 27 to be 0.2 ± 0.8 SNU I find that the discrepancy between the experiment and the standard model to be 2.7σ and while the discrepancy with the low Z model is 1.4σ.

REFERENCES

1. R. Davis, Jr., D. S. Harmer, and K. C. Hoffman, Phys. Rev. Letters 20, 1205 (1968); R. Davis, Jr., Proc. 11th Int. Conf. on Cosmic Rays, (Acta Physica Academy of Science, Budapest, 1970), Hungaricae 29, Suppl. 4, pp. 371-374; R. Davis, Jr., Conf. Rept., The Astrophysical Aspects of the Weak Interactions (Academy of Nazionale Dei Lincie, Cortona, Italy, 1971); R. Davis, Jr., J. C. Evans, V. Radeka, and L. C. Rogers, "Neutrino 1972," Europhysics Conf., ed. A. Frenkel and G. Max (Balatonfured, Hungary, June 1972), Vol. I; R. Davis, Jr. and J. C. Evans, Proc. 13th Int. Cosmic Ray Conf. (Denver, 1973), plus data reported at this conference.
2. J. N. Bahcall, W. F. Huebner, N. H. Magee, Jr., A. L. Merts, and R. K. Ulrich, Astrophys. J. 184, 1 (1973).
3. J. N. Bahcall, N. Cabibbo, and A. Yahil, Phys. Rev. Letters, 28, 316 (1972).
4. R. K. Ulrich, Astrophys. J. 188, 369 (1974).
5. J. N. Bahcall and R. K. Ulrich, Astrophys. J. 170, 593 (1971).
6. D. Ezer and A. G. W. Cameron, Astrophys. Letters 1, 177 (1968); J. N. Bahcall, N. A. Bahcall, and R. K. Ulrich, Astrophys. Letters 2, 91 (1968); G. Shaviv and G. Beaudet, Astrophys. Letters 2, 17 (1968).
7. S. M. Chitre, D. Ezer, and R. Stothers, Astrophys. Letters 14, 37 (1973).
8. A. J. R. Prentice, M. N. R. A. S. 163, 331 (1973); P. Demarque, J. G. Mengel, and A. V. Sweigart, M. N. R. A. S. 165, 19P (1973).
9. R. K. Ulrich, Astrophys. J. 158, 427 (1969); P. Demarque, J. G. Mengel, and A. V. Sweigart, Astrophys. J. 183, 997 (1973), Nature Phys. Sci. 246, 33 (1973); J. W. Roxburgh, Nature 248, 209 (1974).
10. J. W. W. Dilke and D. O. Gough, Nature 240, 262 (1972); R. T. Rood, Nature 240, 178 (1972); D. Ezer and A. G. W. Cameron, Nature 240, 180 (1972); R. K. Ulrich and R. T. Rood, Nature, 241, 111 (1973).
11. R. T. Rood and R. K. Ulrich, in preparation.
12. J. N. Bahcall and R. K. Ulrich, in preparation.
13. R. H. Dicke and M. Goldenberg, Phys. Rev. Letters 18, 313 (1967).
14. A. P. Ingersoll and E. A. Speigel, Astrophys. J. 163, 375 (1971); G. A. Chapman and A. P. Ingersoll, Astrophys. J. 175, 819 (1972), 183, 1005 (1973); R. H. Dicke, Astrophys. J. 180, 293 (1973).
15. H. L. Johnson, Astrophys. J. 116, 640 (1952).
16. W. A. Fowler, Nature 238, 24 (1972).
17. T. Sakurai, Publ. Astron. Soc. Japan 24, 153 (1972).

18. D. Labs and H. Neckel, Solar Phys. 19, 3 (1971).
19. R. W. Kavanagh, Cosmology, Fusion and Other Matters, ed. F. Reines (University of Colorado Press, Boulder, 1972) p. 169.
20. J. N. Bahcall and R. M. May, Astrophys. J. 152, L17 (1968), ibid 155, 501 (1969).
21. M. Gari and A. H. Huffman, Astrophys. J. 174, L151 (1972), ibid.
22. J. N. Bahcall and R. Sears, Ann. Rev. Astron. and Astrophys. 10, 25 (1972).
23. J. N. Bahcall, N. A. Bahcall, and R. K. Ulrich, Astrophys. J. 156, 559 (1969).

GALACTIC NEUTRINO SOURCES AND COSMIC RAYS

W.S. Pallister and A.W. Wolfendale
Physics Department, University of Durham, Durham, England

ABSTRACT

The scope of the present paper is three-fold: to review the status of the detection of cosmic ray muon neutrinos of atmospheric origin, to make an estimate of the cosmic ray background in the Brookhaven Solar Neutrino Experiment and to briefly examine the possibility of detecting neutrinos from supernovae and collapsing stars in the Galaxy.

INTRODUCTION

The neutrinos present at the earth's surface can be divided into two groups; those produced in man-made apparatus (reactors, bombs, accelerators) or generated by radioactive elements in the earth, and those incident on the earth from above, either as extra-terrestrial neutrinos or as secondaries produced in the earth's atmosphere by extra-terrestrial energetic particles i.e. cosmic rays. It is with the second group that this paper is concerned.

Figure 1 gives a schematic view of neutrinos in this group and Figure 2 gives the energy spectra of the major components. Although the electron neutrinos of solar origin have not yet been unambiguously detected presumably they do exist and they will dominate the energy region below 10 MeV or so. At higher energies, knowledge of the various cosmic ray components in the atmosphere, principally the muon component, allows fairly precise energy spectra of ν_μ ($\bar{\nu}_\mu$) and ν_e ($\bar{\nu}_e$) to be predicted and, as is well known (see the next section) the interactions of the so-called atmospheric muon neutrinos have been detected.

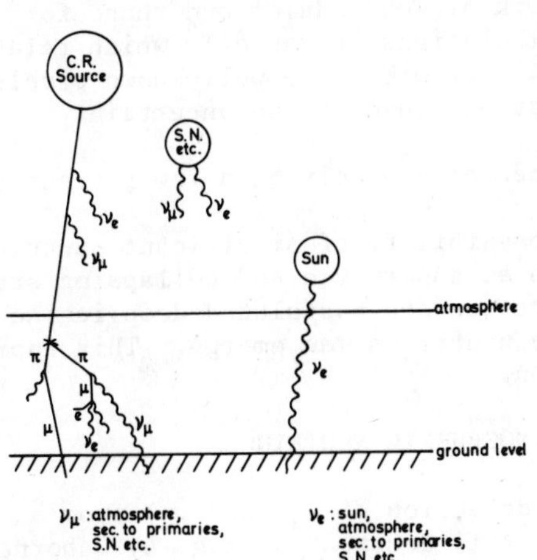

Fig. 1. Cosmic ray neutrinos

As indicated in Figure 1 there should also be a contribution to the flux of ν_μ and ν_e from secondary interactions produced by the primary cosmic rays as they pass through the interstellar medium. Many measurements indicate that at some tens of GeV primary cosmic ray energy the average path length between source and earth is ~ 5 g cm^{-2} (see Shapiro[3] for a recent summary) and what evidence there is suggests that the path length may fall somewhat with increasing energy. Insofar as the attenuation length of the primary particles in the atmosphere is $\sim$ 120 g cm^{-2} the contribution from the

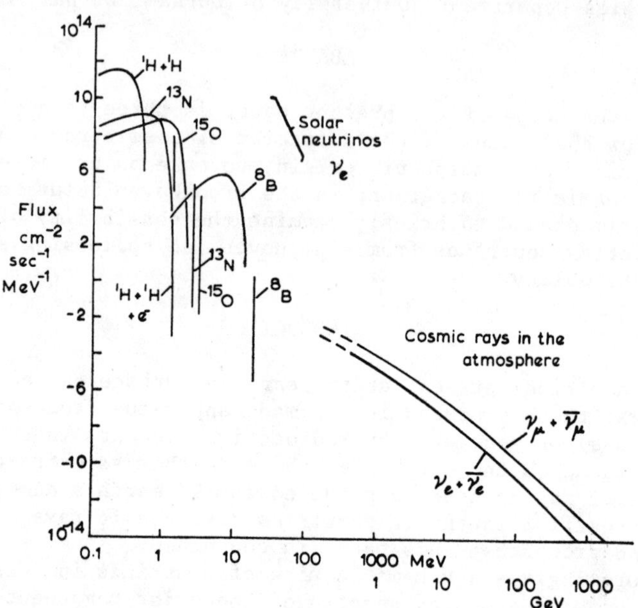

Fig. 2. Energy spectra of cosmic ray and solar neutrinos. The cosmic ray fluxes are from the work of Wolfendale[1] and those for solar neutrinos come from the calculations of Bahcall[2] which relate to a solar core temperature of 14.9 x 10^6K. The well-known problems with the solar model cause the latter fluxes to be uncertain.

interstellar interactions will accordingly only be a few percent of the 'atmospheric neutrino flux'.

Of greater interest is the possibility of significant neutrino fluxes from discrete objects such as supernovae and collapsing stars in general. Here, energetic processes are postulated deep in the stellar interiors from which only neutrinos can emerge. This topic is considered in the final section.

INTERACTIONS OF ATMOSPHERIC NEUTRINOS

The situation regarding the detection of ν_μ (and $\bar{\nu}_\mu$) from atmospheric interactions has been well documented (e.g. by Osborne et al.[4]) and there have been no very recent developments. The situation is summarised in Figure 3. This shows the weighted mean frequency of detected neutrino-induced muons from the experiments in the Kolar Gold Fields[5] (denoted K.G.F.) and the East Rand Proprietry Mine[6] (denoted C.W.I.) compared with expectation. The expectations were calculated assuming a linear increase in total neutrino-nucleon cross-section of magnitude σ_{tot} = (0.8 ± 0.2) $E_\nu . 10^{-38}$ cm^2 GeV^{-1} nucleon^{-1} (and one third this value for the antineutrino cross-section) up to a plateau value above some energy E_o. The form of the relation for the cross-section came from the CERN measurements of Budagov et al.[7]; more recent measurements, extending to higher energies, have been made at N.A.L. and these indicate $\sigma_{tot} \simeq 0.7\ E_\nu\ 10^{-38}\ cm^2\ GeV^{-1}$ $nucleon^{-1}$ and $E_o \gtrsim 100$ GeV. Uncertainties in the various contributory factors: cosmic ray neutrino spectrum fraction of energy taken by the muon, and coefficient of energy in the cross-section relation mean that the (already) small difference in Figure 3 between the weighted mean muon frequency and prediction (for $E_o \gtrsim 100$ GeV, and even $E_o \to \infty$) is not significant. What can be said is that the cosmic ray data give no evidence for a component of neutrinos of extra-terrestrial origin over and above the contribution from the atmosphere. Specifically, one can say that an extra-terrestrial component of muon neutrinos (ν_μ and $\bar{\nu}_\mu$) has a flux of less than about 1/3 of that indicated in Figure 2 (typically, $\lesssim 10^{-7}\ cm^{-2}\ s^{-1}\ MeV^{-1}$ at 10 GeV).

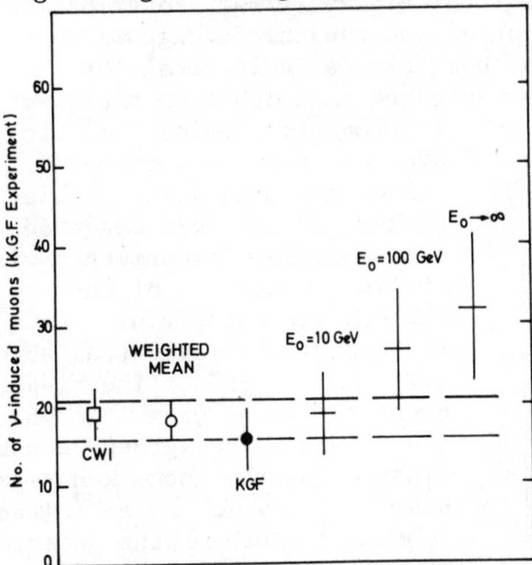

Fig. 3. Comparison of observed and expected frequencies of ν-induced cosmic ray muons. See the text for details.

BACKGROUND IN THE SOLAR NEUTRINO EXPERIMENT

The most serious attempt so far to detect extra-terrestrial neutrinos is the Brookhaven Solar Neutrino Experiment[8] (abbreviated to B.S.N.E.). When the experiment was designed the expected rate of detected 'events' (radioactive ^{37}A atoms) was much higher than the predicted background at the chosen location for the experiment (Homestake Mine, South Dakota). However, the measured rate of production of ^{37}A atoms has turned out to be so low that the background poses a severe problem. This is the reason for examining the matter in some detail.

The interaction sought is $^{37}Cl\ (\nu_e, e^-)\ ^{37}A$ but ^{37}A atoms can also be produced by $^{37}Cl\ (p,n)\ ^{37}A$ and $^{37}Cl(\nu_\mu,\mu^-)\ ^{37}A$ and the ν_e interaction, where the ν are of cosmic ray origin. The first mentioned process is the most important. The protons underground arise from a number of sources:

recoils from neutrons, produced in turn by α-particles,
'stars' produced on absorption by nuclei of negative muons, and
'stars' produced by the nuclear interactions of fast muons (some protons come from the stars directly and others are recoils from neutrons produced in the stars).

The α-particles from the rock walls have been reduced to negligible proportions by flooding the volume surrounding the underground detector[9] thus leaving the muon stars (and the cosmic ray neutrino contribution).

The relative flux of stopping muons falls rapidly with depth and the resulting protons only have significance for depths shallower than about 50 hg cm^{-2}; at the depth of B.S.N.E. their contribution is quite negligible. The calculation of the expected background as a function of depth underground has been discussed in detail by Wolfendale et al.[10] and Figure 4 reproduces a graph from their work. The important points to note are that there is very good agreement between the predicted background and that observed (in subsidiary experiments with smaller quantities of the B.S.N.E. detector fluid: C_2Cl_4) for depths shallower than about 1.2 x 10^3 hg cm^{-2}. The biggest question is the extent to which the predicted background can be trusted for the unchecked region down to 4.4 x 10^3 hg cm^{-2} (the 'nominal' depth of the detector). There are essentially two components in the calculation of the depth-dependence of the background at these depths; the variation of muon intensity and the variation of muon interaction cross-section. The muon intensity is known rather accurately at the depths in question (accuracy ∿ ±10%) and the difference in cross-section between 1.2 x 10^3 hg cm^{-2} and 4.4 x 10^3 hg cm^{-2} is not large. To first approximation, the yield of protons (per incident muon) increases as the mean muon energy $\bar{E}_\mu$ to the power 0.7 (reference [10]) and $\bar{E}_\mu$ varies only from 170 GeV to 300 GeV over the relevant depth range; little additional error is thus expected in the proton yield.

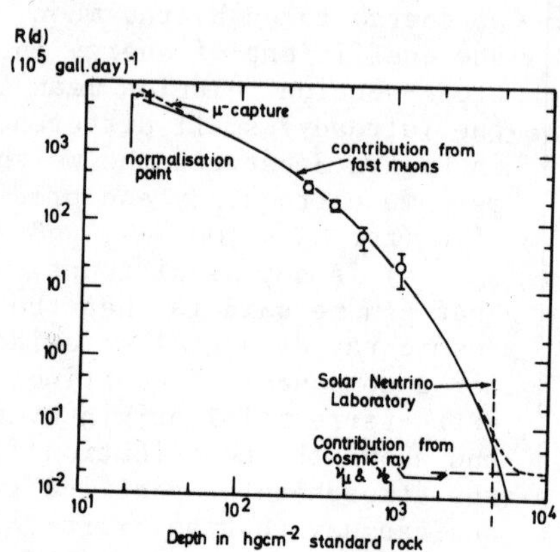

Fig. 4. Predicted background in B.S.N.E.[10] due to various interactions. R(d) is the rate of production of ^{37}A atoms in the full-scale tank.

The derivation of the contribution to the background from cosmic ray neutrino interactions is more difficult [10] but at the B.S.N.E. depth the contribution is not large so that the fractional error in the total background estimate is not serious.

Before making a final estimate of the total background it is necessary to determine the actual depth of the experimental location. The nominal depth is 4.4 x 10^3 hg cm^{-2}; calculations by Wolfendale and Young[12] using details of the rock strata, however, gave a somewhat greater depth of 4.62 x 10^3 hg cm^{-2} standard rock (i.e. corrected to $\frac{Z}{A}$ = 0.5, $\frac{Z^2}{A}$ = 5.5). More recently, the apparatus of Lande et al.[13] used for the anti-neutrino search, and operated at the same location, has been used to give a vertical muon intensity and thus an effective depth using a best-fit depth-intensity relation.

Although the data are still preliminary and a variety of small corrections have yet to be applied they do add somewhat to the precision of depth location. The corresponding depth is now 4.56×10^3 hg cm^{-2}.

Allowing for the various uncertainties the expected background for the B.S.N.E. depth is $(8.0 \pm 2.4)\ 10^{-2}$ ^{37}A atoms day^{-1}.

Although the contribution of cosmic ray neutrinos to the background at the B.S.N.E. location is small, it would of course be more important at greater depths. In view of the difficulty in calculating its magnitude it is rather necessary that independent estimates be made before any new solar neutrino experiment project is mounted at a greater depth.

SIGNIFICANCE OF THE B.S.N.E. RESULTS

Although the physical significance of the solar neutrino data in terms of relevance to solar models is beyond the scope of the present article (it is treated in detail elsewhere [14]) some remarks about the relevance of the measured rates of ^{37}A production with respect to the expected background and with respect to one another are in order. Figure 5 shows the results reported by Davis and Evans[9] at the Denver Conference. Also indicated are the calculated overall mean rate with and without the 'exceptional' Run 27.

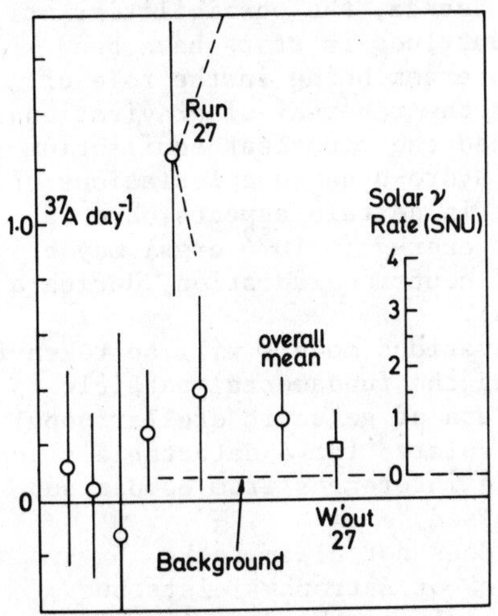

Fig. 5. Measured rates in the B.S.N.E. (Note added later: two later runs[15] give ^{37}Ar production rates close to the background and lend further support to the possibility of run 27 being anomalous).

The high count in Run 27 has caused considerable interest and speculation. There are three possibilities for its explanation:

(i) that it represents an upward fluctuation in the muon background.

(ii) that it is merely a normal fluctuation in the solar neutrino count and background together.

(iii) that it is due to a S.N. or collapsing star.

The first possibility has been examined in some detail. It might be thought that the local interaction of a very energetic muon could produce a large number of ^{37}Ar atoms and thus a large signal and that in view of the correlation of the 'counts' the fluctuations might be non-Poissonian. Our calculations do not support this view and there is no evidence that the background should deviate from the normal Poisson statistics.

The difficulty with examining the second possibility is the problem that the mean solar neutrino rate is not known. If it is the overall mean of the six measured rates (less the muon background) including Run 27, then this Run does not stand out as a significant peak: the goodness of fit for the whole set is 25%. This explanation is obviously the most conservative one.

If, on the other hand, the solar neutrino rate is assumed to be much less than the background then the rates of Table 3 must be compared with the predicted muon background of 0.08 ^{37}Ar atoms day^{-1}. Excluding Run 27 the goodness of fit gives 80%, including it gives $\sim$ 15%. Thus, in these circumstances there is the suggestion of an excess.

To reiterate, the most likely explanation is that Run 27 represents a normal statistical excursion, in which case the overall mean rate has a one standard deviation limit of $\sim$ 0.35^{37}A day^{-1} above background (and problems with solar models are perhaps not too severe). However, the possibility of it being anomalous cannot be ruled out; in parenthesis it can be remarked that a rather detailed analysis of the precision of the derived rate seems desirable - the errors indicated may possibly be an overestimate.

GALACTIC NEUTRINO SOURCES

1. Introduction. During the past decade, the possibilities of energy transport and deposition by neutrinos in stars have been explored by many authors, the main interest being in the role of neutrinos as a possible mechanism for the reversal of gravitational implosion in stars which have concluded their nuclear equilibrium evolution. The conclusions of these hydro-dynamic calculations of stellar collapse, although discordant in certain aspects, are in agreement that a very large amount of energy ($\sim 10^{52}$ ergs) may be expelled from the star in the form of neutrino radiation, during a comparitively short time ($\sim 10^{-2}$ secs).

In the present work, results of various models will be taken in conjunction with hypotheses concerning the fundamental particle properties of the neutrino, and the fate of galactic stellar populations, and detection probabilities calculated for a detectable signal using the B.S.N.E. detector. Possible inferences from comparison with the published data will be discussed.

This part of the analysis given does not claim to be authoratitive, in that the authors are not Astrophysicists but cosmic ray physicists with experience only in the detection of atmospheric neutrinos and the background effects which bedevil attempts at detecting extra-terrestrial neutrinos and which have been referred to already. The treatment is similar in some respects to that given by Domogatsky and Zatsepin[16] and Chudakov et al.[17].

The form of the review is to follow a discussion of supernovae and collapsing stars in general as neutrino sources with the effect of possible neutrino oscillations.

2. The Neutrino Detection Probability. As mentioned earlier neutrino detection in the B.S.N.E. is based upon the $^{37}Cl(\nu_e, e^-)^{37}A$ reaction (threshold 0.8 MeV), for which the cross-section is of order 10^{-42} cm^2, and the subsequent decay of the ^{37}A with a 35.1 day half life.

If a neutrino burst occurs at a distance 'r' from the detector, then the observed number of counts is given by:

$$N = \frac{n\,\sigma\,N_\nu}{4\pi r^2}$$

where N_ν is the number of neutrinos expelled isotropically from the star, n the number of ^{37}Cl atoms, and σ is the effective reaction cross-section. For a detectable signal of at least N counts, the sensitive radius within which the collapse must occur is given by:

$$r_s = \left(\frac{n\,\sigma\,N_\nu}{4\pi N}\right)^{\frac{1}{2}}$$

and if the distribution of sources in the galaxy is approximated as being uniform, the probability of an explosion within the sensitive volume (assuming the galaxy to be a disc, of radius r_g) in which a minimum of N counts are detected is therefore $\Phi_1 = r_s^2/r_g^2$. The probability of the genesis of a neutrino burst anywhere in the galaxy during a particular detection run,

$$\Phi_2 = \frac{\text{exposure time}}{\text{Average time between bursts}}$$

The probability of any run giving a minimum signal of N counts is thus $P = \Phi_1\Phi_2$.

3. Supernovae as Sources. If the neutrino bursts are associated only with supernova explosions, a good estimate of Φ_2 is:

$$\Phi_2 = \frac{100\text{ days}}{26\text{ years}} \simeq 0.01$$

for an exposure time of 100 days using the frequency of supernovae (1 per 26 ± 10 years) estimated by Tammann[18] from a comprehensive survey of observed supernova events as a function of the type of galaxy.

The results of various hydrodynamic calculations on neutrino emission for different mass stars have been taken, and values of Φ_1 obtained, from which an estimate of the detection probability for a minimum of 50 counts per run has been calculated. The results are shown in Table 1. It should be stressed that the calculations adopted were made before the 'discovery' of the neutral current interaction. Inclusion of this interaction will presumably cause a reduction in the predicted neutrino fluxes.

4. Allowance for Neutrino Energy Spectrum and Supernovae Distribution in the Galaxy. The cross section for the reaction $^{37}Cl(\nu_e, e^-)A^{37}$ is very sensitive to the energy of the neutrino, due to the presence of superallowed transitions, and the dependence is approximately of the form $\sigma_\nu(E_\nu) \propto E_\nu^{3.5}$ in the energy range $0.9 < E_\nu < 100$ MeV (Bahcall[22]).

The form of the neutrino spectrum is clearly of importance if accurate detection probabilities are to be calculated, when the former expression for the sensitive radius must be replaced by:

Table I Basic data for neutrinos from Supernovae and Collapsing Stars

(r^{50} is the radial distance from the sun from which SN will give more than 50 detected neutrinos. P_{50} and P'_{50} are, respectively, the probabilities of detecting more than 50 neutrinos within 100 days from SN and Collapsing Stars. The P_{50} values are calculated under simplifying assumptions; more accurate calculations raise them by ~ 2 - see the text).

Author	Star Mass (M_o)	Energy loss by $\nu_e + \bar{\nu}_e$ (ergs)	Average energy of ν (MeV)	Number of ν emitted N_ν	σ_ν (cm^2)	r_{50} (kpc)	S.N. P_{50}	Collapsing Stars P'_{50}
Ivanova et al.[19]	10	0.8×10^{52}	18	2.8×10^{56}	1.5×10^{-41}	1.2	6×10^{-5}	6×10^{-3}
Arnett[20]	2	1.6×10^{52}	18	5.6×10^{56}	1.5×10^{-41}	1.7	1×10^{-4}	1×10^{-2}
	4	2.0×10^{52}	18	7.0×10^{56}	1.5×10^{-41}	1.9	2×10^{-4}	2×10^{-2}
	8	2.5×10^{52}	18	8.8×10^{56}	1.5×10^{-41}	2.1	2×10^{-4}	2×10^{-2}
	32	2.7×10^{52}	18	9.5×10^{56}	1.5×10^{-41}	2.2	2×10^{-4}	2×10^{-2}
Wilson[21]	1.25	5.6×10^{51}	9	3.9×10^{56}	1×10^{-42}	0.4	7×10^{-6}	7×10^{-4}
	2.13	1.5×10^{52}	10	9.4×10^{56}	2×10^{-42}	0.8	3×10^{-5}	3×10^{-3}

$$r_s = \left(\frac{n}{4\pi N} \int_{0.81\ \mathrm{MeV}}^{E_{max}} N_\nu (E_\nu)\ \sigma\ (E_\nu)\ dE_\nu \right)^{\frac{1}{2}}$$

Taking the form of the spectrum given by Wilson[21] for a star of 2.13 M_o and shown in Figure 6 the integration has been carried out numerically with the result that the r_{50} increases by a factor 1.9.

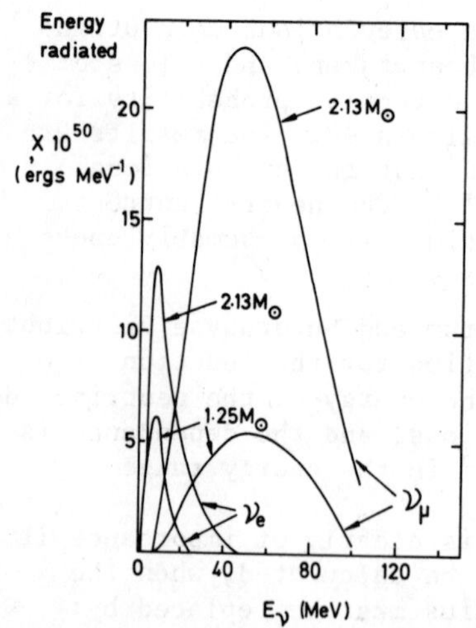

Fig. 6. Neutrino Energy Spectra from the calculations of Wilson[20] for stars of two masses.

The effect of non-uniformity of the distribution of supernovae on the detection probability has been estimated using the surface density of supernovae remnants as a function of radial distance from the centre of the galaxy estimated by Ilovaisky and Lequeux[23]. This radial distribution has been normalised appropriately and used to derived new values of Φ_1. The corrected values of Φ_1 are about one half of those used earlier. The nett result of applying corrections to allow for the neutrino energy spectrum and the distribution of SN in the Galaxy is to raise the values of P_{50} given in Table 1 by $\sim$ 2.

5. Collapsing Stars. Although the hypothesis of neutrino deposition in stars was conceived as a mechanism for supernovae, further calculations indicate that only in a minority of cases (stars $\sim 2\ M_o$) is collapse of the star and neutrino burst followed by any explosion or ejection of matter and thus identification as a supernovae. If these results are correct, then the frequency of ν-bursts will be much higher.

Assuming that, as an upper limit, all stars with mass $> 1.5\ M_o$ must collapse when their supply of nuclear fuel has been expended, and that there is no continuous process by which the star may reduce its mass below the above limit during the quasi-equilibrium period, the problem is to estimate the number of stars of mass $\gtrsim 1.5\ M_o$ ending their nuclear evolution each year. Using astronomical data (Zeldovich and Novikov[24]) to determine the number of stars per unit mass range, and knowing the equilibrium evolution time as a function of mass,

$$t \sim 10^{10}\left(\frac{M_o}{M}\right)^2 \text{ years}$$

the birth rate and hence death rate, (if the stellar population is believed stationary), for all stars of mass $> 1.5\ M_o$ is estimated as ≈ 4 per year. Φ_2 is increased by a factor of 10^2 and the new detection probabilities are thus as shown in Table I.

6. Neutrino Oscillations. If the rest mass of the neutrino is different from zero, and there is a characteristic mass difference between the electron and muon neutrinos $\nu_e \rightleftharpoons \nu_\mu$ oscillations may exist with an oscillation length which is given by:

$$\ell = \frac{E}{m\Delta m} \quad \text{(Pontecorvo}^{25}\text{)}$$

Providing the oscillation length is much less than the galactic radius, approximately half of the sterile muon neutrinos will be detectable using the ^{37}Cl reaction. The oscillation length must be $\gtrsim 10^7$ cm for the neutrino to escape from the star before the 'lepton flip' occurs otherwise the high energy electron neutrino produced will be absorbed. This sets the limits of:

$$10^{-21}\ \text{eV}^2 < m\Delta m < 10^{-6}\ \text{eV}^2$$

for possible detection.

The detection probabilities have been calculated, assuming half the muon neutrinos are detectable, in conjunction with and independent of the collapsing star hypothesis and the results are shown in Table 2.

Some doubt has been thrown[16] on the validity of the results of Arnett considering the muon neutrino as an instantaneous energy sink and neglecting opacity due to the reactions:

$$\nu_\mu + \mu \rightarrow e + \nu_e$$

$$\nu_\mu + \mu \rightarrow \nu_\mu + \mu$$

which have no threshold, in contrast to the reaction:

$$\nu_\mu + n \rightarrow p + \mu^-$$

where the threshold is $m_\mu c^2$. The P-values are thus to be regarded as upper limits.

Table II Basic data for neutrinos from Supernovae and Collapsing Stars

Author	Star Mass M_o	Energy loss $\nu_\mu + \bar{\nu}_\mu$ (ergs)	Average energy of ν_μ (MeV)	Number of ν_μ emitted N_ν	$\bar{\sigma}_{\nu_e}$ (cm^2)	r_{50} (kpc)	S.N. P_{50}	Collapsing Stars P'_{50}
Arnett[20]	2	2.6×10^{51}	35	4.7×10^{55}	1.5×10^{-40}	1.0	4.5×10^{-5}	4.5×10^{-3}
	4	7.6×10^{51}	35	1.4×10^{56}	1.5×10^{-40}	1.8	1.4×10^{-4}	1.4×10^{-2}
	8	5.6×10^{53}	35	1.0×10^{58}	1.5×10^{-40}	15.3	5×10^{-3}	0.5
	32	7.6×10^{54}	35	1.4×10^{59}	1.5×10^{-40}	55	1.0×10^{-2}	1.0
Wilson[21]	1.25	3.0×10^{52}	45	4.2×10^{56}	3.5×10^{-40}	4.8	10^{-3}	0.1
	2.13	1.2×10^{53}	45	1.7×10^{57}	3.5×10^{-40}	9.6	2.5×10^{-3}	0.25

GALACTIC NEUTRINOS - CONCLUSIONS

It has been shown (Table I) that from Supernovae alone a detectable flux of SN would only be expected in the B.S.N.E. once every 10^4 runs or so. If the majority of stars of more than 1.5 M_o collapse and if the models are correct then this frequency would increase by about one hundred. Even higher frequencies appear for collapsing stars if neutrino flip is allowed and a rate of one in ten runs would be possible. If Run 27 were to be definitely attributable to a neutrino burst then the last mentioned situation would appear likely; however, at the present stage it is of course premature to draw this conclusion.

What is clear is that continued operation of the B.S.N.E. project is highly desirable and, in view of developments in neutrino interaction theory further calculations of expected neutrino fluxes are needed.

ACKNOWLEDGEMENTS

The authors are grateful to the Science Research Council for continued support. Drs. R. Davis, J.C. Evans, J.L. Osborne and E.C.M. Young are thanked for helpful discussions.

REFERENCES

1. A.W. Wolfendale, Essays in Physics (Academic Press, London and N.Y.), 2, 131 (1970).
2. J.N. Bahcall, N.A. Bahcall and G. Shaviv, Phys. Rev. Lett., 20, 1209, (1968).
3. M.M. Shapiro, Proc. Roy. Soc. (in the press) (1974).
4. J.L. Osborne, A.W. Wolfendale and E.C.M. Young, Neutrino '72, Balaton-füred (Omkdk-Technoinform), 2, 77.
5. M.R. Khrishnaswamy et al., Proc. Roy. Soc., A323, 489, (1971).
6. F. Reines et al., Phys. Rev., D4, 80, (1971).
7. I. Budagov et al., Lett. Nuovo Cim., 2, 689, (1969); Phys. Lett., 29B, 524, (1969); Phys. Lett., 30B, 364, (1969).
8. R. Davis, D.S. Harmer and K.C. Hoffman, Phys. Rev. Lett., 20, 120J (1968).
9. R. Davis and J.C. Evans, Proc. 13th Int. Conf. on Cosmic Rays, Denver, 3, 2001 (1973), and modified version.
10. A.W. Wolfendale, E.C.M. Young and R. Davis, Nature, 238, 87, 130 (1972).
11. R. Craig, J.L. Osborne, A.W. Wolfendale and E.C.M. Young, J. Phys. A., 1, 61 (1968).
12. A.W. Wolfendale and E.C.M. Young, unpublished (1972).
13. K. Lande, G. Bozoki, W. Frati, C.K. Lee and E. Fenyves, Proc. 13th Int. Conf. on Cosmic Rays, Denver, 3, 2000, (1973).
14. R. Ulrich, these Proceedings (1974).
15. R. Davis and J.C. Evans, private communication, (1974).
16. G.V. Domogatzky and G.T. Zatsepin, Proc. Internat. Sem. on Neutrino Physics and Astrophysics, Moscow, 2, 198, (1969).
17. A.E. Chudakov, O.G. Ryajskaya and G.T. Zatsepin, Proc. 13th Int. Conf. on Cosmic Rays, Denver, 3, 2007, (1973).
18. G.A. Tammann, Astron. and Astrophys., 8, 458, (1970).
19. L.N. Ivanova et al., Proc. Internat. Sem. on Neutrino Physics and Astrophysics, Moscow, 2, 180, (1969).
20. W.D. Arnett, Canad. J. Phys., 45, 1621, (1967).
21. J.R. Wilson, Ap. Journal, 163, 209, (1971).
22. J.N. Bahcall and R. Davis, Phys. Rev. Lett., 26. 662 (1971).
23. A.S. Ilovaisky and J. Lequeux, Astron. and Astrophys., 18, 169, (1972).
24. Y.B. Zeldovich and I.D. Novikov, Relativistic Astrophysics, (Univ. Chicago Press), 1, 465, (1971).
25. B.M. Pontecorvo, Proc. Internat. Sem. on Neutrino Physics and Astrophysics, Moscow, 2, 169, (1969).

NEUTRINO COSMOLOGY+

Sidney A. Bludman
Department of Physics, University of Pennsylvania
Philadelphia, Pa. 19174

ABSTRACT

Cosmological data is reviewed suggesting that the universe may be open and dominated by neutrinos and gravitons rather than by baryons. The thermal history of the Lepton Era is investigated incorporating the effects of neutral currents, additional neutrinos, and a small neutrino mass. In the canonical version of Big Bang cosmology (equal numbers of neutrinos and antineutrinos), the neutrino number and energy density is, like that of photons, gravitationally insignificant unless the neutrino has a small mass (~10 ev).

The neutrino sea can be cosmologically significant if it is degenerate (so that the net leptonic or muonic charge is non-zero) with $\sim 7 \times 10^5$ neutrinos (or antineutrinos) per cc. This density homogeneously spread out is still so low that even the most energetic cosmic ray protons will not be stopped, even if neutral currents exist with the usual weak strength. If these degenerate neutrinos have a small mass ~0.5 ev, they will condense into degenerate neutrino superstars of the size and mass of galactic clusters.

If neutral currents make the $(e\nu)(e\nu)$ coupling 5 x greater than what it is in V-A theory, nucleosynthesis commences a little earlier than conventionally assumed. This increases the cosmological He^4 abundance predicted only slightly from Y = 0.27 to Y = 0.29.

An appendix reviews the effect of neutral currents on neutrino processes in stars.

There are excellent reviews of the subject of cosmological neutrinos by de Graaf two years ago at Balaton [1] and two years earlier at Cortina.[2] I will therefore confine myself to what appear to be the significant changes in the situation in the last two years: (1) New values for the Hubble constant and the deceleration parameter; (2) I will review the implications of these changes for the energy density of neutrinos and for the mass limit of neutrinos; (3) I will speak about the impact of neutral currents on these conclusions. While this completes the subject matter of neutrino cosmology, the Appendix contains some brief remarks about the way neutral weak currents effect neutrino processes in stars.

I. BIG BANG COSMOLOGY

In this section we review those aspects of cosmology that are

+Supported in part by the U.S.A.E.C.

necessary to understand the Lepton Era of the universe. The presently observable data to be explained by an evolutionary cosmology are:

(1) The darkness of the night sky and Hubble's Law $cz = H_o d$ relating galaxy red-shift z to distance d.

(2) The microwave background radiation which is apparently black-body at temperature $T_{\gamma o} = 2.7K$.

(3) The abundance of elements, in particular the large (27%) abundance of He. If the galactic luminosity has always been $L/M \approx (L/M)_{\odot} = 0.2$ ergs $sec^{-1}gm^{-1}$, this amount of helium cannot have been produced in stars but calls for the universe to have once been much hotter (10^{10} K) than it now is.

1. EINSTEIN EQUATIONS AND ADIABATIC EXPANSION

These facts are to be explained by some Friedman cosmology in which the hot universe is homogeneous, isotropic and expanding at a decelerating rate. In Robertson-Walker coordinates, Einstein's field equations read

$$(\dot{R}/R)^2 + kc^2/R^2 = 8\pi G\rho/3 \tag{1.1}$$

$$2\ddot{R}/R + (\dot{R}/R)^2 + kc^2/R^2 = -8\pi Gp/c^2 \tag{1.2}$$

where p and ρ are the pressure and mass density and k = 1, 0, -1 according as spatial geometry is elliptical, flat, or hyperbolic. In the elliptical case, the universe is closed with volume $V = 2\pi^2R^3$. In the flat case, $V = 4\pi R^3/3$ and static solutions are possible.

These field equations obey the Bianchi identity

$$d(\rho c^2R^3) + pdR^3 = 0 ,$$

which can be interpreted as the conservation of energy in an adiabatic expansion. Note that the energy $E \sim \rho c^2R^3$ is constant only for a matter-dominated universe in which the pressure is p = 0.

At the present epoch $t = t_o$, (1.1) and (1.2) read

$$H_o^{\,2} + \frac{kc^2}{R_o^{\,2}} = \frac{8\pi G}{3}\,\rho_o$$

$$H_o^{\,2}(1-2q_o) = -\frac{8\pi G}{3}\,p_o ,$$

where $H_o \equiv (\dot{R}/R)_0$, $q_o \equiv (-R\ddot{R}/\dot{R}^2)_0 > 0$ are the present values of the Hubble constant and the deceleration parameter. In units of $\rho_{CR} \equiv 3\,H_o^{\,2}/8\pi G$, the density is $\Omega = \rho/\rho_{CR}$ so that at present $1 - \Omega_o = -k(c/R_oH_o)^2$ and $\Omega_o > 1, =1, <1$ according as k = 1, 0, -1. The critical density $\rho = \rho_{CR}$ is that density necessary for closing the universe (k=1).

The present size R_o of the universe is fixed by Ω_o

$$(R_o H_o/c)^2 = 1/|1-\Omega_o|, \ k \neq 0 \ . \tag{1.3}$$

Two limiting cases are distinguished

(i) Matter-dominated Universe, $p = 0$

$$\rho R^3 = \text{const} \quad , \qquad q_o = \frac{1}{2}\Omega_o \ ; \tag{1.4}$$

(ii) Radiation-dominated Universe, $p = \frac{1}{3}\rho c^2$

$$\rho R^4 = \text{const} \equiv Q, \ q_o = \Omega_o \ . \tag{1.5}$$

Unless it is full of gravitons and neutrinos so far undetected, the universe is at present and has been as far back as z = 100 matter-dominated. The early stages $R \ll cH_o^{-1}$ of the universe are, in any case, radiation-dominated with

$$R^2 = (\frac{32\ \pi G Q}{3})^{1/2} t \tag{1.6}$$

$$\rho = (3/32\pi G)\ t^{-2} = 4.5 \times 10^5\ t^{-2}\ \text{gm})\text{sec.} \tag{1.7}$$

2. INTEGRATION TO OBTAIN PRESENT AGE

The field equation (1.1) can be integrated

$$t(R) = \int_0^R \frac{dR}{\sqrt{\frac{8\pi G\rho R^2}{3} - kc^2}}$$

to give for the present age in dimensionless units

$$H_o t_o = \int_0^{(|1-\Omega_o|)^{-1/2}} \frac{dx}{\sqrt{\Omega(x)\,x^2 - k}} \ , \ x = H_o R/c \tag{1.8}$$

In the two limiting cases

(i) Matter universe

$$t_o H_o = \begin{cases} q_o(1-2q_o)^{-3/2}\ (\sinh\psi_o - \psi_o)\ , & \Omega_o = 2q_o < 1 \\ q_o(2q_o-1)^{-3/2}\ (\theta_o - \sin\theta_o)\ , & \Omega_o = 2q_o > 1 \end{cases}$$

where $\cosh\psi_o \equiv q_o^{-1} - 1 \equiv \cos\theta_o$.

(ii) Radiation universe

$$t_o H_o = (1 + \sqrt{\Omega_o})^{-1} \quad , \qquad \Omega_o = q_o .$$

The age-density curves for matter- and radiation-dominated universes are plotted in Fig. 1 in dimensionless units.

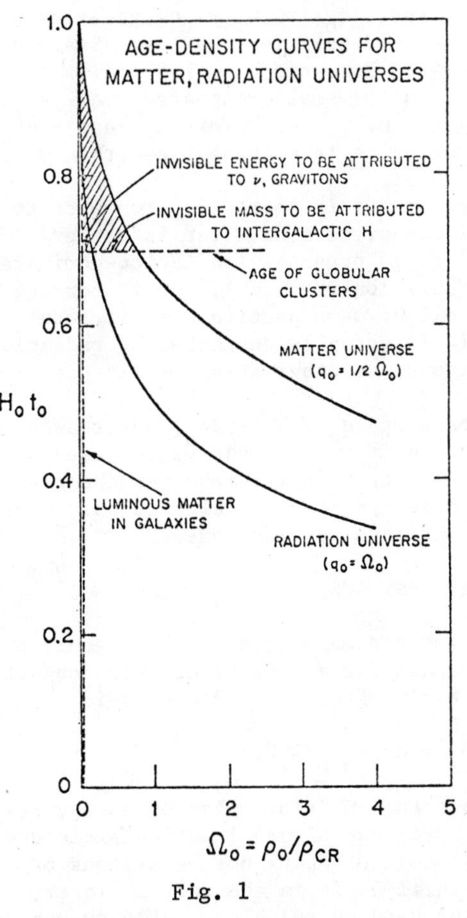

Fig. 1

For a given present density ρ_o, a radiation universe needs to be younger than a matter universe because the radiation universe expands faster than the matter universe. Fig. 1 then relates the three quantities Ω_o, q_o, t_o. We will now review the present data on Ω_o and q_o and on t_o to see whether the universe is open or closed and dominated by matter (as it appears to be) or dominated by radiation (in the form of neutrinos, gravitons or other relativistic particles).

3. PRESENT DECELERATION AND MASS DENSITY

The latest value of Hubble's constant is[3]

$$H_o = 50 \text{ km sec}^{-1} \text{ Mpc}^{-1}$$

whence

$$H_o^{-1} = 19.5 \times 10^9 \text{y}$$

$$cH_o^{-1} = 2 \times 10^{28} \text{cm} = 6000 \text{ Mpc}$$

$$\rho_{CR} \equiv 3H_o^2/8\pi G$$

$$= 0.55 \times 10^{-29} \text{gm/cc}$$

$$= 3000 \text{ ev/cc.}$$

The Hubble constant keeps changing and is now probably known to about 50%. Sandage's latest value[3] of the apparent deceleration (uncorrected for galaxy evolution) is

$$q_o = 0.94 \pm 0.4$$

The density of luminous matter in galaxies is[4] $\rho_o = 1.5 \times 10^{-31}$gm/cc or $\Omega_o = 0.028$. This figure is being revised upwards as diffuse x-rays reveal hot gas clouds between the galaxies in galactic clusters. Nevertheless the mass accounted for is probably considerably less than the critical mass necessary for the universe to be closed, unless[5] galaxies are surrounded by gigantic halos of stars.

The apparent deceleration implies $\Omega_o = 2 \pm 1$ if the universe is presently matter-dominated and $\Omega_o = 1 \pm 0.5$ if the universe is presently radiation-dominated. If the universe is believed to be closed, considerably less matter remains to be accounted for if this missing mass is in the form of radiation (gravitons or neutrinos) than if it is in the form of non-relativistic matter.

Evolutionary corrections decrease q_o if the young galaxies seen far out are brighter than the old galaxies near us. Beatrice Tinsley[6] says: "q_o should be revised downward by ~ 1.5 if its apparent value is near 2, and by less for smaller apparent values, to ~ 0.5 if the apparent value is near 0.5. ... Unless the apparent value is greater than 2, the true value is probably less than 0.5, corresponding to an open world model."

If the true deceleration parameter is corrected downwards to $q_o \approx 1/2$, then $\Omega_o = 1 \pm 0.5$, if the present universe is matter-dominated or $\Omega_o \approx 0.5 \pm 0.2$, if it is presently radiation-dominated. These evolutionary corrections[6] are controversial, but if correct suggest that the universe may well be open particularly if, contrary to visible appearances, it is actually dominated by radiation in the form of neutrinos, gravitons, or other relativistic particles.

Because in any Friedman cosmology $q_o < \Omega_o < 2q_o$, any reasonable q_o quoted is inconsistent with $\Omega_o = .028$, the value based on visible matter in galaxies. $q_o \approx 1$ or even 1/2 requires the presence of huge mass in intergalactic hydrogen or galactic halo stars, or lesser mass in the form of neutrinos or gravitons.

4. PRESENT AGE

A small value ($\Omega_o \sim .028$) for the mass density also requires that the universe must have expanded for a long time. The longest age of the universe based on the age of globular clusters is

$$t_o \geqslant 12 \times 10^9 \text{ y or } H_o t_o \geqslant 0.7.$$

In Fig. 1 the shaded regions show the invisible mass or energy required to reconcile Ω_o and t_o. From the figure less invisible energy is called for if it is in the form of neutrinos, gravitons or other relativistic particles than if it is in the form of intergalactic hydrogen, galactic halo, dead stars, black holes or other nonrelativistic matter. If the missing mass is in the form of relativistic particles, the universe is presently still radiation dominated and only $\Omega_o = 0.2$ is needed to agree with the age limit and corrected q_o.

Our tentative suggestions are:

(i) The universe is probably open if the age is long and evolutionary corrections are large.

(ii) The present universe may still be radiation-dominated since less energy is missing if it is hidden in gravitons and neutrinos than if it is contained in intergalactic H or galactic halo.

II. PHYSICAL EVENTS IN THE LEPTON ERA

1. CANONICAL BIG BANG THEORY

Figure 2 shows the conventional thermal history of the big bang universe, assuming a present mass density $\rho_o \approx 10^{-31}$ gm/cc.

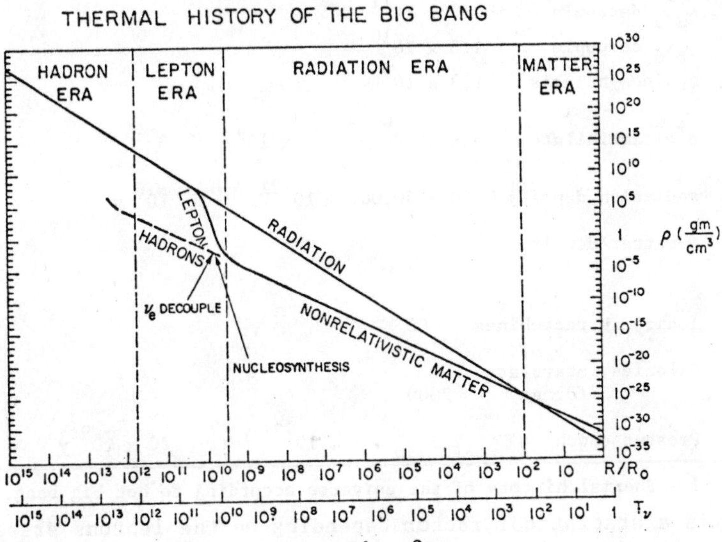

Fig. 2

The key events are tabulated in Table 1.

The Lepton Era in which we are interested begins with the annihilation of hadrons at $T = m_\pi c^2/k = 1.7 \times 10^{12}$K and ends with the disappearance of electrons and positrons at $t = m_e c^2/k = 5.9 \times 10^9$K. Whether the Present Era is matter-dominated or not, the Lepton Era is radiation-dominated so that:

(1) Hadrons are negligible; in particular we may set the next baryon number B = 0.

(2) Space curvature, deceleration and all other aspects of general relativity are unimportant.

This means that our following conclusions do not depend on the present state of the universe or on any details of cosmology, but only on the assumption that the universe was once as hot as 10^{12}K and adiabatically expanded as a relativistic gas.

According to Eq. (1.7), the expansion time

$$t = 650\ \rho^{-1/2} \tag{2.1}$$

$$= 230\ \varkappa^{-1/2}\ T_9^{\ -2}\ . \tag{2.2}$$

In arriving at Eq. (2.2) we have written the total energy density

$$\mathcal{E} = \rho c^2 = \varkappa A T^4,$$

		T_γ (^oK)	ρ (gm/cc)	t (sec)
HADRON ERA	$\pi\pm$ annihilate	1.7×10^{12}	3×10^{14}	4×10^{-5}
LEPTON ERA	$\mu^+\mu^-$ annihilate	1.2×10^{12}		
	$\nu_\mu \bar{\nu}_\mu$ decouple	1.2×10^{11}		
	$\nu_e \bar{\nu}_e$ decouple	1.8×10^{10}		
	nucleosynthesis	1.3×10^{9}		
	e^+e^- annihilate	5.9×10^{9}	5×10^{4}	3
RADIATION ERA	radiation density = matter density	1200 - 40,000	10^{-21}	10^6 y
MATTER ERA	Ionized H recombines	4000		
	Galaxies, stars are formed	2000		
	Present epoch	3	10^{-31}	20×10^9 y

Table I. Thermal history of the universe according to hot big bang.

where $\varkappa$ is a statistical factor depending on the leptons present and on their statistical weights g. The number and energy densities of photons, relativistic electrons or positrons, and chiral neutrinos separately are given in Table II. The species i known to be present at various periods within the Lepton Era are given in Table III along with their contributions to the total energy density (2.3): $\varkappa$ reduces from 25/4 to 18/4 when $\mu^\pm$ annihilate and to 11/4 when $e^\pm$ annihilate.

Shvartzman et al.[7] raised the question of how many unknown neutrinos or other relativistic particles there might be that were

	Photons g=2	Electrons g=2	Neutrinos g=1
$n = \frac{g}{2\pi^2\hbar^3}\int \frac{p^2 dp}{e^{\beta E} \pm 1}$	$20\ T^3$	$15\ T^3$	$7.5\ T^3$
$\varepsilon = \frac{g}{2\pi^2\hbar^3}\int \frac{E\,p^2 dp}{e^{\beta E} \pm 1}$	aT^4	$\frac{7}{8}\ aT^4$	$\frac{7}{16}\ aT^4$

Table II. Number and energy density of photons, non-degenerate relativistic electrons or positrons, and non-degenerate chiral neutrinos or antineutrinos in CGS units. $a=7.56 \times 10^{-15}$ erg cm^{-3} deg^{-4} = 4.75×10^{-3} ev cm^{-3}deg^{-4}.

	$\mu^{\pm}$		$\nu_\mu \bar{\nu}_\mu$		$\nu_e \bar{\nu}_e$		$e^{\pm}$		γ		
after $\pi^{\pm}$ annihil.	$2\cdot\frac{7}{8}$	+	$2\cdot\frac{7}{16}$	+	$2\cdot\frac{7}{16}$	+	$2\cdot\frac{7}{8}$	+	1	=	$\frac{25}{4}$
after $\mu^{\pm}$ annihil.			$2\cdot\frac{7}{16}$	+	$2\cdot\frac{7}{16}$	+	$2\cdot\frac{7}{8}$	+	1	=	$\frac{18}{4}$
After $e^{\pm}$ annihil.			$2\cdot\frac{7}{16}$	+	$2\cdot\frac{7}{16}$			+	1	=	$\frac{11}{4}$

Table III. Contribution $\kappa_i aT^4$ of each particle pair to the total energy density κaT^4. The statistical factors 7/8 and 7/16 from Table II are multiplied by 2 to include particles and antiparticles.

once in thermal equilibrium with our universe. Each such species would contribute to κ raising the temperature at which helium synthesis takes place and thus the primordial He abundance. Fowler[8] shows that κ must be $\sim$1 so that no more than about 10 such additional species of neutrinos or relativistic particles can exist. Empirically the lepton spectrum cannot increase rapidly with energy even if the hadron spectrum does.

The important events in the Lepton Era are:

(1) $\mu^{\pm}$ annihilation at $T^{\mu}_{dec} = 1.2 \times 10^{12}$K, soon followed, if there are no neutral currents, by $\nu_\mu + \bar{\nu}_\mu$ decoupling at $T^{\nu_\mu}_{dec} = 1.2 \times 10^{11}$K.

(2) $\nu_e + \bar{\nu}_e$ decoupling at $T^{\nu_e}_{dec} = 1.8 \times 10^{10}$K when the neutrino-electron collision rate $\tau^{-1} = \sigma nc = (6.7 \times 10^5\ t_q^{-5}\ \text{sec})^{-1}$ (in the absence of neutral currents), falls below the cosmic expansion rate. After decoupling $\nu_\mu + \bar{\nu}_\mu$ and $\nu_e + \bar{\nu}_e$ cool adiabatically as the universe expands.

(3) Deuterium synthesis which commences at $T_{syn} = 10^{10}$K when the photon density is low enough to push the equilibrium $n + p \rightleftarrows d + \gamma$ to the right, removing neutrons from equilibrium. The deuterium abundance is measured. In any case, deuterium build-up leads very quickly to He synthesis.

(4) $e^{\pm}$ annihilation at $T = 5.9 \times 10^9$K. Until this happens the γ, $e^{\pm}$ and decoupled ν have the common temperature T_ν. When $e^{\pm}$ annihilation takes place, the entropy

$$S = \frac{E+pV}{kT} = V\,\frac{\mathcal{E}+p}{kT} = V\,\frac{4}{3}\,\frac{\kappa a T^3}{k}$$

is conserved. If $\nu_e + \bar{\nu}_e$ have already decoupled, the $e^{\pm}$ entropy is dumped into the γ gas, so that

$$S = V\,\frac{4}{3}\,\frac{11}{4}\,aT_\nu^{\;3} = V\,\frac{4}{3}\,1\,a\,T_\gamma^{\;3}$$

makes

$$T_\gamma = \left(\frac{11}{4}\right)^{1/3} T_\nu \text{ for } kT_\nu < m_e c^2 .$$

If $\nu_e + \bar{\nu}_e$ are still thermally coupled to γ when $e^{\pm}$ annihilate then their common temperature is raised by $(9/2)^{1/3}$.

The radiation density ν_μ

$$\mathcal{E}_\gamma + \mathcal{E}_\nu = aT_\gamma^4 + 4\cdot\frac{7}{16}\, aT_\nu^4 = [1 + \frac{7}{4}\,(\frac{4}{11})^{4/3}]\, aT_\gamma^4$$

$$= 1.45\, aT_\gamma^4$$

is, because of the 4 neutrino species present, raised by 45% above what it would be for a photon gas. The present neutrino temperature, number density and energy density, derived in this way from $T_{\gamma o}$ the present relict photon temperature, are given in the left half of Table IV.

	Canonical Big Bang Theory ($\mu = 0$)		Degenerate Neutrino Theory ($\mu \neq 0$)
	γ	ν	deg ν or $\bar{\nu}$
T_o	$2.7K = 0.17 \times 10^{-3}$ ev	$1.9K = 0.1 \times 10^{-3}$ ev	$\lvert\mu\rvert < 45\, kT_\nu$ $= 7 \times 10^{-3}$ ev
n_o	$20\, T_\gamma^3 = 392$/cc	$4.(7.5\, T_\nu^3) = 176$/cc	$< 7 \times 10^5$/cc
$\mathcal{E}_o$	$aT_\gamma^4 = 0.25$ ev/cc $= 4.4 \times 10^{-34}$ gm/cc	$4\cdot\frac{7}{16}\, aT_\nu^4 = 0.11$ ev/cc	$< \mathcal{E}_{cR}$
$\mathcal{E}_{star}$	10^{-3} ev/cc	10^{-2} ev/cc	
$\mathcal{E}_{vis}$		1.5×10^{-31} gm/cc = 80 ev/cc	
$\mathcal{E}_{cR}$		0.5×10^{-29} gm/cc = 3000 ev/cc	

Table IV. Present temperature, number and energy density of photons and of neutrinos on the canonical (μ=0) big bang theory, and on the degenerate neutrino theory ($\mu \neq 0$). The right hand column gives upper limits for the Fermi energy, number and energy density of degenerate neutrinos in the big bang theory without neutrino charge neutrality.

2. EFFECT OF NEUTRAL CURRENTS

Neutral currents, if they exist, permit mechanisms $\nu_\mu + \bar{\nu}_\mu \rightleftarrows \nu_e + \bar{\nu}_e$ for the muon neutrinos to remain in equilibrium after muons have disappeared. If the strength of the diagonal coupling $(e\nu)(e\nu)$ is G_D rather than the universal Fermi value G_F then τ^{-1} is multiplied by $(G_F/G_D)^2$ and $T_{dec}^{\nu_e}$ by $(G_F/G_D)^{2/3}(11/4\varkappa)^{1/6}$. $T_{dec}^{\nu_e}$ is thus insensitive to the existence of additional lepton species, but if $G_D \geqslant 5\, G_F$, the $\nu_e + \bar{\nu}_e$ (and also $\nu_\mu + \bar{\nu}_\mu$) would remain in equilibrium until the $e^- + e^+$ annihilates, raising the temperature of both γ and ν seas. This would accelerate the time scale leading to d and He^4 synthesis, so that the neutrons would be frozen into He^4 when the temperature is a little higher than conventionally. This would only increase the fraction of cosmologically produced He^4 from Y = 0.27 to Y = 0.29.

3. COSMOLOGICAL SIGNIFICANCE OF CANONICAL BIG BANG NEUTRINOS

The present values of photon and neutrino temperatures, number and energy densities are given in Table IV assuming $\mu = 0$ and $m_\nu=0$. These energy densities are large compared with the energy densities out of starlight $\mathcal{E}_{star}$, but negligible compared with the energy density of luminous matter $\mathcal{E}_{vis}$ and the critical energy density $\mathcal{E}_{cR}$.

The number of big bang neutrinos is, however, very large so that if they had a mass $m_{\nu_e} = m_{\nu_\mu} \approx 16$ ev, they would contribute significantly to the deceleration of the universe[9]. (If these massive neutrinos had both chiral states thermally coupled, then $m_{\nu_e} = m_{\nu_\mu} \leq 8$ ev.) These limits on the neutrino mass are many orders of magnitude better than those set in laboratory experiments.

4. DEGENERATE NEUTRINOS: A VARIANT OF CANONICAL BIG BANG THEORY

We must now distinguish a second version of the Big Bang theory.[10,11] In the canonical version, each species of conserved fermions has equal numbers of fermions and antifermions (net charge zero) $Q = B = N_e = N_\mu = 0$ or $\ll N_\gamma$. Since the number density of fermions and antifermions is given by

$$n_\pm = \frac{g}{2\pi^2\hbar^3} \int \frac{p^2 dp}{e^{\beta(E-\mu)} + 1}$$

where g = statistical weight, $\beta = 1/kT$ and μ = chemical potential, charge neutrality $n_+ = n_-$ requires $\mu = 0$. Since the photon has $\mu = 0$, charge neutrality will obtain if all fermions and antifermions are created in pairs out of photons. When $\mu = 0$, the number and energy density of photons, relativistic electrons and neutrinos are as given in Table II.

In another version of the Big Bang theory[10,8], for neutrinos $\mu \neq 0$ so that the chemical potential or Fermi energy of neutrinos determines the excess number of neutrinos ($\mu > 0$) or of antineutrinos ($\mu < 0$). In this theory μ appears as an additional parameter referring, like the net baryonic charge B, to the neutrino number L "before the Big Bang". This theory is called that of degenerate neutrinos or degenerate antineutrinos. (This is misleading since at $\mu = 0$ fermions are still degenerate; the classical Boltzman distribution is realized only when $\mu \ll 0$ for fermions and $\mu \gg 0$ for antifermions.)

The number and energy densities of completely degenerate neutrinos of Fermi energy μ are

$$n_\nu = \frac{4\pi}{3} \frac{|\mu|^3}{(hc)^3} = 2.19 \times 10^{12} \ |\mu(e\nu)|^3 \ \text{cm}^{-3}$$

$$\mathcal{E}_\nu = \frac{\pi \mu^4}{(hc)^3} = 1.64 \times 10^{12} [\mu\ (\text{ev})]^4 \quad \text{erg cm}^{-3}.$$

If the universe is not closed by degenerate neutrinos, $\mathcal{E} < \mathcal{E}_{CR}$ from which the limit $|\mu| < 45\ kT_\nu = 7 \times 10^{-3}$ ev and $n_{\nu o} < 7 \times 10^5\ cm^{-3}$ in the right side of Table IV are obtained. An order of magnitude better limit on $|\mu|$ is obtained[11] from the observed limits on primordial helium. Nevertheless even though the number of degenerate neutrinos is one thousand times greater and their energy about ten times greater than in canonical big bang theory, if homogeneously distributed, they remain undetectable. Even the most energetic cosmic ray protons ($E > 10^{20}$ ev) would be unaffected in number, range or energy by interaction with the neutrino sea. Neutral currents would permit the elastic processes $\nu + P \to \nu + P$, $\bar{\nu} + P \to \bar{\nu} + P$ which degrade all protons, changing, in principal, the cosmic ray power-law spectrum. The present neutrino density is so low that as long as the neutrino coupling is of weak order of magnitude, the mean free path for any neutrino processes exceeds the present size of the universe.

III. NEUTRINO SUPERSTAR

We have seen that, even if degenerate, cosmological neutrinos are undetectable in scattering processes, if they are homogeneously distributed about us. We will now consider the possibility of the neutrinos having a small mass $m_\nu \sim 1$ ev, which will lead to their gravitational condensation. Although $m_\nu \neq 0$ is an initially unattractive possibility, only in its cosmological effects could a neutrino mass reveal itself, and in this way the neutrino would link the very small with the very large.

If neutrinos are degenerate and have a small mass[12,13] they will condense into what Markov[14] called a superstar. If most of the universe is hidden in the form of such neutrinos, the visible matter acts only as test particles revealing the gravitational potential of the generally denser neutrinos. These neutrinos are nonrelativistic, in fact they became nonrelativistic when the temperature dropped to $T_\nu \approx 10^4$K. Such a star forms a nonrelativistic polytrope of adiabatic index 5/3 or polytropic index 3/2. While the total mass M and radius R depend on the central density, all such polytropes obey the homology law

$$MR^3 = 91.9\ m_\nu\ (\hbar^2/Gm_\nu)^3$$

$$= 7.2 \times 10^{51}\ (m_n/m_\nu)^8\ \text{gm-cm}^3, \tag{3.1}$$

and show central peaking of density

$$\rho_c/\bar{\rho} = 5.99\ , \tag{3.2}$$

where $\bar{\rho} = 3M/4\pi R^3$ is the mean density.

In the nearby Coma cluster of galaxies the virial mass[15]

$$M = 4 \times 10^{49} \text{ gm} , \tag{3.3}$$

while the radius, although very hard to define, is roughly[15]

$$R \sim 2 \times 10^{24} \text{ gm} . \tag{3.4}$$

These two figures in Eq. (3.1) give immediately

$$m_\nu \approx 0.5 \text{ ev.}$$

Two further remarks can be made:
(1) If, as Rood suggests[15] the distribution of missing mass follows that of the visible galaxies, except near the center of the cluster, then the neutrinos and galaxies have the same mean square velocity $\langle v^2 \rangle$. This means that except near the cluster center, the neutrinos and galaxies are very far indeed from thermally equilibriating. In fact, the observed central peaking of galaxies in Coma is much larger than (3.2).
(2) If the cosmic neutrinos have, over the cosmos, a mean density $\bar{\rho} \leqslant \rho_{cR}$ but are all swept up into superstars, then the separation between superstars is $\gtrsim 60 \cdot$ Mpc. This corresponds closely to the mean separation observed between galactic clusters, as it should if the missing mass necessary to close the universe is mostly in galactic clusters.

IV. SUMMARY

To summarize: The latest data[3] on Hubble constant and cosmological deceleration and on the age of the universe from globular clusters is most simply reconciled if there is much hidden mass ($\rho \sim 0.2 - 0.4\ \rho_{cR}$) in the universe, particularly if this mass is hidden in neutrinos or gravitons. The density of such relativistic particles is, in canonical Big Bang theory, as insignificant as the density of photons. If the neutrinos are degenerate, corresponding to a neutrino charge for the universe which is nonvanishing like the baryonic charge, their density could be cosmologically significant. Even in this case, however, they could not be revealed by the degradation of energetic cosmic rays. If the degenerate neutrinos had a small mass, however, they would gravitationally cluster and form a nucleus about which clusters of galaxies would condense. In this case, the local density of neutrinos would considerably exceed the cosmic average which is $\leqslant \rho_{cR}$.

APPENDIX: EFFECTS OF NEUTRAL CURRENTS ON STELLAR PROCESSES

We must distinguish two kinds of neutrino processes important in late stages of stellar evolution:
(1) The purely leptonic neutrino interaction $(e\nu_e)(e\nu_e)$ which was so important in the Lepton Era is also important as a cooling mechanism in white dwarfs and in presupernova cores. The degenerate carbon-oxygen cores of evolved stars of intermediate mass are liable to

explode by carbon flash unless they are kept cool by neutrino emission.
(2) The semileptonic neutrino interactions $n \to p+e^-+\bar{\nu}$, $p+e^- \to n+\nu$ (Urca processes). These processes, proceeding through conventional charged currents are important in dynamic situations like the collapse and pulsation of neutron stars.

The effects of weak neutral currents, if they exist, are to supplement (1) and (2) with:
(1') $(ee)(\nu_\mu\nu_\mu)$ and $(\mu\mu)$ $(\nu_e\nu_e)$ interactions tending to symmetrize between muon and electron neutrinos.
(2') $\nu+\eta \to \nu+\eta$, $\nu+\eta \to \bar{\nu}+\eta$ processes which, if they proceed via the isosinglet state of the nucleon η, lead to coherent neutrino scattering[16] proportional to the square of the atomic number A.

Leptonic Neutrino Processes. On the conventional charged current-current theory, the $(e\nu_e)(e\nu_e)$ interaction leads to $\nu_e+\bar{\nu}_e$ production from plasmon annihilation, electron-positron annihilation, photoneutrino production and neutrino bremmstrahlung. Neutral currents tend to symmetrize between ν_e and ν_μ, allowing the production of $\nu_\mu+\bar{\nu}_\mu$ along with $\nu_e+\bar{\nu}_e$, but they also alter the matrix elements by an amount depending on the Weinberg angle θ_W. In Table V, derived from Dicus[17], we tabulate the amount R by which the energy loss rates due to various processes must be multiplied to allow for Weinberg weak neutral currents.

$\sin^2\theta_W$ =	1	0.35	0
(A)	8.5	1.6	
(B)	4.5	1.2	1/2
(C)	1.8	0.8	

Table V. The amount R by which energy loss rates on Weinberg theory exceed those calculated on conventional theory for (A) plasmon annihilation, nondegenerate pair annihilation; (B) extremely degenerate pair annihilation, photoneutrino emission; (C) nonrelativistic photoneutrino emission.

Denoting $C_A = 1/2$, $C_V = 1/2 + 2\sin^2\theta_W$, Dicus obtains

$$R = C_V^2 + (C_V-1)^2$$

for (A) plasmon or nondegenerate pair annihilation processes;

$$R = \frac{1}{2}[C_V^2 + (C_V-1)^2] + C_A^2$$

for (B) extremely degenerate pair annihilation or relativistic photoneutrino emission;

$$R = \frac{1}{6}[C_V^2 + (C_V-1)^2] + \frac{10}{6}C_A^2$$

for (C) nonrelativistic photoneutrino emission.

The results calculated are especially simple in the limit $\theta_W \to 0$: the energy loss rate is doubled by allowing $\nu_\mu + \bar{\nu}_\mu$ production along with $\nu_e + \nu_e$ production, but $C_A = C_V = 1/2$ and the matrix element for each process separately in Weinberg theory is 1/2 what it would be in V-A theory. Consequently, despite the $\bar{\nu}_\mu + \nu_\mu$ emission, all loss rates are reduced by $2 \cdot (1/2)^2 = 1/2$. If $\sin^2\theta_W \approx 0.35$, as seems to be preferred experimentally, then the loss rate by plasmon annihilation is augmented by 80%, but the loss rate by annihilation of degenerate electron-positron pairs is augmented by only 20% . These processes are probably the important ones in ultradense matter. So if neutral currents exist, neutrino cooling in white dwarfs and presupernova will be enhanced. This makes it a little bit more likely that presupernova of intermediate mass can collapse before suffering explosive carbon ignition.

Semileptonic Neutrino Processes. In the collapse and pulsation of neutron stars, electron degeneracy tends to inhibit most leptonic neutrino processes while density changes enhance URCA processes off baryons. Coherent isoscalar neutrino scattering[16], if it exists, will be important in Fe (A=56) which occurs in fully evolved white dwarfs and in the crust of neutron stars. Neutral currents will therefore help to make neutrino transport an important energy transfer mechanism in hot neutron stars, where neutrinos are generated in the hot core and absorbed in the stellar envelope.

Recent developments in neutrino physics thus do not change lepton cosmology significantly, but appear potentially important in prepulsars and in supernova blow-off.

References

1. T. de Graaf in Proc. Cortona Meeting on Astrophysical Aspects of the Weak Interactions (Roma, 1971).
2. T. de Graaf in Proc. 1972 Conference on Neutrinos (Technoinform, Budapest 1972).
3. A. Sandage, Hale Observatories Report (1972); Ap. J. 183, 743 (1973).
4. J. Oort in La Structure et l'evolution de l'universe (Brussels, 1958).
5. J. P. Ostriker, Ap.J. (to be published).
6. B. Tinsley, Ap. J. 178, 319 (1972).
7. V. F. Schvartzman, Sov. Phys. JETP Letters 9, 184 (1969); R. A. Sunyaev and Ya. B. Zeldovich, Comm. on Astrophys. and Space Science 1, 159 (1969).
8. W. A. Fowler in Proc. Cortona Meeting on Astrophysical Aspects of the Weak Interactions (Roma, 1971).
9. R. Cowsik and J. McLelland, Phys. Rev. Lett. 29, 609 (1972).
10. S.Weinberg, Phys. Rev. 128, 1457 (1962).
11. R.V. Wagoner, W. A. Fowler and F. Hoyle, Ap. J. 148, 3 (1967).
12. R. Cowsik and J. McClelland, Ap. J. 180, 7 (1973).
13. G. Marx and A. S. Szalay in Topical Seminar on Weak Interactions, (Trieste, 1973), IC/73/71.

14. M. A. Markov, Phys. Letters 10, 122 (1964).
15. H. J. Rood, T. L. Page, E. C. Kintner and I. R. King, Ap. J. 175, 627 (1972).
16. D. Z. Freedman, National Accelerator Laboratory Pub-73/76-THY, Batavia, Ill. (1973).
17. D. Dicus, Phys. Rev. D6, 941 (1972).

Conference Summary

R. P. FEYNMAN

California Institute of Technology, Pasadena, California 91109

Summary Talk given at

International Conference on Neutrino Physics and Astrophysics

April 28, 1974

Philadelphia, Pennsylvania

As is conventional in summary talks, I'm not going to summarize what happened during today, the last day of the conference. The excuse usually is, of course, that you've just heard it. But I have just discovered the real reason, for the summarizer didn't attend todays sessions, because he is busy preparing his summary.

To summarize a conference is always a silly thing; to have to say everything over again that the other people said. That's a waste of time, so I decided to try to summarize what we know about neutrinos. But the only thing we know about them (except for the intrinsic properties which are so simple) is that they interact somehow. I shall therefore summarize our knowledge of weak interactions and in particular how the results of this conference have expanded our knowledge of weak interactions. OK?

I've also done something that will bore most of you but may be pleasant to some of you. I've decided to explain our ideas in as simple a manner as possible. This means cutting corners and being therefore slightly inaccurate, and using ideas to explain a point that are not absolutely necessary for the point to be true, but make it easier to understand or remember. In this way, people who are not completely in the field like the astrophysicists, will understand some of the ideas. So you'll excuse me then if everything is done all over again in a simple way.

There are certain things besides the things of todays' sessions that I won't cover. In particular, in our struggles we make up a number of possibilities involving new particles of various kinds. There are very many different possibilities for these new particles, and for each there are special techniques that people have invented that would help us find them. I will not summarize that, because there are so many particular suggestions for experiments to look for this or that particular thing that to make a summary of such things is hopeless.

So I would like to say where we stand in our theories of weak interactions. We have a conventional theory of weak interactions invented by Marshak and Sudarshan, published by Feynman and Gell-Mann and completed by Cabibbo - I call it the conventional theory of weak interactions - the one which is described as the V-A theory. In describing the theory of interactions I have to discuss not only the coupling of neutrinos to the weak interaction at one end of the terminal so to speak, but also the terminal end involving the hadrons. In describing that I'll use the simplest minded model. I'll not speak in terms of abstract currents, but we'll simply use the little quark view of the hadrons. This is a possibly inconsistent model that requires much further study but that is a study of strong interactions and belongs in another conference. Nevertheless I'll cheat by using this inconsistent view which says that at low energies you can look at a proton as made out of three quarks that move nonrelativistically, but when you look at them in detail as we do in these high energy neutrino experiments we see their momenta are distributed longitudinally over a wide range, which makes it look like they're moving relativistically even in the center of mass system. I will not discuss the paradoxes, such as why the quarks do not come apart, and all the other strong interaction problems. So if you'll excuse me for cutting corners there then I can explain everything very simply. It's always easy to explain things if you have the wrong explanation.

Now the theory I'm talking about of course is that weak interactions are represented as the product of two currents, and the current is the sum of several parts (see Fig. 1). That immediately suggests the possibility of interaction through an intermediate particle, analogous to electromagnetic interaction via an intermediary photon. This intermediary vector heavy boson is sometimes called a W meson. It is always possible that this current-current interaction is a second order action of a direct first order effect; that a current is coupled to some vector boson, W, and then the vector boson goes and is coupled to the other current. We now, from this particular conference, have evidence shown by Mr. Sciulli that if there is such a boson its mass has to be bigger than 10 GeV, or we would have seen effects in the high energy scattering of neutrinos at NAL. This is high enough that in the rest of the talk we can talk about it as if there is a direct interaction between currents. The current is a combination of vector and axial and it involves different kinds of particles, you all know this stuff so I'll go very quickly. You know that there is a part of the current called the lepton current which has two pieces, one between electrons and their neutrinos, and the other between muons and their neutrinos. The third piece of the current couples to hadrons. We describe it as a coupling to quarks. The charge change means it must be an up-type quark coupled, we assume via V-A, to either a d or s quark. Cabibbo's suggestion is that it is a "c" quark which is neither a pure d nor a pure s but has an amplitude to be a d and an amplitude to be an s, normalized to one. It has been written as $c = d \cos\theta_c + s \sin\theta_c$ where θ_c is Cabibbo's angle. It is only understood empirically. When I make calculations for this talk, I shall take $\sin^2\theta_c$ as .046, but we don't know it that well. From the u,d term we have reactions which involve no strangeness changes, and from the u,s term we have strangeness changes of the hadrons (equal to their change in charge, $\Delta S = \Delta Q$).

That's the theory and now we would like to talk about how we stand relative to experiment (Fig. 2). In doing this I will describe what we've discovered and what we've heard reported in this conference. We've heard reported that the conventional theory is OK when compared to low energy. There is no exotic problem - there is no problem with double β decay - or in any of the intrinsic properties of μ. Everything is alright, so we can summarize everything by saying so far so good. (I will remind you that there are a few peculiarities in the theory of strong interactions which were not mentioned in this conference because it is not a neutrino problem. There is a $\Delta I = 1/2$ rule for meson decay into other mesons which is not understood. The $\Delta I = 1/2$ rule for baryon decays is possibly now understood as being a consequence of the quark model. There are also the p-wave baryon hadronic decay rates which people are still going crazy trying to explain. Otherwise the numbers fit very nicely but vaguely to our vague understanding of matrix elements.)

Now we turn to high energy and I'm really beginning to summarize our conference. We are interpreting these experiments, as letting us test whether the proton can be looked upon as being made of parts.

At very high energy I shall use a picture of the hadrons called the quark-parton model. The word partons just means parts, one kind of part that a proton might be made out of are quarks. That's called the quark-parton idea: We have evidence reported here at this conference about whether the pieces out of which a proton is made might be quarks.

I drew a picture (see Fig. 2) of the typical collision between a neutrino and a proton. Everybody uses the most convenient coordinate system for them (in the laboratory, it's the proton at rest). I found the easiest one for me is that the momentum transferred from the leptons to the hadrons (by means of the virtual intermediate meson W) called "q" is purely spacelike (i.e., carries no energy). The proton coming in from the right is socked by the incoming neutrino at some angle α and a μ^- goes off so that they give a momentum which is purely spacelike and very large. This momentum is then given to the proton. In this system the proton is moving, highly relativistically from the right. In this naive model (which is instigated by field theory) it is represented as being made out of some parts which are idealized particles. These idealized particles, the partons, are pure point-like objects with very simple electromagnetic and weak interaction point-like local couplings. That is just a consequence of field theory ideas. The question is, what field theory? That question is just the same as the question, what kinds of particles are the parts, what are the partons, are they quarks? As a part of our model we go further than field theory when we go on, and we say that when a particle is moving very relativistic the various parts share this total longitudinal momentum in various proportions (while the transverse momenta remain small). Thus in this picture the various parts inside have different proportions, x, of the total momentum, with certain probabilities. The transverse momentum in this picture is supposed to be limited, there is some evidence that this is the case. You might even believe you understand that; we should think of the proton in a system where it is at rest with the stuff inside moving back and forth with only finite internal momenta, when you transform to the moving frame the transverse momentum is not changed and is still finite. It sounds good and I'll go along with that but that's the naiveté of this particular model. That's where probably the thing is wrong if it's going to be wrong. However, with that view, we have the following observations. We can understand what happens very easily. The W meson is a vector and therefore can be polarized in three ways, either a right-hand circular or left-hand circular, or having a zero component of angular momentum (z is the direction of motion). That is, the z-component of angular momentum is either +1, -1 or 0. I'll call those the right circular, the left circular and scalar cases (although it's not scalar, it's the vector polarization with pure time component). These cases do not interfere in this experiment of "deep inelastic scattering." Now the amplitude to produce these different kinds of polarization depends on this angle α with the usual $1 - \cos\alpha$ or whatever it is you get when you do those things in space with different angular momenta. When converted back into the laboratory, those factors become respectively the following probabilities (these cases don't interfere in this experiment, where we measure neither the spin of the proton nor the character of the hadronic products). The left circular polarization is produced, let us say, with probability one, then the right spinning one will be produced by a probability $(1-y)^2$; [(1-y) is the energy in the μ coming off in the laboratory, divided by the neutrino energy]. The scalar polarization case is produced with probability 1-y. That is all you have to know, the rest is easy.

You can always remember which way the 1-y goes, the following way: Neutrinos couple only when they're spinning to the left (in fact there are no neutrinos that aren't spinning to the left) and when things are very relativistic like the μ, it is also coupled spinning to the left. Thus when the angle α is 0 (which corresponds to (1-y) = 0 the neutrinos coming in and the μ's coming out are both spinning to the left so the spin has been flipped, by one unit left. This angular momentum is taken out by the W meson which therefore spins purely left, in this case that (1-y) = 0). Now what happens when a W meson hits a piece of the proton, a parton, let's take it to be a quark, a particle of spin 1/2 coupled via V-A. It knocks it from forward to backward, it just goes directly backwards and again we have this same problem of angular momentum. If it's a relativistic quark and so coupled only if it is spinning to the left, then when it turns around to go the other way it's still spinning to the left so the angular momentum of the W must be to the left. Thus quarks only couple (relativistically) to the left spinning W. In the same way antiquarks (which V-A couples only to the right) are coupled to right spinning W's only. Scalar W's do not couple relativistically to spin 1/2 quarks. So if it's a quark, in other words, it spins to the left, it's hit by the left spinning W. The only kind of quark you can do that with is a down quark (or strange, but this is much smaller and we shall, in first approximation neglect the Cabibbo angle). It has to be a down quark because the charge has to be conserved; you go from a neutrino to a μ^- so the W is positive and must raise the charge of the quark. You go from a down quark to an up quark so this component (W^+, left spin) only tests how many down quarks there are in the proton. Therefore, if we could select left-hand spinning W's we would directly measure the number of down quarks inside a proton and that's what we do because we can select it by looking at the y-dependence (flat) and that's all there is to it. Now the right spinning W^+ can't couple with the quark, it can only couple with the antiquark which is spinning to the right, and so we find that the coefficient of the $(1-y)^2$ piece is the amplitude of finding the opposite reaction and therefore the anti-up quarks that are in a proton (if any). Therefore, the cross section for scattering a neutrino for a proton looks like this. It's proportional to the number of down quarks times the number 1 and the number of anti-up quarks times the number $(1-y)^2$ that's where all the formulas come from. If d(x) is the number of down quarks of momentum fraction x and $\bar{u}(x)$ the number of anti-up we have the cross section for neutrinos on protons $\sigma_{\nu P}$ as proportional to $\sigma_{\nu P}=2x[d(x)+(1-y)^2\bar{u}(x)]$.

The unit is $G^2s/2\pi$ where G is the Fermi constant and the s of the reaction is, in the lab, approximately the neutrino energy times twice the proton mass M. The quantity x is controlled by q the momentum transfer (hence the momentum of the W), by the rough conservation of energy of the quark; since the W carries no energy, only momentum q, the quark must be turned around with no change in energy, hence it must have momentum q/2 before and -q/2 after. Thus a particular momentum, and hence x, is selected in these experiments; in detail, $x = -q^2/2M\nu$ where ν is the energy lost by the neutrino in becoming a muon in the lab system.

There is nothing that the W will couple to when it has scalar polarization of 0 angle momentum if the partons have spin 1/2. That they have spin 1/2 has been checked by the electron scattering measurement. The so-called Callan-Gross relationship is again a conservation of angular momentum, if the electrons were scattering from objects of spin 1/2. Now of course the cross section for scattering depends on how many quarks carry the fraction of momentum x, the number of down quarks. But the fraction x is simply a matter of how much momentum W has. And you can get the same momentum W, that is the same x by using different angles, α. And that means that in the laboratory there are some quantities that correspond to the momentum of the W and its relation to the momentum of the proton which involves x, and some other quantities that involve α that are variables in the laboratory. Therefore the scattering is going to depend on only certain combinations of the variables in the laboratory. That's called the scaling principle, it was discovered by Bjorken. If we use the right variables, the famous $-q^2/2M\nu = x$ and y the energy of the hadron over the energy of the initial neutrino; only those variables are involved and the scattering should just depend on those in the particular way indicated. By the way, what we're doing is very easy to understand, we're trying to find the momentum distribution of some parts - you have a swarm of bees and they're flying around, you scatter radar from them and you look at the reflected light and the change in frequency depends on how fast the bee was moving that it scattered from. So we scatter neutrinos and pick up the μ's that come back and the change in energy in those neutrinos to make the μ's gives us the speed that the pieces were moving inside the proton. And that's all there is to it. And this scaling business is only this, that if you measure what the distribution of the momenta of the parts is by using scattering at a particular angle and energy then you can change the experiment and use different frequency radar waves and different angle of deflections and you should get the same momentum distribution. Thus the internal consistency of the idea can be tested. Do we get the same distribution when we use different angles, and so on? Of course if the energy is too low, the answer should not be the same distribution, because during the time of the scattering the velocity is changing, but when we get to high energy we are supposed to have gotten a universal function which is called the great idea of scaling. That's the way it's supposed to work and now let's see how it works. By the way, it's very quick if we know what neutrinos do on protons we can figure out what anti-neutrinos do on protons, either starting over again and changing the rights and lefts or just imagining it's probably correct if you just change the d to $\bar{d}$ and u to $\bar{u}$ and sure enough it gives the formula for scattering anti-neutrinos on protons so you see that if we had protons alone we could determine the functions $d(x)$, $\bar{u}(x)$, $\bar{d}(x)$ and $u(x)$ all independently, in principle, from data. If we had neutrons we could make a consistency check. Because the case of neutrons differs from protons only in that the down quarks are replaced by up quarks and vice versa, so that if $u(x)$ represents the number of up quarks inside a proton it also represents the number of down quarks inside a neutron. Therefore, just like neutrinos scattering from the proton measures the down quarks that are in the proton, then neutrinos scattered from neutrons measure the down quarks in the neutron, but that is the same as the up quarks in the proton. Thus the cross section on neutrons $\sigma_{\nu N}$ is given by $\sigma_{\nu N} = 2x[u(x)+(1-y)^2\bar{d}(x)]$. There therefore would be a direct connection or consistency check between $\sigma_{\nu N}$ and $\sigma_{\nu P}$ if we could separate neutron and proton scattering. It's very nice that we have so many connections, the scaling laws, the necessity to have this mathematical form as a function of y, (a constant plus $(1-y)^2$) and a consistency between the neutrinos on neutrons and on protons. However so far we don't have separate scattering from neutron and proton. We only have scattering from iron, which is a mixture, about 50-50. (I make corrections when I give my theoretical predictions later on, for the fact that it's 26 protons, 29.85 neutrons). Let us define $q(x)$ as a measure of the number of non-strange quarks, ups and downs together. Let $q(x) = x(u(x) + d(x))$ so that $q(x)$ represents the amount of momentum in the range dx carried by non-strange quarks. Then the scattering of neutrinos from iron will be (from averaging $\sigma_{\nu N}$ and $\sigma_{\nu P}$)

$$\sigma_{\nu Fe} = q(x) + (1-y)^2\bar{q}(x) \quad .$$

For the anti-neutrinos we get

$$\sigma_{\bar{\nu} Fe} = \bar{q}(x) + (1-y)^2 q(x) \quad .$$

I hope you'll excuse me for taking all your time to show you what you know very well, but for me it's rather bad to have everything kind of hidden in some kind of f_1, f_2 and f_3 and various linear combinations with strange positivity conditions when all that they represent is the idea that u, d, $\bar{u}$, $\bar{d}$ and s and all these other numbers have to be positive since you can't have a negative number of quarks inside. I find this a simple way of understanding everything, and incidentally, in the same way, we can easily understand that there are certain sum rules. For example, the Adler sum rule just says that the total number of ups minus downs minus down bars plus up bars has to be 1 so that the total isospin of all of the components adds up correctly.

To summarize then, by scattering neutrinos from iron we can measure $q(x)$ and $\bar{q}(x)$, the momentum distribution of the non-strange quarks and of the non-strange antiquarks in the proton.

But furthermore, we have data from SLAC in which we measured scattering by electrons; they did the analogous experiment with electrons, and the big challenge was: do we see the same quarks by electrons as we see by neutrinos? It's a really dramatic moment to find out whether we really interpreted the scattering by electrons correctly in thinking of protons as made out of parts and quark parts at that. So we try to look at the partons with some other tool, and the neutrino is that tool. What we're trying to find out in these experiments, then, is; do we see the same distribution of partons? Electrons measure of course, ups and up bars and down and down bars and so on, but with coefficients that depend on the charge squared, because the probability of scattering an electron from an up quark is proportional to the square of the charge which is 4/9 and for down quarks 1/9 (see Fig. 2). If you never did believe that "nonsense" that quarks have nonintegral charges, we have a chance now, in comparing neutrino to electron scattering, to finally discover for the first time whether the idea that quarks have non-integral charges is physically sensible, physically sound; that's exciting. We don't have the same functions in σ_{eP} as we do in $\sigma_{\nu Fe}$ but electron scattering has also been done with neutrons, σ_{eN}. If you take the sum of electron scattering from protons and neutrons (which is like iron) you get 5/9 (from 4/9 and 1/9) times the number of quarks and the number of antiquarks, plus 2/9 the number of strange quarks.

$$\text{Elect, P + N} = \frac{5}{9}\,(q(x) + \bar{q}(x)) + \frac{2x}{9}\,(s(x) + \bar{s}(x)) \quad .$$

The same quantity $(q(x) + \bar{q}(x))$ is involved in summing up the neutrino and anti-neutrino cross section in iron. It should depend on y as $1 + (1-y)^2$ times a function of x. This function should be nearly the same function of x that we obtained in the electron scattering if we can neglect the functions $s,\bar{s}$ due to the strange quarks. We don't expect many strange quarks in the proton. We would not only have the same distribution but could check the 5/9. That's what I want to pay attention to, I want to point out in the summary what we're doing - what it meant when something agreed - it meant that we are checking the charges on the quarks. (You probably all knew that. I've got to make a summary, and now you know everything because you said it, and you heard it but how else can you make a summary if you don't just say the same thing over again.) By the way, the presence of $s + \bar{s}$ in the electron case just means that the electron data has to be higher; in other words, the neutrino cross section experimentally must be lower than an amount predicted, by neglecting that term, from the electron data. But it shouldn't be very much lower because every expectation is that the number of strange quarks in a proton is much smaller than the number of non-strange ones.

I was challenged to give an answer theoretically, as to what we think should happen in the experiment that Barish and Sciulli were doing. I replied that I had no difficulty, if the theory were right, with the sum of the neutrino and anti-neutrino for iron, as you just saw, because that involved the same numbers that we had already measured via electrons. But what happens when we do either one or the other ($\sigma_{\nu Fe}$ or $\sigma_{\bar{\nu}Fe}$) separately? It requires that you get out a crystal ball and guess as to how big $\bar{q}$ is compared to q. This (see Fig. 3) is the crystal ball I used to estimate how many $\bar{q}$'s there are compared to q's, the relevant size and the behavior of $\bar{q}$ and s and $\bar{s}$. I would like to explain what I did to fill my envelope (you remember the three envelopes Sciulli described that I was supposed to be opening as a result of the NAL experiment; a numerator which was the number of events, a denominator which was the neutrino flux, and the theoretical estimate of the ratio for each case, neutrino and anti-neutrino). I have to tell you more or less what I did and how I did it. This q(x) is the momentum of non-strange quarks and is, therefore, not the number. You have to divide by x to get the number, so q(x)/x is the number of quarks, $\bar{q}(x)/x$ is the number of antiquarks and the net number of quarks in the proton $\int(q(x) - \bar{q}(x))dx/x$ has to be three in order to get the conservation of all the quantum numbers (charge, isospin, and strangeness, or if you prefer simply baryon number). That's called the Llewellyn Smith sum rule. Now we also know to make any sense at all (i.e., to make this integral converge) the number of quarks and antiquarks must approach each other at $x = 0$. I write this integral as $\int(q+\bar{q} - 2\bar{q})dx/x$. Now we do have the data for $q+\bar{q}$ supposing of course that SLAC data is right, that we are interpreting it correctly and neglecting the small $s+\bar{s}$. Then we have this function $q+\bar{q}$ drawn in Fig. 3 From it we have to subtract some kind of a $2\bar{q}$ thing, divide by x and integrate and get a number as big as 3. It's almost impossible. This number $q+\bar{q}$ is so small experimentally you have to squeeze all the difference (of $q+\bar{q}$ and $2\bar{q}$) into small x to make use of that singularity near $x = 0$ to make the q and $\bar{q}$ different enough that you get some contribution of the integral over as much volume as you can. If you do a rough guess that it costs something to insist that a function will change rapidly, I try the slowest possible rates of change in the functions. (That's not proving anything, it's just guessing, that's all I was doing.) Then I say $2\bar{q}$ will disappear quickly, and get out of there so that I can get a big integral. Thus I have to have the $\bar{q}$ fall fast and the q then rise so that the sum is more or less constant (or whatever precise behavior it has for small x, it doesn't really make much difference to the answer, the curve for $q(x) + \bar{q}(x)$ was extrapolated to small x). It turns out that in order to get that integral as big as 3, $\bar{q}(x)$ has to fall very rapidly, in fact so fast that the area below the $\bar{q}(x)$ is about 5% of the area under the quark curve $q(x)$. Thus from this point of view, about 5% of the momentum would be in the antiquarks. You can't prove it has to be 5% because there are other shapes $\bar{q}(x)$ could have; you can have it come down faster and then go off with a very low value for a very long while. But no matter what you say, in order for this model to work, to be consistent, it's necessary that the number of antiquarks for any reasonable x's, not small but reasonable, like above 0.10, have to be a very small fraction of the number of quarks or we can't get a big enough input. So I drew this picture for $\bar{q}(x)$ of Fig. 3 predicted from that. (It has been noted by a number of people that this is the case and I'm not trying to say that I invented anything. We've been waiting for this experiment for many years. It was sometimes noted that this rapid variation in $\bar{q}$ was necessary and I've heard the expression "bizarre behavior" to describe this; yes indeed this is the behavior expected but I do not consider it bizarre). Perhaps we can summarize the result (that $\bar{q}$ is small) in the following way. In the rest system, as the nonrelativistic theory says, there are just three quarks and practically no pairs, for some obscure reason. When you look at this in the highly relativistically transformed system you would see just three quarks and no antiquarks. But mathematically you can show this is not quite possible and you have to have a number of pairs, for any field theory that has forces, has to make pairs. But the evidence that $\bar{q}(x)$ is so small shows that we make very few pairs of quarks and antiquarks (as measured by their momentum contribution) (although why, we do not know).

This also suggests s and $\bar{s}$ are very small. For the only way to get strange and antistrange quarks in the proton is when we're making the very few quark and antiquark pairs, because the normal quark model is supposed to be two ups and a down, nothing else. That is, the normal quark model has a slight complication, there are antiquarks and there are strange quarks that are probably produced by more or less the same mechanisms. Since there isn't going to be many antiquarks, I would guess it is only going to be consistent if there aren't very many strange quarks either. As a kind of a guess, I estimated that the number of strange quarks is less than the number of antiquarks which expresses the number of pairs, actually if you want to make a numerical guess it's about half, because there are two kinds of quarks, ups and downs in $\bar{q}$ and there's only one kind in s. Therefore probably s and $\bar{s}$ are more or less equal and less than quark bar, which means that with the experiments that we have, it has practically no effect - 1% corrections due to the uncertainty in this s as long as you are willing to say they are less than $\bar{q}$ and that the $\bar{q}$ is about .05. So that we can get rid of our uncertainties by that guesswork and now I compare the results to experiment (see Fig. 4).

Gargamelle did an experiment with a neutrino energy of 1-5 GeV and we have very good agreement with all those numbers and ideas. They got the charge to be 5/9 and the integral to be 3 and everything to be just great and we should be very happy because we're getting somewhere. But if it hadn't worked, all the theorists would have said it was because the energy is too low. It is therefore not fair, if it does work, to take any credit. Nevertheless it is interesting to ask why is it right at such energies. I don't know, but I would like to make some comments. I would like to say that the methods described by Mr. Von Krogh are quite sound. It looked like there is a certain hocus pocus using x and x' to get the thing, and so on, but I think it's an interesting idea which I would like to explain. You know that the W meson is supposed to be coupled in a combination V-A to the quarks in the proton. Suppose, just for a moment, it was coupled only with V, and we do the experiment with neutrinos. Now, we can do an experiment with electrons that is exactly coupled with the same V, and therefore (using isotopic spin you can change some charge numbers around but it's exactly the same vector), we should be able to predict perfectly exactly what happens with the neutrino experiment in terms of what happens with the SLAC experiment with electrons. Even if scaling isn't right, never mind scaling, it's a complete identity - because we're scattering with neutrinos or with electrons the proton sees the same thing, it sees something coming in giving it momentum with a vector coupling, V. I'm only repeating to emphasize it to you as a part of the summary, excuse me, it's the same with everything, I've got nothing new, it's not mine, everybody else said all this. It is right, therefore, to compare it directly to SLAC data. But people (Bloom and Gilman) have studied SLAC data, massaged it gracefully, and have discovered that there is a way of remembering all the SLAC data which is to use a funny variable x' instead of x. As Mandula explained to me, there was no more reason to have to define the original x as $-q^2/2M\nu$ rather than as $-q^2(2M\nu+M^2)$ the expression for x'. I just didn't know what combination to use. In the limit of high energies they are the same thing, but at low energies it turns out there's a certain particular way of fiddling about, so that the SLAC data can be remembered; in terms of a parameter x', the function looks more universal. Alright, that just summarizes the data. Therefore using the same summary system for neutrinos is a perfectly legitimate thing, if the coupling were exactly vector. It is really vector and axial and we have no legal right to suppose that the axial coupling is going to imitate precisely the same kind of off-scaling behavior if you wish, or scaling using x', as the vector. However, perhaps, we have discovered that the axial and vector behavior might not be too different. It was reported in this meeting, the axial form factor, at least, goes down with momentum just like the electric one. The vector coupling, goes as $(1-q^2/m^2)^{-2}$, for the axial you can't tell the form, but it goes down at about the same speed and the form and the mass m could easily be the same. It is just about the same value as the vector case. So it is possible that we have something to learn, that someone should now sit down and make an analysis, possibly a little more organized than what Mr. Von Krogh did,as to what he wants to assume and where all the variables go in order to make the best guess as to how the axial curve behaves, from knowledge of the way the vector curve behaves. One could try to put the things together and sort of put them in a more formal way, because it seems to me the world is a little simpler than we have a legal right to expect. Someone might guess that simplicity as a reasonable sort of semi-phenomenological guess, devising a system for comparing electron data with neutrino data at any energy, and therefore being able to use any energy at all in this comparison and be able to make more sense out of the connections. That's a suggestion for improving our way of understanding low energy. In other words, I take this as being a possibility that something is simple that we didn't expect to be simple.

On the other hand, all the objections to the low energy are cured. if we go to high energy and there are two experiments at high energy from NAL reported here. One, which is called 1A, was described by Mr. Cline, and the other one, number 21, is reported by Mr. Sciulli. I'll discuss first the one described by Mr. Sciulli (Fig. 4). The first remark to make is that all those distributions as a function of x that you saw were all consistent with the shape of the distribution at SLAC. Good thing, electrons see the same distribution as neutrinos do. That is very impressive because we've increased the energy by a factor of five and we look at a much higher momentum transfer, yet the partons still look like points, and it's the same points in the same distribution. Thus the idea that we used to use to interpret SLAC is very sound. As we all know, you may convert that idea if you want to be more formal, from the statement about particles inside to a statement about singularities of a certain commutator on the lightcone. That is, at the lightcone the commutator has a certain jab, and that jab is very sharp and continues to stay sharp as we go up in energy. That's very impressive.

If I may have a moment to think historically, it's very much like the days of Rutherford in which there was an atomic model in which there was a smear of charge, but by scattering α's he found there was a hard sharp center of charge. So in the same way, by scattering these things on a proton, we have found out that although at first we might have thought that the charges were somehow smeared around in that space, and nature was smeared in space, it turns out no, nature has local couplings and sharp point-like insides. We still see it as pointlike even though we've got a 40-50 GeV^2 momentum transfer. This statement that the mass of the W boson has to be bigger than 10 GeV is also equivalent to saying the quarks look like points down to energies of this same order of 10 GeV. This happens also to be as close as we know that electrons look like points from tests of quantum electrodynamics. 10 GeV is the number that we know for electrons. By the way, if a hydrogen atom were magnified so it is 200 feet in diameter, then the proton is a centimeter across and we know that an electron is a point to as near as the eye can see which is a 200th of an inch (the visual acuity of the eye) and we now know that inside the proton there are points to the same acuity. It is always good to get your models magnified but not over-magnified, atoms look now 200 feet in diameter.

The distributions in y were not tested as sensitively as the distributions in x but they were perfectly consistent with the distributions in y to be expected, nearly flat for neutrinos, nearly $(1-y)^2$ for antineutrinos. That's the end of that summary, I just summarized everything that he showed. Now comes the interesting question, about this charge squared; we have the right shape but do we have the right size? Measuring the size means measuring whether the quantum numbers, the square of the charges on the quarks as seen internally is really $(2/3)^2 + (1/3)^2$. This is the first (high energy) test of that, as electrical quantities, not just adding up some quantum numbers. In order to do that we have to have the absolute total cross sections for neutrinos. The total cross sections for the neutrinos and antineutrinos is an integral of the expressions in Fig. 2. They are therefore expressible directly in terms of the total momentum carried by the non-strange quarks and antiquarks. The total cross section, from this point of view, measures directly, presumably, the total momentum carried by the non-strange quarks and antiquarks, $\int(q+\bar{q})dx$ (without using any SLAC data or numbers from other experiments). It is the momentum of the nonstrange quarks, the things that are coupled to the neutrinos, that is what is being measured directly. Combining that with the data from SLAC will give us the mean square charge of the quarks (correcting for the strange quarks which I will neglect here), which should be 5/18, 4/9 and 1/9 averages 5/18, which is .278. I have been asked if I would give the detailed numbers of Sciulli instead of just his slide because a number of people couldn't get the numbers off the slide very well when he showed it. In Fig. 4 you find the values for the cross-sections. These are the result of opening those envelopes and dividing those numbers, they have not been changed in any way since that moment, so nobody's been prejudicing anybody. They (Sciulli and Barish) have asked me to please present the result for the total cross sections so here we are. (I would also like to say of course the slight hesitation and worry that Mr. Sciulli had is because the number itself you may later find changed by a few percent. However, in the errors that are given here there is a systematic error that's been guessed at large enough to include many things they might have forgotten in their hurry to make the calculations. Probably when the data is more carefully analyzed these numbers might move around a few percent one way or the other, and it is possible that the systematic errors may be somewhat reduced. These are not final official figures, in that sense, but you can guarantee the final figures will be in the range of the errors given here, and the errors themselves may be smaller.) The energy of the neutrinos from the π neutrinos in the experiment was on the average 37.5 GeV (the range was ± 9 GeV). The cross section for those neutrinos on iron was $33 \pm 8 \times 10^{-38}$ cm^2, and so forth (see Fig. 4). For antineutrinos the cross section is 12.5 ± 2.5. The sum of the two is $45.3 \pm 9 \times 10^{-38}$ cm^2. For neutrinos from k-mesons, which have an average energy of 108 GeV (of range ± 15 GeV), the cross sections were 146 ± 39 of the same units of 10^{-38} cm^2. And the antineutrinos gave 41 ± 18. The numbers marked "theory" are those I've computed using the SLAC data. (I used .30 for the sum of the integral of νW_2 on the proton and on the neutron. This is a slightly different answer than Mr. Bloom quotes (he got .28) because I wasn't quite as prejudiced as exactly how to go near small x. I also corrected for the inequality of the number of neutrons and protons in iron. There is a 1% uncertainty because of my lack of knowledge of s, the number of strange quarks. Of course we predict the sum of σ_ν and $\sigma_{\bar{\nu}}$ much better than we do each alone because each alone requires that uncertain estimate of .05 for how much momentum is in quark bars and how much in quarks.) You see that the theory fits very nicely with the exception of the neutrinos from kaons on iron, which is distinctly outside our experimental error. You may remember that in Sciulli's curves the highest energy point for neutrino scattering on iron was higher than the curves, and this is the way it looks numerically. As we noted, from $\sigma_\nu + \sigma_{\bar{\nu}}$ we can obtain the total momentum carried by the quarks. (It turns out accidentally, the way numbers go, that in these units, you just take 1/2 the slope of the cross-section curve versus energy. Theoretically the cross section should be (some slope) x (the energy in GeV) x 10^{-38} so you just divide this cross section by the energy to get the slope and half the slope turns out to be the total momentum carried by the quarks. My predictions for the slope was .80 for neutrinos, .30 for antineutrinos.) So this experiment with the π neutrinos at 37.5 GeV tells us the momentum of nonstrange quarks is .61 ± .10. The total momentum of the proton is one so the rest of the momentum is carried by god-knows-what, called gluons. We don't know what carries the rest of the momentum but it's not at all crazy that something does. If all the momentum were in the quarks we would be rather nervous because if you have quarks interacting, all held together somehow, and they're more or less relativistic as the distributions show (because if a non-relativistic system were looked at, in motion like this, then the distribution of the momenta as a function of x would have a very narrow peak) then there is some delay in interaction across the face of the thing. One particle interacts with another but not instantaneously across the distance so one has lost its momentum before the others have picked it up, so that temporarily there's a missing momentum. This is usually described as being the momentum in the field, the momentum in the "photons" that are being exchanged. Thus on the average the momentum of the interacting parts of an object cannot be the total momentum of the whole thing. It is not impossible that the total quark momentum be less than one, in fact it just tells us that there's more to it than just quarks. There is something else, to hold the quarks together. At present we have no experimental evidence on what it is in detail. It is called "gluons," it holds things together, that is all we know. The higher energy neutrinos give .86 ± .20, the average of these two numbers (with the weights from the errors) gives you .65 ± .08. I compare that another way, SLAC data taken directly with the errors given for the integrals (and multiplying by 9/5) gives the mean momentum .54 ± .07. This number from SLAC data can be compared with this number from neutrino data. Insofar as they are equal, the sum of the square charges of the quarks is 5/9. One more way (it's all the same data, but just to make sure you get the idea) the mean charge squared which should be 5/18 = .28 comes out .25 ± .04 for the 37.5 GeV neutrinos, .18 ± .05 for the 108 GeV neutrinos and .23 ± .04 for the average. These errors do not include the error that might be in SLAC data which amounts at this scale to another ± .04. The 108 GeV data is still outside the experimental error by 1-1/2 standard deviations and what could we say about that. One and a half standard deviations is not worth a theory. (Another fact, someone told me, is that in spite of statistical theory, in high energy physics data is three standard deviations off at least 10% of the time.)

If you want to take it more seriously, you can ask why does the neutrino cross section rise at these higher energies? One nice suggestion is that new channels are being opened in the hadrons; there are various alternative suggestions for why. Another suggestion is that something is wrong with the β decay coupling theory. The intermediate boson effects would make the neutrino cross section go down, not up. Some new gradient coupling on the neutrino side would be required. It is easy to find out if the neutrino coupling or the hadron behavior is at fault. All we have to do is compare data at the same q^2 and ν with one energy neutrino and another energy neutrino, because the hadron end is affected by q^2, ν but not by the neutrino energy directly (except via the polarization of the W). Such data would test whether it is the β decay theory or if it's an error or the hadron machinery and Mr. Cline described various regions of the x,y world that you have to look at to explore various kinds of failure of hadron physics.

Now the other experiment (1A) that was done at NAL and described by Mr. Cline, reminds us that everything might not work so well theoretically. In those experiments it appears that the stuff does not scale, that the y distribution observed from antineutrino scattering which should look nearly like $(1-y)^2$ according to the theory, does look like that for low energy neutrino scattering but does not for high energy neutrino scattering. It looks like there is more scattering at large y. This is not inconsistent at all with the experiment (21) described by Mr. Sciulli because that experiment is, at the moment, not sensitive to high y. I would advise you, however, again, not to worry, just wait. I don't say which way it's going to come out, I just say you'll get it resolved one way or the other. By the way, this cannot explain the rise of the cross section with energy (in experiment 21) because they found a rise in cross section in a region where they weren't sensitive to this effect. We shall find out more in two ways; first of all, Mr. Cline said, and I quote, "we are still not sure whether this is real," in other words, he wants to make some more checks. The other is, experiment 21 has lots of high y data for which they were unable to measure the exact momentum of the muon, but you can calculate how many you're supposed to get if the distribution is really $(1-y)^2$ and how many you would get if it's not and they would be very distinctly different if this result of experiment 1A were right. The data is all taken, it just has to be analyzed and we just have to wait about a month as I understand it, to get a number. (It, of course, will be equivocal, it will be half way in between, as usual, with an error that will allow you to have either answer. That's my prediction, it's going to be equivocal.) That is the situation with the conventional theory. I would now like to describe a new subject, namely difficulties with the theory and attempts to correct the difficulties.

Now that we think it looks good experimentally, we worry about it not looking good theoretically. I'm going to try to include, for completeness, all the things that are not satisfactory (see Fig. 5). (In speaking of difficulties of our present day theories I should also have to talk about difficulties of the naive quark hadron model. That involves manifold difficulties that you are very well aware of; the statistics problems involving the three quarks, whether they're slow or fast, why they don't come out and whether the partons really have limited transverse momenta and if so, how, etc., but those all belong to the strong interaction problem and so I will not discuss them. But to summarize this meeting we do have to discuss weak interaction theory.)

The first thing is that it's only a first order theory, if you try to do anything in higher order it gives complete nonsense. You can say it diverges, or unitarity fails or something, you cannot extend it directly to higher orders, you get divergent difficulties. Furthermore, we have one measured phenomenon which should be interpreted a second order effect of the weak interaction theory. This is the mass difference of k-long and k-short. And experiment says the result is not only convergent but also small. It's like the days of the Lamb effect. In 1928 electrodynamics gave infinity for higher order processes, everybody knew infinity was not right but it took a good experiment to give it the value that wasn't infinite, namely the Lamb shift. Then everybody started trying to get the experimental numbers. So what we should do here, since we have an experiment sensitive to second order - we should try to get that number.

There is now evidence for neutral currents and that is not expected in the conventional theory. That shows an incompleteness of the theory.

The third point is the Cabibbo angle. It's alright, it's very nice, and it fits everything but it's very mysterious. What on earth could determine that number, how does that come out, it's a funny world. It is not a complete theory, we don't like parameters in a theory.

Another thing that the theory gives us no clue about is the fact that there is a muon and electron doubling. Why is there a doubling? I mean that question in the sense of, are there other μ's and ν's at higher energy? We don't know, we don't understand that at all, and the theory is incomplete about that. We just wrote down two terms because we saw two things, so if we write down three terms are they connected?

We also have a CP violation.

I see no reason to include the solar neutrino problem in here because it is only 1-1/2 standard deviations off of some solar models and the principle is, you don't make a new theory because of 1-1/2 standard deviations.

Now people have struggled to fix this theory, in particular to fix the divergences, they start there. The way they thought to do it was to insist that the field theory be convergent. (This was described here by Ben Lee.) To do that then the idea is to make up some new diagrams to cancel the divergences. It turns out to be possible if the theory has deeply imbedded in it certain symmetries. Those are called gauge theories (see Fig. 6). They have one extra thing that comes out of them that is beautiful, they permit one to try to make a symmetry between weak interactions and electromagnetic interactions and connect those two. This wasn't on the list as an incompleteness of conventional theory because we don't know whether they are connected, but it's a nice thing - a possible relation of electromagnetic and neutrino interactions. The question is, of course, does this attempt to get rid of the divergences help us with any other thing? I ask this because it is always my experience that if you are on the right track you get more out than you put in. After you get the right idea a few things come out that you didn't put in. Whey they made the gauge theory they had wanted to make it fit with electrodynamics. They wanted to make it converge. So they put the two things in, the symmetry and the method of convergence, by force to make those two things happen, but nothing else with the exception of the neutral current came out. No understanding of the Cabibbo angle, no understanding of the small kaon mass difference, and so forth. We do not have any clues as to any of the fundamental difficulties of our theory by this step. There are very very many gauge theories, people try different kinds of symmetry, a very nice idea, it might be right, but one must remember that there is no yield of new things.

One particular example of such a theory is the Salam-Weinberg theory. Here we add to a gauge theory the principle that there is a relation to quantum electrodynamics, and the particular method used by Salam-Weinberg implies that there are neutral weak currents. (I'll repeat much of what Madam Gaillard said.) The problem is, how are these neutral currents coupled to the hadrons. The charged weak current is, as we have said, coupled, is the up quark u multiplied by a linear combination that we will call the Cabibbo quark $c = d \cos\theta_c + \sin\theta_c$. Now, if you add neutral currents in order to get convergence and symmetry, then the natural way to have it coupled is with the Cabibbo quark and the anti-Cabibbo quark $(\bar{c}c)$ plus or minus (depends on your theory) the up quark and the anti-up quark $(\bar{u}u)$. Those are the only kind of quarks that the charge current couples to so that when you twist it to find the corresponding neutral combination, this is the combination you get. But the trouble now is that the term $\bar{c}c$ expanded in terms of s and d contains a term $\bar{s}d$ and that means that there would be neutral currents coupled substantially to a strangeness changing component. However, we know experimentally (from K° decays) that such a term must be very small, if it exists at all. So the theory is dead? No, the result is evaded - the theory perpetually evades experiment by inventing new things. It is not predicting these new things, I have to explain, but it invents them to evade the obvious results of the nice theory it would be if we didn't have to fix it up all the time. An evasion in this particular case is to invent a quark called c' which is another linear combination, nothing new. Just the orthogonal combination of known things $c' = d \sin\theta - s \cos\theta$. Then the neutral current could contain $\bar{c}c$ and $\bar{c}'c'$. If you add them the cross terms in $\bar{s}d$ disappear. You get no $\bar{s}d$ term and you have solved the problem (of the absence of leptic decay of the neutral K°). But now you turn around and you remember symmetry and you ask what charge current goes with that symmetric one $\bar{c}c + \bar{c}'c'$. Now you're in trouble, if you keep the symmetry and you want the charged current coupling to work right, then the $\bar{u}c$ coupling implies something has to go with the c'. Let us call it u'. It has to be another thing to be symmetric, so you have to invent another kind of quark. That quark, u', therefore is a fourth type, quarks must have four types, and you mumble SU_4 instead of mumbling SU_3. This means a new quark type. It turns out, of course, that u' has to have an electric charge of + 2/3, and the conservation of this new quark type inside the baryon system (just like the conservation of strange quarks gives us the strangeness number rules for interactions), implies a new rule for strong interactions; some new quantum number (the net number of these u' quarks inside) would be conserved. That new quantum number is called charm. So what they mean by a "charmed" particle is not a new particle but new hadronic states which have a new quantum number (because the hadrons have one or more of these new quarks inside). We have no evidence for any of these new hadron states, or the charm quantum numbers. How to look for these states was discussed in considerable detail in the meeting but I won't summarize that here. Whether there is or is not charm, doesn't effect our deep inelastic scattering experiments, but I won't say any more.

There is one important problem that the idea of charm does solve (I think it was this that suggested charm in the first place). The k-short, k-long mass difference is, as we have said, of second order in G but divergent, proportional to the square of a "cut-off mass," in the conventional theory. In the usual gauge theories it is convergent (for everything is) the "cut-off mass" being about the mass of the W-bosons. The problem is that this, though finite, is much too large compared to experiment - a "cut-off mass" more like the low GeV region is indicated. In a system including charm there is cancellation among diagrams where u' replaces u; the "cut-off mass" now becoming like the difference of mass of u' and u - or, more precisely, the difference of mass of charmed hadrons states and normal hadron states. From this we conclude that the charmed hadron states cannot be more than 1 or 2 GeV above their normal counterparts.

That brings us, however, to one of the great successes of the Salam-Weinberg model, which is to predict neutral currents (see Fig.9). But I should like to follow the advice of Mr. Mann. Neutral currents should be studied in their own right. That means the experimenters should say, alright, we have neutral currents, let's find out what their properties are. (Rather than just comparing them to the theory of Salam and Weinberg.) The reason is that there are many and various kinds of symmetry gauge theories which involve properties for the neutral currents different from that of Salam-Weinberg. There have, historically, also been many other suggestions of neutral currents (for example, $\Delta I = 1/2$ rule for non-leptic decays made trouble in the beginning and people suggested neutral currents to explain that).

So it could happen that these guys thought of neutral currents and that there are, in fact, neutral currents, but for two entirely different reasons. There are so many particles that if you find one of them it is no longer possible to say that it confirms a theory because there are so many possibilities. So it is better to think of the analysis of neutral currents in a more general way. To outline the possibilities; the neutral coupling is probably a lepton current times a hadron current. The first question is what kind of a lepton current, in fact what kind of lepton, presumably a neutrino coming in, but what comes out? Let's start by guessing it's a neutrino again which comes out, and being left-handed they would have to couple various ways, the simplest one is the V-A, but there are other possibilities here. It is easy to do experiments which get rid of a few of them. Mr. Sakurai pointed out we already have data which very strongly indicates (the y distribution falling instead of rising) that it's not s and p coupled, but I don't want to start that whole hassle again. The point is I would like to hear a discussion of the properties of things, not in terms of a particular model, but in terms of possibilities determined by experiment because there are too many theoretical possibilities. If I take, as best guess, the example that the lepton current is V-A neutrino-neutrino then next I have the problem of the hadron current. That could be many things, the most natural guess and the one that is consistent with the guess of the Salam-Weinberg, is again that it is some linear combination of V and A. I choose to write it as a linear combination of V-A and V+A. To what extent does it couple with V-A and to what extent to V+A? It's only a mathematical expansion, we leave it to experiment to determine the coefficient. Now let's take the V-A part. That's the part that screws the right way according to the charged current coupling. It is neutral, so it could be coupled to a u and $\bar{u}$ quark or a d and $\bar{d}$ quark or an $\bar{s}s$ (not $\bar{s}d$ to avoid $K^+ \to \pi^+\nu\nu$). Let's not confuse things by adding charm and so forth, because now we're only doing experiment with uncharmed particles anyhow (there being, presumably, few charmed quarks in the proton). So we have unknown coefficients α,β,γ to couple those in (see Fig. 7). In the same way with the V+A term we have unknown coefficients, α',β',γ' so we have six unknown coefficients to discover. The simplest guesses a theorist would make without knowing anything about gauge theory symmetries among various currents is to try some nice isospin or SU_3 symmetry. For instance, perhaps $\alpha = \beta = \gamma$, that's a nice possibility, but we don't have to be prejudiced about it. If you do that then the isospin change here is 0. Another neat possibility is to have pure isospin 1 just like the isospin of the non-strangeness changing charged current. That would make $\alpha = -\beta$ and $\gamma = 0$, but they could be anywhere in between. There are many possibilities, and it might be wise to think of the various experiments as simply determining these coefficients. There is already some evidence that the isospin change $\Delta I = 1$ is small because the Δ production for protons by neutrinos is very small (if the current were pure I spin 0, $\alpha = \beta$, the Δ could not be formed at all).

Various experimental quantities should be calculated in terms of these parameters, to see what values of the parameters might agree with all experiments. For example, the "neutral current" differential cross-section σ_ν^N (x,y) for neutrinos on iron (N = Z) leading to neutral leptons (presumably neutrinos) is $(\alpha^2+\beta^2)\sigma_\nu^c + (\alpha'^2+\beta'^2)\sigma_{\bar{\nu}}^c$ where σ_ν^c, $\sigma_{\bar{\nu}}^c$ are the cross sections on iron for neutrinos and antineutrinos making muons (if we disregard strange quarks). This is true in the scaling region for any x,y. Experiments reported here for R_ν, the ratio of total neutral to total charged cross sections for neutrinos, (which should be $(\alpha^2+\beta^2) + (\alpha'^2+\beta'^2)\rho$ where ρ, the ratio of total cross section for charge currents for $\bar{\nu}$ and ν, is known to be near .38) are not entirely consistent, but the value is roughly 0.2. For antineutrinos $R_{\bar{\nu}}$ is $(\alpha^2+\beta^2) + (\alpha'^2+\beta'^2)\rho^{-1}$ and is experimentally, possibly 0.4. When more accurate values of these ratios and other measurements become available we shall be able to get a clearer idea of these parameters, and might be led to a simplifying guess.

Now I should like to discuss what should happen next with these high energy experiments (see Fig. 8). What happens if we ultimately discover real deviations from scaling? If we do find that, then the first problem is whether it is in the weak theory, perhaps an effective mass of W or something else in a weak theory. I have explained how you can separate that from a hadron effect very neatly (in principle) by using the same q^2, ν and two different energies of neutrinos. If the scaling failure is hadron physics, it may be new channels are opening up inside, for example, charm. And don't forget the famous fact that the rate that e^+e^- makes hadrons is a factor of three too large, according to the naive quark parton model. That looks like something important; although perhaps the logic of the model expectation is wrong, it shouldn't have been as small as we thought (I doubt this explanation). Alternatively perhaps we see here that some new channels are opening or something. At any rate it is very worthwhile admitting the possibility that something's going to go wrong. It's interesting that it hasn't gone wrong yet in any of our deep inelastic reactions (except with the clue of Mr. Cline) which are in the negative q^2 region. This may have many other explanations, new particles which exist but which are not found in the proton, and so on, that I'm surprised that when we heard of theories of some of our new particles we didn't hear more about them as being produced in this e^+e^- reaction. For here we have an actual experiment which is making something. I'm sorry I said e^+e^- into hadron, the experiment doesn't measure that, it measures e^+e^- into things that look like hadrons and they are not absolutely sure in many situations how much is hadrons and how much is something else, so we had better leave it that way.

We also have to worry about the naiveté of the parton model. The parton model got its ideas from field theory and the idea of the Foch wave function, which says that you can describe the wave function of a system by giving the amplitude to find all the different ideal particles, the quanta of the field, inside. That is where it began, but it added the idea that the perpendicular momentum is limited - an idea that was suggested by some experience with high energy hadron collisions. This idea is not really consistent with the ideas of field theory, and anyone who studies field theory deeply can prove again and again, (it has often been done) that the naive parton theory with this limited transverse momentum $K_\perp$ is not valid. In other words they say scaling cannot hold, for the scaling is derived by assuming that the $K_\perp$ is limited. It may be that field theory doesn't hold, but the theorem (possibly-it hasn't been proved yet) is that if a field theory holds scaling doesn't hold, so it's always good to look for deviations experimentally. The deviations from having larger transverse momenta in the wave function (and some hint of that is indicated already by high transverse momentum hadronic experiments) indicate that there should perhaps be logarithmic deviations in scaling.

We heard a report by Mr. Zee on the characteristic behavior of the type of logarithmic deviations expected from one particular type of field theory. There are very few types of field theories that we even think that we can calculate. They all diverge or blow up or do something terrible. But there is a special combination, which is colored quarks with colored glue, which people believe is controllable. It is called "asymptotically-free" theory. With it they make estimates on how the logs will appear and they've been able to prove that the moments of x (that is, the average $(\text{momentum})^n$ for integral n, the integral of x^n times νW_2 over the integral of νW_2) should vary with inverse powers of the logarithm of q. In the particular case of the mean x, it should vary inversely as roughly the square root of $\ell n\, q^2$. These people also tried to invert these formulas to try to find out how the function νW_2 actually had to behave. I don't believe that the inversion was right, I think they made too many assumptions about the uniformity of the convergence, but that's a personal opinion. However, it has nothing to do with it. The right and easy way is to test directly the consequence that they're mentioning, that the mean value of some power of x varies inversely as an appropriate power of $\ell n\, q^2$. The way experiments are, with their errors and so forth, clearly the lowest power is the most sensitive to test it out on, and so we'll take the mean x. It should, according to them, vary as $(\ell n\, q^2)^{-1/2}$ (.48, or whatever that is, not exactly 1/2). The trouble is we don't know q^2 over what! If it were q^2 over 1 GeV then it turns out that this is a rather substantial variation over some possibly observable range. So it is a good idea, and I would suggest, (and I think Gell-Mann would agree with me here because he's always saying the same thing), take the data directly and integrate to find the mean x and plot it against $\ell n\, q^2$ to show whether it is constant or has a tendency to fall with $\ell n\, q^2$.

We are now stepping out. You see we made an important step, and this meeting culminates that step in a way. It is telling us it really looks good that the protons are made out of parts, quark-like parts, and that the field-theory type ideas or parton-like ideas or current-commutator ideas, are on the right general track (in spite of the e^+e^-) in describing the hadron particles. Therefore we now must move on and improve the naiveté of these parton models, and so on, which were good as crutches to get a good idea what could be happening, but do not yet represent solid theory.

We should now start to think about trying to find out, for example, if asymptotic field theory is on the right track. And they say that this mean x should drop with the power of $\ell n\, q^2$ more or less. Let's see if it's right. Or let us see, if instead, scaling behavior is actually exact.

I think this meeting is at a good moment. As it finished, it is apparently agreeing with an idea, (with some hints of difficulty in Cline's distribution as a function of $(1-y)^2$), culminating otherwise apparently nicely. It is remarkable that results agree so well when we vary the energy so much. The nearest competitive theory which gives different predictions, by the way, had a mean square charge of .5 (that's the Han-Nambu theory of quarks, assuming the SLAC energy is high enough to break them up!), .5 instead of .28, that's definitely excluded. Our experimental errors of 15 or 20% appear rather remarkably small in such comparisons. Finally therefore, I think of this meeting this way; that nothing is ever a given moment in history, we are always making steps, but here we are kind of completing one, a test of the naive quark-parton model. So maybe we ought to start moving in a new direction to develop a more complete theoretical picture so we can say whether scaling should work, or where it should break down. At the same time, this meeting stands at the opening up of an entirely new phenomenon - that of the neutral currents.

During the talk a series of view graphs were shown (Fig. 1 to Fig. 8 below). The content of them is reproduced here, although with the text most of the material in them is often unnecessary and without the text, they are unclear. They are reproduced because they form a kind of outline of the material of the talk which may be of use to the reader. In some cases they explicitly contain equations which are only alluded to in the text.

CONVENTIONAL THEORY OF WEAK INTERACTIONS VIA

NAIVE QUARK VIEW OF HADRONS:

INTERACTION = CURRENT • CURRENT

(VIA INTERMEDIARY HEAVY BOSON W?

$M_W > 10.3$ GeV)

CURRENT IS (V - A) $= (\bar{e}\nu_e) + (\bar{\mu}\nu_\mu) + (\bar{c}u)$ ← HADRONIC CURRENT

(PARTICLES, LEFT HELICITY

ANTIPARTICLES, RIGHT) $c = d \cos\theta_c + s \sin\theta_c$

$(\sin^2\theta_c = .046)$

FIGURE 1

CONVENTIONAL THEORY: COMPARE EXPERIMENT

Low Energy O.K. (except ? a) why $\Delta I = 1/2$ for non-leptonic meson decays

b) P-wave non-leptic baryon decays)

High Energy: - Naive Quark Parton View of Hadrons

νP deep inelastic

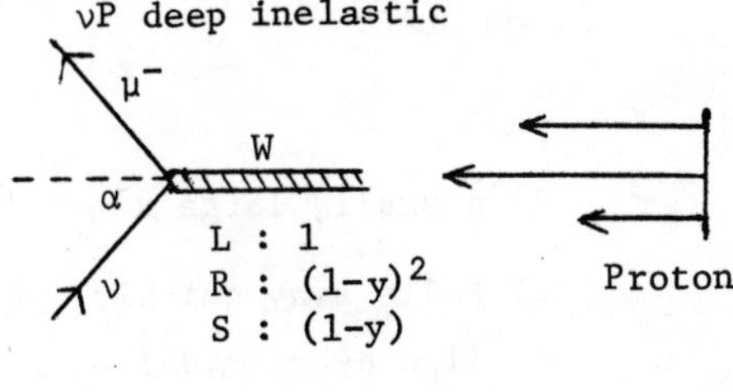

$\nu \to \mu^- \therefore \quad d \to u : L$

$\quad\quad\quad$ or $\quad \bar{u} \to \bar{d} : R$

$$\boxed{\sigma_{\nu P} = 2x[d(x) + (1-y)^2\bar{u}(x)]}$$

(unit is $G^2S/2\pi$)

$$\therefore \sigma_{\bar{\nu}P} = 2x[\bar{d}(x) + (1-y)^2u(x)]$$

$$\sigma_{\nu N} = 2x[u(x) + (1-y)^2\bar{d}(x)]$$

$$\sigma_{\bar{\nu}N} = 2x[\bar{u}(x) + (1-y)^2d(x)]$$

$$\boxed{\begin{aligned}\sigma_{\nu Fe} &= q(x) + (1-y)^2\bar{q}(x)\\ \sigma_{\bar{\nu}Fe} &= \bar{q}(x) + (1-y)^2q(x)\end{aligned}}$$

where $q(x) = x(u(x) + d(x)) =$ Momentum in non-strange quarks

$\bar{q}(x) = x(\bar{u}(x) + \bar{d}(x)) =$ Momentum in non-strange anti-quarks

$$\text{Electric } P/x = \frac{4}{9}(u+\bar{u}) + \frac{1}{9}(d+\bar{d}) + \frac{1}{9}(s+\bar{s})$$

$$\text{Electric } P+N = \frac{5}{9}(q(x) + \bar{q}(x)) + \frac{2x}{9}(s+\bar{s}) = \nu W_2^D$$

$$\sigma_{\nu Fe} + \sigma_{\bar{\nu}Fe} = [1 + (1-y)^2](q(x) + \bar{q}(x))$$

FIGURE 2

THEORY <u>ESTIMATE</u> OF $\int \bar{q}(x)dx / \int q(x)dx$ and $\bar{s}, s$

$$q(0) = \bar{q}(0) \quad (\approx 0.5 \text{ by SLAC})$$

$$\int_0^1 (q(x) - \bar{q}(x))dx/x = \text{Net Number of quarks} = 3$$

$$\therefore \int_0^1 [(q+\bar{q}) - 2\bar{q}]dx/x = 3$$

To get this, with $q+\bar{q}$ from SLAC:

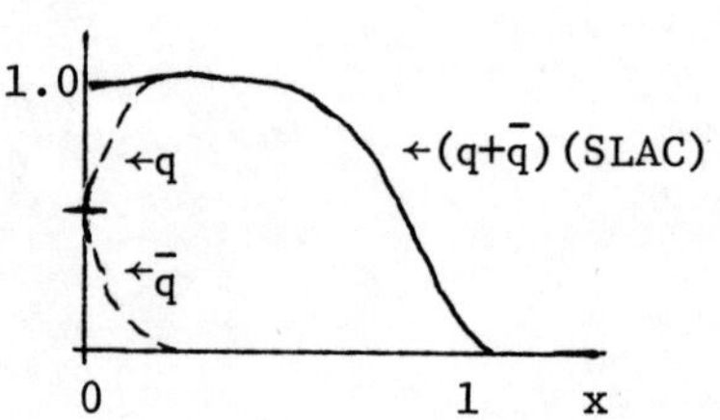

(1) $\bar{q}$ small, large x

(2) Falls away quickly (1/e by $x \sim .05$)

(3) $\int_0^1 \bar{q}(x)dx / \int_0^1 q(x)dx \approx .05$

Also guess $xs \sim x\bar{s} < \bar{q}$

FIGURE 3

EXPERIMENTS

Gargamelle E_ν 1 to 5 GeV Very good agreement. Why?

$$\int \bar{q}\, dx / \int q\, dx \simeq .05$$

NAL (Experiment 21, Sciulli report)

Distributions O.K. - Consistent

(Insensitive to large y) $M_W > 10$ GeV

Total x-section: yields total momentum carried by non-strange quarks

(σ = SLOPE $\cdot E_\nu \cdot 10^{-38}$. Momentum Quarks = (0.5) SLOPE)

Add SLAC: yields mean square charge on quarks

(if theory: 5/18 = .28; SLAC says momentum quarks = .54 ± .07)

Results

$\bar{E}_\nu$ (GeV)		σ in 10^{-38} cm^2 EXP.	THEORY	Momentum quarks	(mean charge squared implied)
π 37.5	σ_ν	33 ± 8	30		
	$\sigma_{\bar{\nu}}$	12.5 ± 2.5	11		
	Sum	45.3 ± 9	41 →	.61 ± .10 ↓	0.25 ± .04
				Average .65 ± .08	0.23 ± .04
	σ_ν	146 ± 39	86		
	$\sigma_{\bar{\nu}}$	41 ± 18	32	↑	
	Sum	187 ± 43	118 →	.86 ± .20	0.18 ± .05

NAL (Experiment 1A, Cline) Not Scale

Large y, antineutrinos not $(1-y)^2$ for large E_ν .

FIGURE 4

DIFFICULTIES AND INCOMPLETENESS OF CONVENTIONAL WEAK THEORY

Higher order diverge ↔ Gauge theories cure.

$K_L - K_S$, Δm is small!

Neutral currents

Cabibbo angle?

μ, e doubling

CP violation

~~Solar Neutrinos~~

DIFFICULTIES OF NAIVE QUARK HADRON MODEL

Manifold: Statistics. Non-Relativistic or Relativistic?

Parton $K_\perp$? No free quarks. Gluons?

etc. etc.

FIGURE 5

NEUTRAL CURRENT THEORIES

Gauge theories → Symmetry

+ Relation to Q.E.D. (e.g., Salam, Weinberg) → Neutral Currents

---

Charge Current $\bar{c}u$ $c = d \cos\theta_c + s \sin\theta_c$

Neutral Current $\bar{c}c \pm \bar{u}u$
(by Symmetry) ↳ contains $\bar{d}s \cdot \sin\theta_c \cdot \cos\theta_c$ $\Delta S = 1$

Experiment says: No neutral $\Delta S = \pm 1$

Evasion Define $c' = -d \sin\theta_c + s \cos\theta_c$

Suppose Neutral Current Contains $\bar{c}'c' + \bar{c}c$

$\therefore$ No $\bar{d}s$.

Problem Charge Current? Symmetry?

$\therefore$ $\bar{c}u + \bar{c}'u'$

u' = New quark type, charge + 2/3.

Implies new hadron states with new "charm" quantum number (no. of u').

Can solve small $K_L - K_S$ mass difference if charmed states not too high.

FIGURE 6

NEUTRAL CURRENTS, GENERAL PROPERTIES

"Study in their own right." A. Mann

Lepton current · Hadron current

V,A or $\not{S}$, $\not{P}$, T

Example lepton current V-A, $\bar{\nu}_\mu \nu_\mu$

$\therefore$ Hadron Current

$$(V-A)\ (\alpha\bar{u}u + \beta\bar{d}d + \gamma\bar{s}s)$$

$$+\ (V+A)\ (\alpha'\bar{u}u + \beta'\bar{d}d + \gamma'\bar{s}s)$$

Examples if $\Delta I = 0$ then $\alpha = \beta$ ($=\gamma$? SU_3)

Astrophysics: Scattering of ν_μ from nucleus is coherent, amplitude as N + Z.

if $\Delta I = 1$ then $\alpha = -\beta$ ($\gamma = 0$)

Astrophysics: Amplitude N - Z.

Total cross-section ratio on Fe, Neutral to charged = R

Neutrinos $R_\nu = \alpha^2 + \beta^2 + \frac{1}{3}(\alpha'^2 + \beta'^2)$, disregarding antiquarks, strange quarks

Antineutrinos $R_{\bar{\nu}} = \alpha^2 + \beta^2 + 3\ (\alpha'^2 + \beta'^2)$

FIGURE 7

WHAT NEXT

High Energy.

Scaling deviations?

Weak Theory M_W? Other?

New Channels or Particles? e.g. charm
(Note" $e^+e^- \rightarrow$ hadrons!)

Naiveté of Parton Model? e.g. $K_\perp$ not small?

Logarithmic Deviations

Example $\bar{x} \sim (\ell n\, q^2)^{-a_1}$

$$a_1 \sim 1/2 \quad .$$

FIGURE 8

		L.C. Number	ISBN
No. 1	Feedback and Dynamic Control of Plasmas (Princeton 1970)	70-141596	0-88318-100-2
No. 2	Particles and Fields - 1971 (Rochester)	71-184662	0-88318-101-0
No. 3	Thermal Expansion - 1971 (Corning)	72-76970	0-88318-102-9
No. 4	Superconductivity in d- and f-Band Metals (Rochester 1971)	74-188879	0-88318-103-7
No. 5	Magnetism and Magnetic Materials - 1971 (2 parts) (Chicago)	59-2468	0-88318-104-5
No. 6	Particle Physics (Irvine 1971)	72-81239	0-88318-105-3
No. 7	Exploring the History of Nuclear Physics (Brookline 1967, 1969)	72-81883	0-88318-106-1
No. 8	Experimental Meson Spectroscopy - 1972 (Philadelphia)	72-88226	0-88318-107-X
No. 9	Cyclotrons - 1972 (Vancouver)	72-92798	0-88318-108-8
No.10	Magnetism and Magnetic Materials - 1972 (2 parts) (Denver)	72-623469	0-88318-109-6
No.11	Transport Phenomena - 1973 (Brown University Conference)	73-80682	0-88318-110-X
No.12	Experiments on High Energy Particle Collisions - 1973 (Vanderbilt Conference)	73-81705	0-88318-111-8
No.13	π-π Scattering - 1973 (Tallahassee Conference)	73-81704	0-88318-112-6
No.14	Particles and Fields - 1973 (APS/DPF Berkeley)	73-91923	0-88318-113-4
No.15	High Energy Collisions - 1973 (Stony Brook)	73-92324	0-88318-114-2
No.16	Causality and Physical Theories (Wayne State University, 1973)	73-93420	0-88318-115-0
No.17	Thermal Expansion - 1973 (Lake of the Ozarks)	73-94415	0-88318-116-9
No.18	Magnetism and Magnetic Materials - 1973 (2 parts) (Boston)	59-2468	0-88318-117-7
No.19	Physics and the Energy Problem - 1974 (APS Chicago)	73-94416	0-88318-118-5
No.20	Tetrahedrally Bonded Amorphous Semiconductors (Yorktown Heights, 1974)	74-80145	0-88318-119-3
No.21	Experimental Meson Spectroscopy - 1974 (Boston)	74-82628	0-88318-120-1
No.22	Neutrinos - 1974 (Philadelphia)	74-82413	0-88318-121-X